The Joy of Mathematica

Second Edition

**Instant Mathematica for Calculus,
Differential Equations, and Linear Algebra**

LIMITED WARRANTY AND DISCLAIMER OF LIABILITY

The Joy of Mathematica®
Second Edition

Instant Mathematica for Calculus, Differential Equations, and Linear Algebra

Alan Shuchat and Fred Shultz

Wellesley College

A Harcourt Science and Technology Company

San Diego   San Francisco   New York   Boston
London   Sydney   Tokyo

Academic Press
A Harcourt Science and Technology Company
525 B Street, Suite 1900, San Diego, CA 92101-4495 USA
http:// www.academicpress.com

Academic Press
Harcourt Place, 32 Jamestown Road, London, NW1 7BY, UK
http://www.hbuk.co.uk/ap/

Harcourt/Academic Press
200 Wheeler Road, Burlington, MA 01803 USA
http://www.harcourt-ap.com

Library of Congress Catalog Card Number: 99-066840

International Standard Book Number: 0-12-640730-4

PRINTED IN THE UNITED STATES OF AMERICA
99 00 01 02 03 04 EB 6 5 4 3 2 1

Contents

Chapter 12   Differentiating Functions of Several Variables   181

Chapter 13   Integrating with Several Variables   189

Chapter 14   Working with Vector Fields   195

Chapter 15   Solving Differential Equations   207

Chapter 16   Working with Vectors and Matrices   235

Part III   Exercises and Labs   261

Chapter 17   Functions   263

Chapter 18   Limits and Continuity   281

Chapter 19   Derivatives in One Variable   301

Chapter 20   Integrals in One Variable   329

Chapter 21   Sequences and Series   357

Chapter 22   Parameterized Curves   389

Chapter 23   Surfaces and Level Sets   407

Preface

If you're already familiar with *Mathematica* and *Joy*, here are the major changes from the first edition of *Joy*, which was published in 1994. If not, you can find more details below.

■ What's New in This Edition of *Joy*?

We have rewritten the *Joy* software and book from the ground up to take advantage of the major changes that have taken place in *Mathematica* and to make *Joy* easier to use for mathematics instruction and self-study.

The *Joy* software now:

- Runs on Windows and Power Macintosh
- Takes advantage of new features in *Mathematica* 4.0 (requires 3.0 or higher)
- Includes a palette for easy entry of common mathematical notation
- Runs faster and operates internally to *Mathematica*

The *Joy* book now:

- Contains ready-to-use exercises and labs for the mathematics classroom
- Includes more coverage of multivariable calculus and differential equations, in addition to single-variable calculus and linear algebra
- Contains materials suitable for both reform and traditional approaches to calculus and other courses

We were pleased to see that many people in disparate disciplines found the original version of *Joy* useful in their work. We believe this edition will be equally useful for that broad audience, while at the same time providing materials that instructors can use without change in math courses.

■ What Is *Mathematica*?

Mathematica is an exciting and powerful computer algebra system and programming language for mathematics. Programs such as *Mathematica* have revolutionized research and teaching in mathematics and the way it is used in science, engineering, and economics. They make it possible for anyone with a personal computer to do many of the tasks that before could only be done laboriously by hand or with mainframe computers that were not available to most people. *Mathematica* enables you to carry out a wide variety of computations with numbers and symbols and to visualize them with striking two- and three-dimensional graphs. For example, you can manipulate algebraic expressions, solve equations algebraically and numerically, calculate exactly with rational numbers and to any number of decimal places with either real or complex numbers, differentiate and integrate symbolically and numerically, graph curves and surfaces, create animations from graphs, and do many other things.

Beginning with Version 3.0, *Mathematica* includes typesetting capabilities for creating traditional mathematical notation as well as the linear notation of earlier versions. It also includes the ability to create palettes of buttons that will enter notation, execute commands, etc.

Since *Mathematica* was first released in 1988, a wide variety of related materials has appeared. There are now books that demonstrate *Mathematica*'s abilities and uses, teaching materials for courses that use *Mathematica*, materials for learning the *Mathematica* language, and computer packages that extend *Mathematica*'s scope. It is no exaggeration to say that *Mathematica* and programs of its kind have changed the face of mathematics and other technical fields forever.

■ What Is *The Joy of Mathematica*?

The Joy of Mathematica is a collection of *Mathematica* notebooks and packages that create menus and dialogs, and a palette of buttons that create common mathematical notation. By pointing and clicking in *Joy*, anyone can access *Mathematica*'s power instantly, without first learning its commands or syntax. *Joy* makes it easier to use and learn *Mathematica* by substituting menus and dialog boxes for typing commands. *Joy* is ideal for all new *Mathematica* users, as well as for more experienced users who want to broaden their knowledge by exploring new topics in *Mathematica*.

Joy's menus and dialogs make *Mathematica* immediately accessible to the user who wants to:

- Learn *Mathematica* easily
- Use *Mathematica* quickly and "out of the box"
- Use *Mathematica* only occasionally

With *Joy*, students who know only the most basic techniques for operating a personal computer can use *Mathematica*. Teachers and other professionals can use *Joy*'s menus to select the most common mathematical operations found in *Mathematica*. *Joy* makes it possible to use *Mathematica* as a tool for solving problems without learning its programming language. On the other hand, anyone can use *Joy* to learn *Mathematica* by observing how *Joy* works.

■ How Do *Joy* and *Mathematica* Work Together?

In *Mathematica*, the usual way to perform a computation is to type a command as input, following certain rules of syntax. *Mathematica* then executes the command and replies with the corresponding output. *Joy* works by creating and typing the commands for you.

A *Joy* menu bar appears on the screen below *Mathematica*'s own menus. The *Joy* menus refer to specific areas of mathematics.

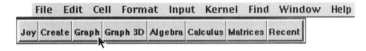

Joy's menus (below) and *Mathematica*'s menus (above).

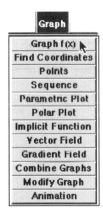

Joy's **Graph** menu.

When you make a selection from a *Joy* menu, a dialog box appears for you to fill in.

Plot the function(s)

$$x^3 + 1, \ 2\,\text{Sin}[3\,x]$$

x – interval : $-5 \leq x \leq 5$

y – interval :   Automatic ▾

☐ **Equal scales**
☐ **Axes at origin**
☐ **Legend**
☐ **Assign name**

graph1

Example   Cancel   OK ◤

Joy's **Graph f(x)** dialog.

Joy uses this information to construct a *Mathematica* command. *Joy* sends the command to *Mathematica*, which executes it just as if you had typed it as input and then displays the output. You can choose whether to display the actual commands that *Joy* creates or to paraphrase them in a way similar to what you would see in a math textbook. You can also choose to do both.

For example, here is the result of filling in *Joy*'s **Graph f(x)** dialog.

▷ Graph

| $x^3 + 1$ | Black | Solid |
| $2\,\text{Sin}[3\,x]$ | Red | . . . . . |

on $-5 \leq x \leq 5.$

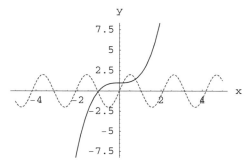

Out[2]= - Graphics -

This is the command that *Joy* creates behind the paraphrase.

```
In[2]:=  Plot[{x³ + 1, 2 Sin[3 x]}, {x, -5, 5}, AxesLabel -> {"x", "y"},
         PlotStyle -> {Black, {Red, Dashing[{.008}]}}]
```

Displaying commands helps you learn *Mathematica*'s commands and syntax gradually as you use *Joy*, if you wish to do so. Paraphrasing is useful, for example, in classes where the instructor prefers not to devote time to explaining the commands.

You can also use *Joy* and *Mathematica* interchangeably. You can type commands directly into *Mathematica*, make selections from *Joy*'s menus, and move back and forth freely between *Mathematica* and *Joy*. You can use *Joy* with *Mathematica* notebooks that others have created. Using *Joy* and *Mathematica* together in this way gives you access to the full power of *Mathematica* while still relying on *Joy* to make many common tasks easier.

■ About This Software

This version of *The Joy of Mathematica* software requires

- Windows 95 (compatible with Windows 98) or a Power Macintosh
- *Mathematica* version 3.0 or higher (optimized for *Mathematica* 4)
- The same memory (RAM) requirements as *Mathematica*
- *Mathematica*'s kernel and front end on the same computer
- One *Joy* user at a time (multiple users require additional licenses)

■ About This Book

This book is a manual for the *Joy* software, an introduction to using *Joy* and *Mathematica* for mathematics and its applications to other fields, and a supplement to mathematics texts in calculus, differential equations, and linear algebra. It is divided into three parts.

Part I: An Introduction to *The Joy of Mathematica*

Chapter 1, A Brief Tour of Joy. This chapter contains the minimum you need to know to get started, regardless of how you plan to use the software or what courses you may be taking.

Chapter 2, More about Joy. This chapter is intended as a reference for when you want to get beyond the basics or are having trouble.

Part II: How to Use *Joy* for … shows how to use *Joy* for the mathematics in single- and multivariable calculus, differential equations, and linear algebra. Each chapter illustrates briefly the menus and techniques you need to employ *Joy* for the topic at hand. They are written in the same "do it now" style as the Tour. These chapters assume you have taken the Tour but are independent of each other.

Part III: Exercises and Labs contains chapters on topics similar to ones in math textbooks. They contain examples that show how to use *Joy* to illuminate mathematical concepts, especially where several calculations or dialogs must be combined. There arc short and medium-length exercises and more extended labs.

Some examples of labs are

- What is the area under the normal probability curve?
- How can we visualize the definition of convergence for an infinite series?
- Can we predict the future growth of the U.S. population from past patterns?
- How should we design a waterslide?
- Does the speed of Halley's comet change as it moves along its orbit?
- What happens when we combine two reflections? a reflection and a rotation?

To instructors: Each chapter in Part III lists the specific "how to" sections that are needed to do the exercises and labs. Each lab begins with an introduction to be read and questions to be answered in advance. The main body of the lab contains specific questions to be answered with *Joy*, along with any necessary instructions for how to use *Joy*. Afterward, there are questions asking students to summarize what the lab has shown and to integrate their results with the work they did before the lab. Some have an optional section with open-ended explorations or additional questions. In order to help students do the labs successfully, you may want to discuss the answers to crucial pre-lab questions before they start their computer work.

■ Getting Started

A good way to begin is to take the *Brief Tour of Joy* in Chapter 1.

The Tour takes about a half hour to go through at your computer, using *Joy* to follow along with the text and to reproduce its examples. It is written in a "do it now" style with pictures of the *Joy* menus, the dialog boxes as you would fill them in, the paraphrases or commands, and *Mathematica*'s output. Then skim Chapter 2 so you'll know what's there if you need it, and move on to the chapters that particularly interest you.

We've tried to make the book easy to use by giving tips and warnings where appropriate. They are shown in the text in this way:

  ♀ A tip for using *Joy* and *Mathematica* more effectively

  ⚠ A warning to be careful

■ Acknowledgments

We would like to thank Wellesley College for its support of this project, and our students and colleagues at Wellesley who used the first edition of *Joy* and tested the second edition. We would also like to thank our colleagues elsewhere, especially Bill Barker and Ray Fisher, for their support and helpful feedback. We also wish to acknowledge reviewers Dean Allison, Jim Delany, and Scott Metcalf. All have made valuable suggestions that led to the improvement of the *Joy* software and book.

We are particularly grateful to our wives, Alix Ginsburg and Mary Shultz, and to our children, for their patient and understanding encouragement of our work.

Alan Shuchat and Fred Shultz
Department of Mathematics
Wellesley College
Wellesley, MA 02481
ashuchat@wellesley.edu
fshultz@wellesley.edu

Installation

The Joy of Mathematica is very easy to install on your PC or Macintosh. Be sure your system meets the following requirements:

- Windows 95 (compatible with Windows 98) or a Power Macintosh
- *Mathematica* version 3.0 or higher (optimized for *Mathematica* 4)
- *Mathematica*'s memory (RAM) requirements
- *Mathematica*'s kernel and front end on the same computer
- One *Joy* user at a time (multiple users require additional licenses)

■ Installing *Joy* on Your Hard Disk

The *Joy* CD-ROM contains the *Joy of Mathematica* folder and also Installation folders for Macintosh and Windows.

Joy of Mathematica

Joy Macintosh Installation   Joy Windows Installation

Power Macintosh Installation

- Drag the *Joy of Mathematica* folder from the CD-ROM to any location on your hard disk.

You're ready to use *Joy*.

Windows 95/98 Installation

Automated installation.

1. Copy the *Joy of Mathematica* folder to the top level of the C: drive, i.e., so that the *Joy of Mathematica* folder is on the C: drive and is not inside any other folder.

2. Open the folder *Joy* Windows Installation on the CD.

3. Double click on the file *JoyInstall* (or *JoyInstall.bat*) on the CD. (Make sure Step 1 is completed first!)

Copyright   JoyInstall.bat   Readme.txt
Notice.txt

Virtually at once you should get the message, "*Joy of Mathematica* installation is complete!" At that point you may remove the CD. Then if you wish, you can move the *Joy of Mathematica* folder to a new location on your PC. *Joy* is ready to use.

You may instead get the message, "Please make sure that the *Joy of Mathematica* folder has been copied to the top level of the C: drive (i.e., not inside any folder), and then try again." This means the installer couldn't find the *Joy of Mathematica* folder where it expected it. This could happen if you run the *JoyInstall* program before copying *Joy*, or if you haven't copied *Joy* to the right place. If this happens, try Steps 1 and 2 again. Instead, you can install *Joy* manually, as shown next.

Manual installation.

1. Copy the folder *Joy of Mathematica* to any location on your PC.

2. For each of the four files listed below, take the following steps to change the "Read-only" property for that file to false:

   a) Find the file.

   b) Right-click on the file. From the resulting popup menu choose Properties ....

   c) In the dialog that appears, click in the box beside "Read-only" to remove the checkmark.

   d) Close the dialog.

3. The four files to change are in the folders *JoyNotebooks* and *JoyPackages*, which are inside the *Joy of Mathematica* folder:

a) in the folder *JoyNotebooks*

       Joy Menu

b) in the folder *JoyPackages*

       JoyPreferences
       JoyStartupPreferences
       JoyStartupPreferences 4.0 PC

When you've made each of these files not "Read-only," *Joy* is ready to use.

■ Using *Joy*

The *Joy of Mathematica* folder contains these items:

Joy of Mathematica.nb

JoyNotebooks   JoyPackages

⚠ These items should remain inside the *Joy of Mathematica* folder and not be renamed.

- To start *Joy*, open *Joy of Mathematica.nb*.
  ☿ In some Windows systems, this file may simply appear as *Joy of Mathematica*.

- Chapter 1, *A Brief Tour of Joy*, gives an overview of how to use *Joy*.

Part I

An Introduction to *The Joy of Mathematica*

Chapter 1
A Brief Tour of *Joy*

This chapter gives an overview of *The Joy of Mathematica* and how to use it. It discusses the most common techniques and menus you will need to know and illustrates how best to use *Joy* for graphing and symbolic and numeric calculations. Several of the examples that we illustrate in this chapter can be approached in more than one way using *Joy*. We've chosen the methods shown here to illustrate a variety of techniques.

■ 1.1 Starting *Joy*

You can start up *Joy* either before or after you've launched *Mathematica*.

- Open *Joy of Mathematica.nb* in the *Joy of Mathematica* folder. In some Windows systems, this file may simply appear as *Joy of Mathematica*.

Joy of Mathematica.nb

  If *Mathematica* is already open, then you can also choose **Open** from *Mathematica*'s **File** menu, navigate to the *Joy of Mathematica* folder, and open *Joy of Mathematica.nb*. If *Mathematica* is not already open, *Joy* will also start up *Mathematica*.

This picture will appear on the screen:

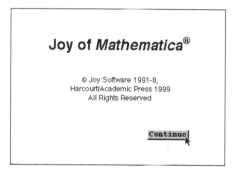

- Click **Continue**.
- Wait for the startup process to finish and for the *Joy* menu to appear.

   This may take a while if *Mathematica* isn't already open or hasn't done any calculations in this session.

■ 1.2 Menus and Dialogs

When *Joy* has been launched, its *menu bar* will appear on the screen, together with a *palette* of buttons. You will also see a blank *Mathematica* window, called a *notebook*. This is the file that will contain *Mathematica*'s input and output. The appearance of the screen varies with different computers.

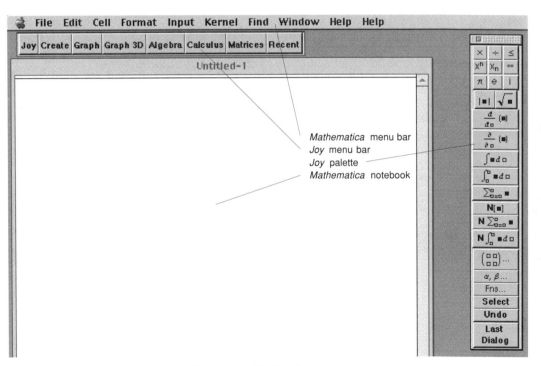

Starting *Joy* (Macintosh version).

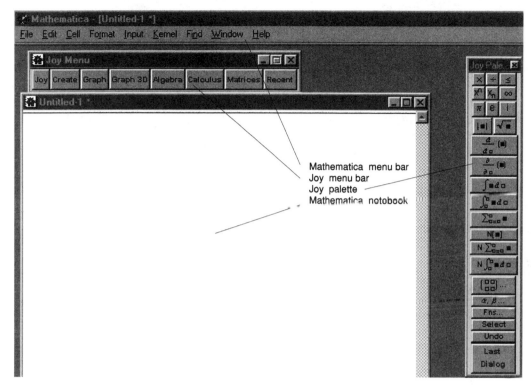

Starting *Joy* (Windows version).

Clicking in the *Joy* menu bar brings up menus containing dialogs for specific mathematical operations. Here's how it works with graphing.

- In the *Joy* menu bar, click **Graph**.

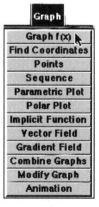

Joy's **Graph** menu.

The **Graph** menu drops down from the *Joy* menu bar. The first selection in the menu is for graphing expressions of the form $y = f(x)$.

- In the **Graph** menu, click **Graph f(x)**.

A dialog box appears, with some information already filled in. Every *Joy* dialog contains a *default example* that shows you the format for filling in the dialog. You enter mathematical expressions in *input fields*, which appear as boxes in the dialog. The default example in the **Graph f(x)** dialog shows how to graph two curves simultaneously.

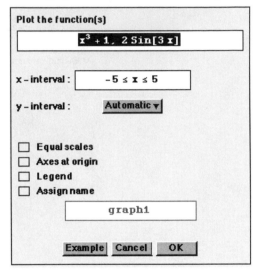

Joy's **Graph f(x)** dialog.

1.2.1 Filling In a Dialog

Instead of using the default example, we'll change the dialog and graph $y = 10(x + 1)\sin 2x$ and $y = x \sin 2x$.

    ☿ Functions in *Mathematica* are always entered with square brackets [ ] and the names of the standard functions are capitalized, e.g., `Sin[2x]`. *Mathematica* uses parentheses ( ) only for grouping parts of expressions, e.g., `10(x+1)Sin[2x]`.

- Type `10(x+1)Sin[2x]`.

  The default functions are automatically selected (highlighted) when the dialog opens. Typing over an expression that is selected always replaces it with what you are typing.

- Type a comma **,** and then type `x_Sin[2x]`.

  The symbol ␣ means that you should press the Space bar. Typing a blank space between `x` and `Sin[2x]` indicates multiplication. You may use an asterisk * instead of a space.

You don't need a space or asterisk if you use parentheses or if the left-hand factor is a number, e.g., `2x` in `Sin[2x]`.

The default entries in a *Joy* dialog will always show when multiple entries are possible, such as these two functions. Always use commas to separate multiple entries. If you wish, you may insert a space after a comma to make it easier to read the entries.

If you make a mistake in typing, such as `sin[x]` or `Sin(x)`, *Joy* will warn you of a possible error. Click **Cancel** to return to the dialog and make any corrections. Section 2.4 gives more information.

Next, here's how to change the graphing interval to $-10 \leq x \leq 10$.

- Drag across the default interval to select it.

   Instead, you can just select the endpoints and change them one at a time.

- Fill in $-10 \leq x \leq 10$.

   You can type $\leq$ by clicking the **Inequality** button in the *Joy* palette.

Inequality button.

- ♡ On the Macintosh, you can hold down the Option key while typing < instead of clicking the button, i.e., you can type OPTION − <. On other computers, you can type <= and *Mathematica* will interpret it as $\leq$.

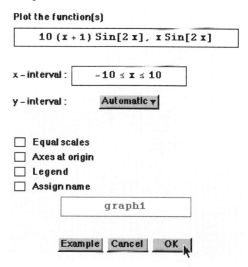

Change the functions and interval.

- Click **OK**.

Whenever you complete a *Joy* dialog, click the **OK** button. This causes *Joy* to take the information in the dialog, construct the appropriate syntax, and send it to *Mathematica*.

⚠ In *Joy* dialogs, you need to click **OK**. Pressing the Enter or Return key or just clicking in another window will not work.

Mathematica now graphs the functions and displays a summary of the information you entered in the dialog. The graphs appear in different colors on a color monitor.

▷ Graph

10 (x + 1) Sin[2 x]	Black	Solid
x Sin[2 x]	Red	. . . . .

on –10 ≤ x ≤ 10.

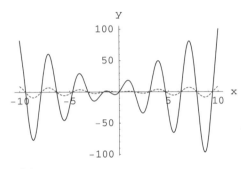

Out[2]= - Graphics -

1.2.2 Using the Last Dialog Button

It is easy to change the *x*- and *y*-values that are used in the graph without retyping the functions. In the preceding graph, the *y*-values were set automatically. Here's how to zoom in around the origin, where the graph of $y = x \sin 2x$ appeared almost flat.

- On the *Joy* palette, click **Last Dialog**.

Last Dialog button on the *Joy* palette.

The **Last Dialog** button always brings back the last dialog and values that you entered. Here it brings back the **Graph f(x)** dialog.

- Change the *x*-interval to – 3 ≤ **x** ≤ 3.
- For the *y*-interval, click the triangle ▼ to open a popup menu.

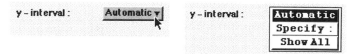

Click the triangle (left) to open the popup menu (right).

The popup menu gives you different ways to choose the *y*-interval.

- Click **Specify**.

  The new choice will be highlighted briefly and the popup menu will close. The **Graph f(x)** dialog now allows you to specify a *y*-interval.

- Change the *y*-interval to – 5 < y ≤ 10.

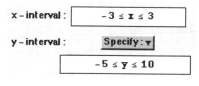

Change the *y*-interval.

- Click **OK**.

Mathematica graphs the curves using the new intervals:

▷ Graph

```
10 (x + 1) Sin[2 x]      Black      Solid
x Sin[2 x]               Red        .....
```

on – 3 < x ≤ 3 and 5 ≤ y ≤ 10.

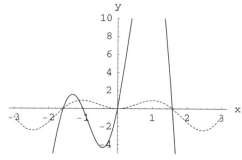

Out[3]=   - Graphics -

1.2.3 Displaying *Mathematica*'s Commands

So far, when you have used a *Joy* dialog the *Mathematica* output has been accompanied by a description like those in a mathematics textbook. *Joy*'s default setting keeps *Mathematica*'s actual commands hidden from view, and you can choose whether to display the commands. For example, you may wish to see the commands if you want to use *Joy* to learn *Mathematica* or if you just want to know what's going on behind the scenes. Here's how to display an individual command along with its paraphrase. Section 2.3.1 shows how to set *Joy* to display commands all the time.

- Click the triangle ▷ that appears next to the description of the output.

```
▷ Graph

10 (x + 1) Sin[2 x]      Black       Solid
  x Sin[2 x]             Red         . . . . .
```

Click the triangle.

The triangle changes to ▽ and the *Mathematica* command appears below the paraphrase.

```
▽ Graph

10 (x + 1) Sin[2 x]      Black       Solid
  x Sin[2 x]             Red         . . . . .

on -3 ≤ x ≤ 3 and -5 ≤ y ≤ 10.
```

In[3]:= **Plot[{10 (x + 1) Sin[2 x], x Sin[2 x]}, {x, -3, 3}, AxesLabel -> {"x", "y"},
        PlotRange -> {-5, 10}, PlotStyle -> {Black, {Red, Dashing[{.008}]}}]**

- To hide the command and display only its paraphrase, click the triangle again.

1.2.4 Creating Expressions with the *Joy* Palette

Here is how to create the function

$$f(x) = 4x^2 - \frac{1}{x^2} + \frac{x-1}{x} .$$

- Choose **Create** ▷ **Function** from the *Joy* menus, i.e., click **Create** in the menu bar and then click **Function** in the **Create** menu.

Create menu.

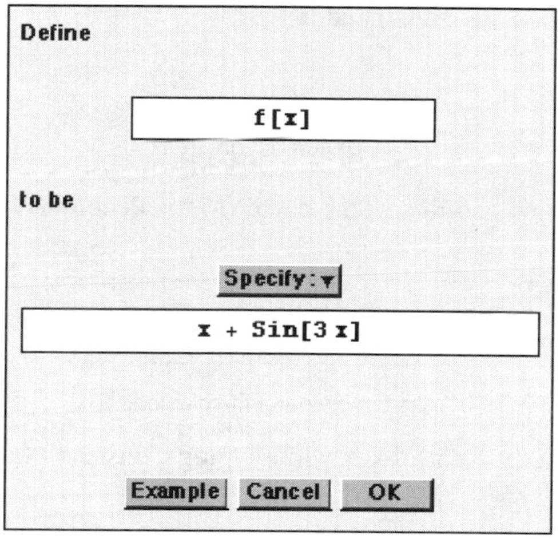

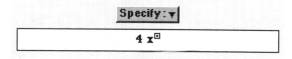

Define

f[x]

to be

Specify: ▼

x + Sin[3 x]

Example Cancel OK

Create Function dialog.

- Keep the name f[x].
- Drag over the expression for the default value of the function to select it.
- Type 4x.
- On the *Joy* palette, click the **Superscript** button.

xⁿ

Superscript button.

This moves the cursor to the exponent. The exponent is indicated by a small square placeholder that is selected (highlighted).

Specify: ▼

4 x

The exponent is selected.

☿ You can hold down the Control key while typing ^ instead of clicking the **Superscript** button, i.e., you can type CTRL – ^.

- Type 2.
- Press the → key to move the cursor forward.

  The → key always moves the cursor forward to the next character. You can also type [CTRL] – ⎵ (hold down the Control key and press the Space bar), which moves the cursor out of a part of an expression.

- Type –1.
- On the *Joy* palette, click the **Divide** button.

Divide button.

This creates a fraction and moves the cursor to the denominator, which is indicated by a highlighted placeholder.

$$4\,x^2 - \frac{1}{\blacksquare}$$

The denominator is selected.

♀ You can hold down the Control key while typing / instead of clicking the **Divide** button, i.e., you can type [CTRL] – /.

- Type x^2 as before.
- Press the → key *twice*, once to move the cursor out of the exponent and again to move it out of the denominator.

  You can press [CTRL] – ⎵ twice instead.

- Type + and then type x – 1.
- Drag across x – 1 to select it.

$$4\,x^2 - \frac{1}{x^2} + \boxed{\text{x – 1}}\ \text{\}}$$

Drag across x–1.

- Click the **Divide** button in the *Joy* palette.

  This creates a fraction with $x - 1$ in the numerator. When the numerator has more than one term, you need to select it before clicking the **Divide** button.

$$4\,x^2 - \frac{1}{x^2} + \frac{x - 1}{\blacksquare}$$

x–1 is in the numerator and the denominator is selected.

- Type **x**.

  This completes the expression $4\,x^2 - \frac{1}{x^2} + \frac{x-1}{x}$.

- Click **OK**.

Mathematica now displays this statement:

▷ Define f[x] to be $4\,x^2 - \frac{1}{x^2} + \frac{x-1}{x}$

Mathematica creates the function but does not display any output.

1.2.5 Using the Recent Menu and Example Button

Here's how to graph $f(x)$ without retyping $4\,x^2 - \frac{1}{x^2} + \frac{x-1}{x}$. The domain for the graph will be $-5 \le x \le 5$.

- In the main *Joy* menu, click **Recent**.

  The **Recent** menu lists the most recent selections you have made from the *Joy* menus and is useful when you choose just a few menu items repeatedly.

Recent menu.

- Click **Graph f(x)**

  This brings back the **Graph f(x)** dialog again. It's easiest to go back to the default example and start fresh.

- In the **Graph f(x)** dialog, click **Example**.

Example button.

  The **Example** button restores the default example and settings of the dialog and often makes entering new data easier. It also reminds you of the format you need to use to fill in the dialog.

- Type f[x] in place of the expression to be graphed.

- Keep the default interval, $-5 \le x \le 5$.

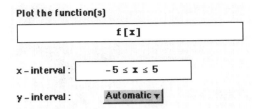

- Click **OK**.

Mathematica now displays the result:

▷ Graph f[x] on -5 ≤ x ≤ 5.

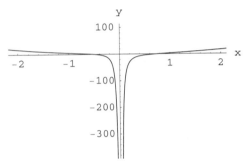

Out[5]= - Graphics -

■ 1.3 Using *Mathematica* Directly

When you use *Joy* to create commands for *Mathematica*, it pastes them into the *Mathematica* window, or notebook. Having *Joy* enter a command is the same as typing it in the notebook. Sometimes, you'll work directly in the notebook. For example, here's how to use *Mathematica* to calculate 2^{1000}, a number that is beyond the range of many calculators.

- If necessary, click in your *Mathematica* notebook below the last output to make a horizontal line appear beneath it.

  A horizontal line appears automatically below any output that *Mathematica* creates. If you click somewhere else, the line disappears. In that case, you must click again to make it reappear.

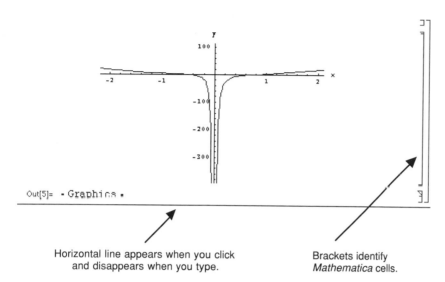

Out[5]= • Graphics •

Horizontal line appears when you click
and disappears when you type.

Brackets identify
Mathematica cells.

Every *Mathematica* notebook is made up of *cells* that contain *input, output,* and other information. Cells are generally indicated by *brackets* on the right-hand side of the notebook. The horizontal line indicates that whatever you type next will begin a new cell below the line.

- Type 2^{1000}.

  One some platforms, you may need to click an extra time before typing. Use the **Superscript** button to enter the exponent. The horizontal line disappears when you start typing.

- Press the Shift and Return keys simultaneously, i.e., press $\boxed{\text{SHIFT}} - \boxed{\text{RET}}$.

  Pressing these keys at the same time causes *Mathematica* to evaluate the expression or execute the command in the cell. Pressing only $\boxed{\text{RET}}$ creates a new line but doesn't cause an evaluation or execution.

  ♡ On Macintosh and some PC keyboards, you can press $\boxed{\text{ENTER}}$ on the numeric keypad instead of $\boxed{\text{SHIFT}} - \boxed{\text{RET}}$.

Mathematica calculates the exact value:

In[6]:= 2^{1000}

Out[6]= 10715086071862673209484250490600018105614048117055336074437503883703510 5
11249361224931983788156958581275946729175531468251871452856923140435984
57757469857480393456777482423098542107460506237114187795418215304647498
35819412673987675591655439460770629145711964776865421676604298316526243
86837205668069376

⚠ If you click elsewhere before pressing [SHIFT] – [RET], this value will not appear. *Mathematica* acts on the input in the last cell where you clicked or typed. So if you click elsewhere, you must click in the input cell and then press [SHIFT] – [RET].

1.3.1 Using the Numeric Button

The number 2^{1000} has 302 digits, so it's hard to read in this form. Here's how to make it easier.

- If necessary, click below the last output to make the horizontal line reappear.
- Click the **Numeric** button on the *Joy* palette.

Numeric button.

A new command, N[ ], appears in the *Mathematica* notebook. Here N stands for *numeric*, and this command will put any expression you type inside the brackets [ ] into numeric form.

N[▣]

- Type 2^{1000}.
- Press [SHIFT] – [RET].

Mathematica expresses 2^{1000} in numeric form:

In[7]:= **N[2^{1000}]**

Out[7]= 1.07151×10^{301}

You can use the **Numeric** button to express any number in decimal form. For example, here's how to do it with π.

- If necessary, click below the last output to make the horizontal line reappear.
- Click the **Numeric** button and then click the π button on the *Joy* palette.

π button.

- Press [SHIFT] – [RET].

The result is:

In[8]:= **N[π]**

Out[8]= 3.14159

Mathematica expresses its output to 6 significant digits by default. Here's how to change the number of digits to, say, 150.

- If necessary, click below the last output to make the horizontal line reappear.
- Click the **Numeric** and π buttons.

  A short blinking cursor appears just to the right of π. The cursor indicates that anything you type will appear at this point.

Short blinking cursor.

- Type a comma **,** and then **150**.
- Press SHIFT − RET .

Mathematica displays the answer:

In[9]:= **N[π, 150]**

Out[9]= 3.14159265358979323846264338327950288419716939937510582097494459230781640628620899862803482534211706798214808651328230664709384460955058223172535940813

Another way. By selecting an expression before clicking the **Numeric** button, you can convert it to numeric form without retyping. Here's how to do that with the value of $f(5)$, where $f(x) = 4x^2 - \frac{1}{x^2} + \frac{x-1}{x}$ is the function you defined in Section 1.2.4.

- If necessary, click below the last output to make the horizontal line reappear.
- Type **f[5]** and press SHIFT − RET .

Mathematica evaluates the function:

In[10]:= **f[5]**

Out[10]= $\dfrac{2519}{25}$

- Drag across the output to select it. Make sure you highlight the entire fraction.

In[10]:= **f[5]**

Out[10]= $\dfrac{2519}{25}$

- Click the **Numeric** button on the *Joy* palette.

The highlighted square ■ in the **Numeric** button N[■] indicates that clicking the button will operate on whatever you have selected.

- Press SHIFT — RET.

The result is:

$In[11]:= \quad \mathbf{N}\left[\dfrac{2519}{25}\right]$

$Out[11]= \quad 100.76$

1.3.2 Annotating Your Work

You can annotate your work by typing directly in your *Mathematica* notebook. Just click where you want your comments to appear and start typing.

- If necessary, click below the last output to make the horizontal line reappear.
- Type (* and then your comment. Finish by typing *).

For example:

$In[11]:= \quad \mathbf{N}\left[\dfrac{2519}{25}\right]$

$Out[11]= \quad 100.76$

(* This is the value of f[5] as a decimal.*)

Annotating your output.

♡ Including (* and *) indicates that this is a comment rather than a command for *Mathematica* to execute. If you didn't mark the comment this way, executing the cell would lead to an error message.

■ 1.4 Working with Notebooks

This section contains basic information you need to know to be able to work with your *Mathematica* notebooks.

1.4.1 Saving Your Work

It is a good idea to save regularly while using any computer program, especially before printing. For example, if you run short of memory you can unexpectedly lose any work you have not saved. You can save your *Mathematica* notebook in the same way that you save files or documents in other applications on your computer.

- From *Mathematica*'s **File** menu, choose **Save** or use the keyboard equivalent for your computer (Macintosh ⌘-S, Windows [CTRL]-S).

  There is no Save command in *Joy*'s menus.

- Choose the disk and folder or directory in which you wish to save the notebook.

  How to do this depends on your computer system.

- Enter the name you wish to give your notebook.
- Click **Save**.

1.4.2 Cleaning Up Your Notebook

Cutting and copying cells. You may wish to remove cells from your notebook if, for example, they contain a mistake. Let's create an intentional mistake. Suppose you want to evaluate $f(10)$.

- If necessary, click in your notebook beneath the last output to make a horizontal line appear.
- Type f(10), with parentheses instead of square brackets, and press [SHIFT] − [RET].

The result is:

In[12]:=　**f (10)**

Out[12]=　10 f

Because you typed parentheses ( ) instead of brackets [ ], *Mathematica* interpreted your input as "multiply *f* by 10."

It doesn't hurt to leave this mistake in your notebook, but here's how to remove it if you want to clean up.

- Click on the bracket that groups the input and output cells together.

  As you position the cursor over the bracket, it changes to ←. Clicking *selects* the cells.

click to select cells

In[12]:=　**f (10)**

Out[12]=　10 f

- Choose **Cut** from *Mathematica*'s **Edit** menu (Macintosh ⌘-X, Windows [CTRL]-X).

Sometimes, you may want to copy cells and paste them elsewhere without cutting them out of your notebook. To copy cells instead of removing them:

- Choose **Copy** instead (Macintosh ⌘-C, Windows [CTRL]-C).

- Click where you want to paste the cells.
- Choose **Paste** (Macintosh ⌘-V, Windows [CTRL]-V).

  ♥ To select cells that are contiguous to your selection but not grouped with it, drag along their brackets.

  If you want to select many contiguous cells that aren't all grouped together, it's easier to click on the bracket for the first cell, scroll to the last cell, hold down the [SHIFT] key, and click on the bracket for the last cell.

Removing cells vs removing information. When you remove cells, you are only changing the appearance of the notebook. *Mathematica* retains the information that was in the cells. For example, you created $f(x) = 4x^2 - \frac{1}{x^2} + \frac{x-1}{x}$ earlier. Now you'll cut out the cells that defined this function but see that $f(x)$ keeps its meaning.

- Scroll up to and select the cells where you created $f(x)$.

  There are two cells, one containing the paraphrase and a blank cell below it that contains the hidden *Mathematica* command. Drag along their brackets to select them both.

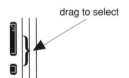

drag to select

$$\triangleright \text{Define } f[x] \text{ to be } 4x^2 - \frac{1}{x^2} + \frac{x-1}{x}$$

- Choose **Cut** from *Mathematica*'s **Edit** menu (Macintosh ⌘-X, Windows [CTRL]-X).

  The cells disappear.

- Scroll down to the last output and click below it to make a horizontal line appear.
- Type f[x] and press [SHIFT] − [RET].

Here is the result (*Mathematica* automatically rearranges the terms):

In[13]:= **f[x]**

Out[13]= $-\dfrac{1}{x^2} + \dfrac{-1+x}{x} + 4x^2$

The definition of $f(x)$ remains in force even though you removed the cells with the definition. It will remain in force until you give this function another meaning or quit *Mathematica*.

  ⚠ Some *Joy* dialogs refer to the **Last Output**. This always means the last output that *Mathematica* created, whether you removed it or not.

1.4.3 Printing Your Work

Printing an entire notebook

- From *Mathematica*'s **File** menu, choose **Print** (Macintosh ⌘-P, Windows CTRL-P).

  There is no Print command in *Joy*'s menus.

- Make any changes you wish to *Mathematica*'s print dialog and click **Print** (Macintosh) or **OK** (Windows).

Printing selected cells. You might want to print only some of the cells in your notebook. For example, here's how to print your earlier computation of π.

- Scroll up and select the input and output cells where you calculated π to 150 digits.

In[9]:= **N[π, 150]**

Out[9]= 3.1415926535897932384626433832795028841971\
6939937510582097494459230781640628620899\
8628034825342117067982148086513282306647\
0938446095505822317253594081 3

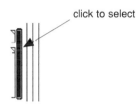

click to select

- Choose **Print Selection** from *Mathematica*'s **File** menu.
- Click **Print** (Macintosh) or **OK** (Windows).

Here's how to print cells that are not contiguous to each other.

- Click on the bracket at the right of the first cell or group that you wish to select.
- Hold down the ⌘ (Macintosh) or CTRL (Windows) key and click on the bracket for the next cell or group of cells you wish to select.
- Repeat until you have highlighted all the cells you wish to select.
- Choose **Print Selection** and print as before.

1.4.4 Closing and Opening Notebooks

Here's how to close your notebook and open a new one, for example, to start a new project or to free up memory. You can open and close these documents the way you would in other applications.

Closing a notebook

- From *Mathematica*'s **File** menu, choose **Close** (Macintosh ⌘-W, Windows CTRL-F4).

Opening a new notebook

- From *Mathematica*'s **File** menu, choose **New** (Macintosh ⌘-N, Windows CTRL-N).

   This opens a blank notebook, just like the one that opens when you start *Joy*. If more than one is open at a time, *Joy* pastes its commands into the frontmost notebook.

Opening an existing notebook

- From *Mathematica*'s **File** menu, choose **Open** (Macintosh ⌘-O, Windows CTRL-O).
- Navigate to the file you want and click **Open**.

   How to do this depends on your computer system.

   ⚠ When you close one notebook and open another in the same session, *Mathematica* retains all the information from the notebook that you closed. For example, if you let $f(x) = 4x^2 - \frac{1}{x^2} + \frac{x-1}{x}$ then $f(x)$ will still have this meaning after you close the notebook, until you give it a new one.

   However, if you quit the *Mathematica* application and reopen it later, all the definitions you have made and all the results of your calculations are "forgotten" *even though they appear in the notebook*. Section 2.2.2 explains how to reestablish your old results in *Mathematica*'s memory if you reopen a notebook in a new session.

■ 1.5 Quitting *Joy*

This concludes the Tour. Chapter 2, *More about Joy*, contains additional important and helpful information about using *Joy*, such as using output names and numbers, getting on-line help, and changing *Joy*'s default settings. It is probably best to read it over briefly at your leisure and refer to it when you need to use the techniques it describes. Later chapters contain topics relating to specific areas of mathematics.

- From the *Joy* menus, choose *Joy* ▷ **Close** *Joy*.

Joy menu.

You can now quit from *Mathematica* or else continue working without *Joy*. Here is how to quit.

- From *Mathematica*'s **File** menu, choose **Quit** (Macintosh) or **Exit** (Windows).

Chapter 2

More about *Joy*

This chapter contains additional information that is helpful in learning how to use *Joy* to work with *Mathematica* and in learning to use *Mathematica* itself. It includes changing *Joy*'s settings, keyboard shortcuts, reopening notebooks, on-line help, and troubleshooting.

■ 2.1 Creating Expressions

This section summarizes how to use *Joy* to create mathematical expressions. It includes using the *Joy* palette or keyboard shortcuts to construct expressions, copying and pasting into *Joy* dialogs, using names and output numbers, and entering expressions directly in your *Mathematica* notebook.

> ♡ If you have accidentally closed the *Joy* palette, you can reopen it by choosing **Joy** ▷ **Palette**.

2.1.1 The *Joy* Palette and Keyboard Shortcuts

You can use the *Joy* palette to create expressions such as $\frac{a+b}{2}$. This format is easier to interpret and less prone to error or misunderstanding than mathematically equivalent expressions such as $(a+b)/2$. For example, it is easy to omit the parentheses in $(a+b)/2$ and enter $a+b/2$ instead, but that would be equivalent to $a + \frac{b}{2}$ rather than $\frac{a+b}{2}$.

You can use keyboard shortcuts for some of the buttons on the *Joy* palette. The following table summarizes the buttons and keyboard strokes for entering basic mathematical expressions.

Expression	Palette	Keyboard
x^n	$\mathsf{x^n}$	CTRL – ^
x_n	$\mathsf{x_n}$	CTRL – _
$\dfrac{x}{y}$	÷	CTRL – /
$x\,y$	×	_(space), *
$\sqrt{x}$	$\sqrt{\blacksquare}$	CTRL – 2

For example, here's how to use the keyboard shortcut instead of the **Divide** button ÷ to enter the quotient $\frac{a+b}{5}$, either in a *Joy* dialog or in your *Mathematica* notebook.

- Type a+b.
- Drag across a+b to select it.

    You need to select a + b first because it has more than one term. Instead of dragging, you can use the **Select** button on the *Joy* palette or press SHIFT – ← repeatedly until all of a+b is selected.

- Type CTRL – / (i.e., hold down the Control key while typing /) and then type 5.

    The shortcut CTRL – / creates the quotient and puts a+b in the numerator.

- Press → to move the cursor out of the denominator.

The **Square root** button $\sqrt{\blacksquare}$ and shortcut work a little differently. For example, there are two ways to enter $\sqrt{u}$ with the button. In both methods, you can type CTRL – 2 (hold down the Control key while typing 2) instead of clicking the button.

- Type u and select it.
- Click the **Square root** button.

    ⚠ If you type u and don't select it, the result is u $\sqrt{\blacksquare}$.

or

- Click the **Square root** button.
- Type u.

    ♡ To edit this expression, e.g., to change $\sqrt{u}$ to $\sqrt{x+y}$, click very close to u and click the **Select** button, drag across it, or backspace over it. If you click too far away, you may inadvertently select the $\sqrt{}$ sign.

The *Joy* palette also contains buttons for entering special characters and expressions. The **Greek** and **Functions** buttons on the palette open subpalettes of buttons.

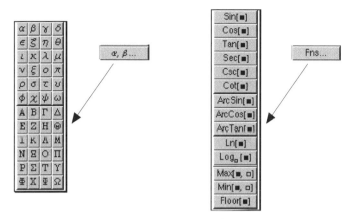

Greek (left) and **Functions** (right) palettes.

The accompanying table shows the keyboard equivalents for entering the symbols on the *Joy* palette. Some keyboard strokes depend on the computer system you are using (e.g., to create π, you can type \[Pi] on all computers and also OPTION -p on Macintosh).

Symbol	Palette	Keyboard
π	π	\[Pi], OPTION - p
Greek letters	$\alpha, \beta \dots$	\[Alpha], ESC -a - ESC , ...
$\leq$	$\leq$	<= , OPTION - <
∞	∞	\[Infinity], OPTION - 5
e (displayed as E or e)	e	E
i (displayed as I or i)	i	I

2.1.2 Navigating within Expressions

The following table summarizes how to use the keyboard or mouse to move around within expressions in *Joy* dialogs or in your *Mathematica* notebook. There is often more than one way to accomplish what you want to do, and the table indicates these choices.

To Accomplish	Do This
Move one character	→ , ←
Move to next placeholder (□)	TAB
Move out of subexpression	repeated → , CTRL – ⎵ (space)
Select subexpression	drag, click **Select**, SHIFT – → or ←
Expand selection	repeat above
Undo typing	click **Undo**, ⌘ – Z (Mac), CTRL – Z (Win)

2.1.3 Copying and Pasting in *Joy* Dialogs

You can frequently avoid retyping expressions that you've entered earlier by copying them from the *Mathematica* window and pasting them into a *Joy* dialog. For example, suppose you've already graphed a polynomial such as $x^3 - 2x - 1$ and now want to find its roots by solving the equation $x^3 - 2x - 1 = 0$. If you haven't defined the polynomial as a function $f(x)$ or given it a name, here's how you can accomplish this without retyping.

- From the *Joy* menus, choose **Graph ▷ Graph f(x)**.
- Graph $x^3 - 2x - 1$ on the default interval $-5 \le x \le 5$.

Here is the result:

▷ Graph x^3 – 2 x – 1 on –5 ≤ x ≤ 5.

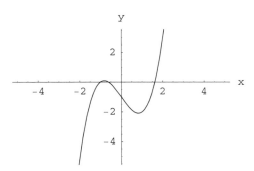

Out[2]= - Graphics -

- Select the expression $x^3 - 2x - 1$ in the description above the graph.

▷ Graph x^3 – 2 x – 1 on –5 ≤ x ≤ 5.

Be sure not to select any blank spaces on either side of the expression. If you do, there may be an error message and you will need to select and copy it again.

- Choose **Copy** from *Mathematica*'s **Edit** menu or use the keyboard equivalent for your computer (Macintosh ⌘-C, Windows CTRL-C).
- From the *Joy* menus, choose **Algebra ▷ Solve**.

  The default equations in the **Solve** dialog are automatically selected.

- Choose **Paste** from *Mathematica*'s **Edit** menu or use the keyboard equivalent for your computer (Macintosh ⌘-V, Windows CTRL-V).
- Type =0.
- Change the variable to x.

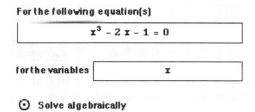

For the following equation(s)

$$x^3 - 2x - 1 = 0$$

for the variables   x

⊙ Solve algebraically

- Click **OK**.

Mathematica now finds the roots.

▷ Solve the equation $x^3 - 2x - 1 = 0$ for x algebraically.

Out[3]= $\left\{ \{x \to -1\}, \left\{x \to \frac{1}{2} (1 - \sqrt{5})\right\}, \left\{x \to \frac{1}{2} (1 + \sqrt{5})\right\} \right\}$

■ 2.2 Reopening Your *Mathematica* Notebooks

You may wish to save your *Mathematica* notebook and reopen it at another time. When you reopen a notebook, you will be able to see all its statements. How you can use these statements depends on whether you are reopening the notebook in the same *Mathematica* session or in a later one. This section describes what happens in these two situations.

2.2.1 Reopening a Notebook in the Same *Mathematica* Session

Mathematica assigns every result an *output number*, which you can use to refer to the result in subsequent dialogs or commands. Many *Joy* dialogs also let you assign an *output name* in addition to the output number. You can choose the name to make it easy to remember to which output it refers, and use it in place of the output number. If you close and later reopen your *Mathematica* notebook, the output names will always appear. Whether you will be able to see any input and output numbers depends on how your copy of *Mathematica* has been set. It's a good idea to assign names to expressions you may want to refer to or use again.

If you reopen your notebook in the same *Mathematica* session as when you created it, all the definitions you have made will still be valid. For instance, if you gave $x^2 - 3x + 4$ the name quadratic and didn't give that name to a different expression in any notebook later in that session, then the meaning of quadratic will still be $x^2 - 3x + 4$ during the session. The output number that *Mathematica* assigned to this expression will also remain valid, although it may no longer be visible. If the output number is not visible, it will be hard to use.

2.2.2 Reopening a Notebook in a Different Session

If you reopen your notebook in a different session, all your definitions and output numbers will be lost. You will be able to see the statement where you assigned the name quadratic, but *Mathematica* will not "remember" that it stands for $x^2 - 3x + 4$. If you only want to read the notebook or print it out, this is not a problem. However, if you want to resume your calculations you will need to recreate any definitions that you will be using again.

Here's how to recreate your definitions when you reopen a *Mathematica* notebook in a new session. The easiest way to reestablish the meanings of all the definitions is to reexecute all the commands in your notebook.

- Choose **Select All** from *Mathematica*'s **Edit** menu (Macintosh ⌘-A, Windows CTRL-A).
- Press SHIFT − RET.

   Each command will now be reexecuted, and input and output numbers will appear on the appropriate statements.

   ⚠ Commands in your notebook that contain output numbers for other commands may not reexecute the way you intend, since the output numbers may change. However, if this was the only notebook you used in the *Mathematica* session where you created it, if you didn't erase any commands before saving the notebook, and if it is the only notebook you have opened in the session where you are reexecuting commands, this procedure will usually work correctly.

   If you have annotated your notebook with comments, be sure to include the comments within (* and *) to keep *Mathematica* from interpreting them as commands to be executed.

You may only wish to reexecute those commands where definitions are made that you will be using in additional calculations. For instance, you may only want to reexecute the command that defined quadratic. Here is how to do that.

- Click anywhere in the statement you wish to reexecute.
- Press SHIFT − RET.

   This recreates the definition.

■ 2.3 Changing *Joy*'s Settings

This section shows you how to change *Joy*'s default settings so that you can control how *Joy* works. Most of these settings are in the *Joy* ▷ **Prefs** dialog.

- From the *Joy* menus, choose *Joy* ▷ **Prefs**.

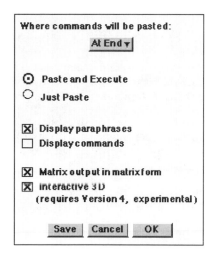

Prefs dialog, default settings.

For all preferences in this dialog:

You can make a change apply to the current session only or save it for future sessions.

- If you want to save a change for future sessions, click **Save**.
  If you change your mind after saving, click **Cancel**.

- If you are finished with the **Prefs** dialog, click **OK**. Otherwise, continue with more changes.

  ⚠ If you click **OK** without clicking **Save**, the change is only valid for the current session.

The following sections contain detailed information about these preferences.

2.3.1 Automatically Displaying *Mathematica*'s Commands

In *Joy*'s default setting each *Mathematica* command that *Joy* creates is hidden from view and a written description appears instead, like those in a mathematics textbook. Section 1.2.3 shows how to display the *Mathematica* commands one at a time. You can choose to display them automatically, if you wish. Here's how to set *Joy* to make this choice.

- If the **Prefs** dialog is not open, choose *Joy* ▷ **Prefs**.

- To display *Mathematica*'s commands, click **Display commands**.
  To turn off *Joy*'s paraphrases, *uncheck* **Display paraphrases**.

  You can display both commands and paraphrases.

- To proceed, follow the instructions at the beginning of Section 2.3.

2.3.2 How *Joy* Places Its Input to *Mathematica*

Joy's default setting is to place the *Mathematica* commands it creates at the end of your notebook. If more than one *Mathematica* notebook is open, the command goes at the end of the current notebook, i.e., the notebook that is frontmost just before you click **OK** in a *Joy* dialog.

You can change the *Joy* setting so that the command goes just after the current location of the *insertion point* or at the end of the cell containing it. The insertion point is the location in the current *Mathematica* window where any typing would go and is marked by a flashing vertical line. Here's how to change *Joy*'s default setting.

- If the **Prefs** dialog is not open, choose *Joy* ▷ **Prefs**.
- Click the popup menu **Where commands will be pasted**.

- If you want the command to appear just after the insertion point, click **In Place**.
  To make it appear after the cell containing the insertion point, click **After Cell**.
  To close the menu without making changes, click **At End**.

You can also decide whether *Mathematica* should automatically execute the command *Joy* pastes into the notebook. The default setting is for it to execute these commands each time.

Here's how to change the setting:

- If you want *Joy* to paste commands without executing them, click **Just Paste**.
- To proceed, follow the instructions at the beginning of Section 2.3.

2.3.3 Interactive 3D Graphics (Version 4 only)

Mathematica 4.0 includes the ability to modify 3D graphics interactively, although on an experimental basis. That is, interactive 3D graphics may work differently in later versions of *Mathematica*. *Joy* supports this interactive mode if you are using Version 4, and Section 10.2 shows how to manipulate 3D graphics in either the interactive or noninteractive mode.

Since the interactive mode is experimental, *Joy* allows you to turn it on and off at will. It is turned on by default. Here's how to change this setting:

- If the **Prefs** dialog is not open, choose *Joy* ▷ **Prefs**.

- Check or uncheck **Interactive 3D** as appropriate.

   This choice is grayed out in Version 3.

- To proceed, follow the instructions at the beginning of Section 2.3.

Changing the setting does not affect existing 3D graphs but determines whether new graphs will be interactive or not.

2.3.4 Matrix Form

Mathematica represents matrices internally as lists of lists, e.g.,

$$\begin{pmatrix} 1 & 2 & 3 \\ 4 & 1 & 6 \\ 7 & 8 & 1 \end{pmatrix} = \{\{1, 2, 3\}, \{4, 1, 6\}, \{7, 8, 1\}\}.$$

In *Joy*'s default setting, *Mathematica* will display matrix output in rectangular form. You can change the setting to display matrices in list form instead.

- If the **Prefs** dialog is not open, choose *Joy* ▷ **Prefs**.
- Check or uncheck **Matrix output in matrix form** as you wish.
- To proceed, follow the instructions at the beginning of Section 2.3.

2.3.5 Opening *Mathematica*'s Palettes

Mathematica has its own palettes of symbols and buttons. *Joy* closes these by default and opens its own palette instead. Here's how to open a *Mathematica* palette.

- From the *Mathematica* **File** menu, choose **Palettes** and select a palette.

Here's how to set *Joy* so that in future sessions it will automatically open the *Mathematica* palettes you chose.

- From the *Joy* menus, choose *Joy* ▷ **Startup Prefs**.
- *Uncheck* **Close *Mathematica* palettes**.

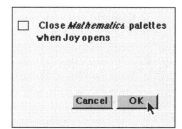 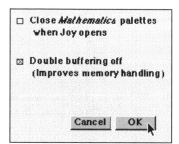

Startup Prefs dialog, Versions 3 (left) and 4 (right) .

- Click **OK**.

  ⚠ When *Mathematica* launches, it automatically opens only those of its palettes that were open in the last session. So if you use *Joy* in its default mode, where these palettes are closed, and later launch *Mathematica* without *Joy*, you will need to open any *Mathematica* palettes manually using **Palettes** in the **File** menu.

■ 2.4 On-line Help

The *Joy* software includes several forms of assistance. The *Joy* palette contains buttons to enter mathematical expressions and to show you the correct syntax for creating them from the keyboard. This section describes *Joy*'s automatic error checking and *Mathematica*'s own help tools.

2.4.1 If *Joy* Diagnoses a Syntax Error

If you make an error in notation in a *Joy* dialog, *Joy* will try to diagnose your mistake and suggest a correction.

For example, if you enter `sin[x]` instead of `Sin[x]` in **Graph ▷ Graph f(x)**, you will see this:

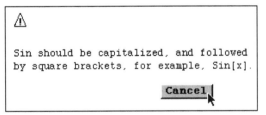

- Click **Cancel** to return to the *Joy* dialog so you can correct the error.

  Some *Joy* error messages include a **Continue** button to use if you believe you filled in the dialog correctly and want to continue.

If you enter an expression with unmatched parentheses or brackets, say, `a(b+c]` instead of `a(b+c)` in **Algebra ▷ Simplify**, this message may appear:

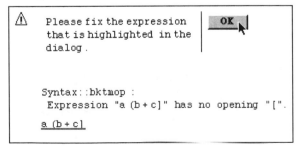

- Click **OK** to return to the *Joy* dialog and fix the mistake.

2.4.2 If *Mathematica* Beeps While You Are Typing

If you enter an incorrect expression in a *Joy* dialog and *Joy* does not catch the error first, or if you enter it directly in your *Mathematica* notebook, *Mathematica* may beep as a warning. If you do not see an error, choose **Why the Beep?** from *Mathematica*'s **Help** menu. This will display a message that you can use to diagnose the problem. For example, you may see this *Mathematica* message:

There are unmatched brackets, parentheses, or list brackets in the expression.

2.4.3 *Mathematica*'s Help Browser

Mathematica includes its own on-line help that you can use for topics that are beyond the scope of this book, such as *Mathematica* commands or options that *Joy* doesn't include. Choose **Help...** under *Mathematica*'s **Help** menu to open the Help Browser. If you know the command or syntax that you want help with, type it in the input field and press the **Go To** button. Otherwise, click **Master Index** and type the beginning of a word that best describes what you are looking for. Click on one of the words that appears and *Mathematica* will display information about it. The appearance of the Help Browser may depend on the version of *Mathematica* you are using.

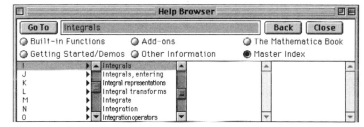

Click **Getting Started/Demos** for help with using the Help Browser and for an introduction to *Mathematica*. The Help Browser also contains the text of *The Mathematica Book* (Stephen Wolfram, 3rd ed., Wolfram Media/Cambridge Univ. Press, 1996; 4th ed., 1999).

■ 2.5 Troubleshooting

In solving problems that can arise, it helps to understand that *Mathematica* consists of two separate programs, the "front end" and the "kernel." Some problems can be traced to the front end and others to the kernel. The front end is the *Mathematica* interface that you see, and normally you do not need to deal directly with the kernel.

2.5.1 Startup Problems

Starting up with the right file. To start *Joy*, double-click *Joy of Mathematica.nb*.

♡ In some Windows systems, this file may simply appear as *Joy of Mathematica*.

Joy of Mathematica.nb

Finding files at startup or later. If certain files are not in their expected locations, *Joy* will display an error message. *Joy* requires that the installation files

Joy of Mathematica.nb

JoyNotebooks   JoyPackages

remain in the *Joy of Mathematica* folder and not be renamed. *Joy* also requires that the *Mathematica* front end and kernel be on the same computer.

If Mathematica can't start or connect to the kernel. This can occur if the front end crashed earlier but the kernel remained open, and you restarted *Joy* or *Mathematica*.

- Quit from the front end by choosing **Quit** (Macintosh) or **Exit** (Windows) from *Mathematica*'s **File** menu.
- To learn if the kernel is open and how to quit from it, follow the instructions in Section 2.5.4 (***If the front end quits and the kernel remains open***).

2.5.2 Undo and Cancel

If you make a mistake while typing, you can click the **Undo** button on the *Joy* palette. You can also press ⌘-Z (Macintosh) or CTRL-Z (Windows).

Undo button.

Undo removes what you have just typed in a *Joy* dialog or *Mathematica* notebook and restores any previous contents. However, it doesn't undo the results of any evaluations of expressions in *Mathematica*.

If you realize you've made a mistake but have already clicked **OK** or pressed SHIFT – RET, you may be able to cancel a command before *Mathematica* finishes executing it. Press ⌘ – **.** (Macintosh) or ALT – **.** (Windows), i.e., ⌘ or ALT and a period simultaneously. You may need to do this several times.

It often doesn't matter whether you cancel a command, since you can just correct it and execute it again. Sometimes, though, it takes *Mathematica* a long time to execute the command and it's more useful to cancel it.

For some commands, canceling may not interrupt *Mathematica* for a very long time. In this case, you can either let *Mathematica* work or you can save the notebook, quit *Mathematica*, and restart.

2.5.3 Operating Problems

If Mathematica beeps for no apparent reason.

- Choose **Why the Beep?** from *Mathematica*'s Help menu.

  Mathematica's message may say there is an internal problem with line breaks. This is harmless. Sometimes making your notebook's window wider will eliminate these messages.

If you type in a Mathematica notebook but nothing appears. On some platforms, you may need to click in the notebook before typing even though a horizontal line appears.

If the OK button doesn't work (Windows and Mathematica 3 only). If the **OK** button in a dialog doesn't work, click the **Example** button or else click **Cancel** and **Last Dialog**. This behavior can occur if you close a *Joy* error message by clicking the close box instead of one of the buttons.

If Mathematica operates slowly. This can happen if your computer doesn't have enough memory (RAM). Section 2.5.5 discusses managing the memory you have more effectively. You may also want to install more physical RAM in your computer. If *Mathematica* slows down mainly when creating graphics, consider adding more video RAM.

Macintosh: If your System Folder contains the extension KeyAccess, make sure you have version 5.0.6.1 or higher. Earlier versions can slow down *Mathematica*.

If there are stray characters that can't be erased or there is "ghosting" in a Joy dialog. This is a harmless cosmetic issue that will disappear when your computer screen contents are redrawn, e.g., when another window covers part of the dialog. Here is one way to redraw the screen immediately.

- Click the **Last Dialog** button to restore an open dialog to its prevous state.

- If the problem remains, drag your notebook window over the part of the screen that needs redrawing and release the mouse button. Then drag it back to where it was.

If the Joy palette disappears. This happens if you inadvertently close the palette.

- Choose *Joy* ▷ **Palette**.

If the Joy menu bar disappears. This happens if the menu bar is covered by other windows or (Windows, *Mathematica* 3 only) if you inadvertently close the menu bar.

To check if the menu bar is present and make it visible:

- In *Mathematica*'s **Window** menu, choose *Joy* **Menu** if it is present.

To reopen the menu bar if it has been closed (*Windows, Mathematica 3 only*):

- Reopen *Joy of Mathematica.nb*.

If blue error messages appear within a Joy dialog. You can clean these up by clicking the **Example** button.

If Mathematica says it can't find the JISFontMapping file. This seems to be harmless, but deleting *Mathematica*'s preferences may eliminate the message. *Mathematica* will create fresh preferences files automatically.

- See Section 2.5.4 for how to delete the preferences (***If Mathematica crashes frequently***).

2.5.4 If *Mathematica* Quits Suddenly or Hangs Up

Any computer program may crash if it runs out of memory or if a file is temporarily corrupted. Restarting the program often resolves the problem. Sometimes you may need to restart your computer. Here are some types of crashes and how to handle them.

If Mathematica hangs up during a graph (Macintosh and Mathematica 3 only). This can happen if no printer is selected. Select a printer in the Chooser (**Apple** menu).

If the front end and kernel both quit.

- Restart *Joy*. If this doesn't help, then you should restart your computer.

If the kernel hangs up and the front end is open. Sometimes *Mathematica* seems to hang up in the middle of executing a command.

- Try to cancel the command as shown in Section 2.5.2. You may need to do this several times.

  If *Mathematica* doesn't cancel the command, then the front end may no longer be able to communicate with the kernel and you will have to quit them separately.

- From *Mathematica*'s **File** menu, choose **Quit** (Macintosh) or **Exit** (Windows). This will make the front end quit.
- Quit the kernel as shown immediately below and restart *Joy*.

If the front end quits and the kernel remains open.

Macintosh: The kernel is open if *MathKernel* appears in the **Applications** menu in the upper-right corner of the screen.

- Choose *MathKernel* from the **Applications** menu.

- Press ⌘-OPTION-ESC.

    A message appears asking if you want to force *MathKernel* to quit.

- Click **OK**.
- Restart *Joy*.

Windows: The kernel is open if *Mathematica Kernel* appears in the task bar at the bottom of the screen.

- Choose *Mathematica* Kernel from the task bar.
- From the Kernel's **File** menu, choose **Exit**.
- Restart *Joy*.

If the kernel quits and the front end is open. A message may appear saying that the kernel has quit. Another symptom is that the *Joy* menu buttons and some of the palette buttons no longer work.

- From *Joy*'s menus, choose *Joy* ▷ **Close *Joy***.
- From *Mathematica*'s **File** menu, choose **Quit** (Macintosh) or **Exit** (Windows).
- Restart *Joy*.

If Mathematica crashes frequently. One likely cause is that *Mathematica* has insufficient memory. Try this first, and then see Section 2.5.5.

- Before opening *Joy* or *Mathematica*, close any open applications.

Another possible cause of frequent crashes, especially if they occur soon after starting, is that *Mathematica*'s preferences files for the front end may have become corrupted. Deleting them may alleviate this problem. *Mathematica* will create fresh preferences files automatically.

⚠ These instructions are for *Mathematica*'s default installation. If this is not your own computer, check with your instructor, the owner, or the system adminstrator, since these files may have been customized for your location. You may be asked for the *Mathematica* password.

- If *Joy* and/or *Mathematica* is running, quit them.
- Follow these steps for your computer and then restart *Joy*.

Macintosh:

- Open the Preferences folder within the System Folder.
- Find the *Mathematica* folder and rename it Old *Mathematica*.

Windows:

- Open \ProgramFiles\Common Files*Mathematica*\3.0 (or 4.0 for that version).
- Find the FrontEnd folder and rename it Old FrontEnd.

2.5.5 Managing *Mathematica*'s Memory

Some *Mathematica* commands require a lot of memory (RAM). Exact (rather than numeric) integration, three-dimensional graphics, and exact solutions of differential equations are particularly memory-intensive. To protect against running out of memory and crashing, you should save your work regularly.

If *Mathematica* seems to be slowing down, it may be getting short of memory. There are some things you can do to increase the amount of memory that is available while you are working:

Clear the clipboard. If you have recently cut or copied something that takes a lot of memory, you can clear most of the clipboard by copying a single character from your notebook.

Close notebooks you are not using. Notebooks that you are not using take up memory unnecessarily.

Save a notebook you want to continue using. If that doesn't work, close and reopen the notebook. Saving a notebook often releases memory, even when the notebook stays open. This is particularly true when the notebook contains 3D graphics. If you need to close and reopen the notebook, all your previous work will still be in *Mathematica*'s memory so you won't need to redo any calculations, but the memory used to display the notebook will be freed up. Here's how to do that.

- Choose **Close** from *Mathematica*'s **File** menu, or press ⌘-W (Macintosh) or [CTRL]-F4 (Windows).
- To reopen a notebook, choose **Open** from *Mathematica*'s **File** menu, or press ⌘-O (Macintosh) or [CTRL]-O (Windows).
- See Section 2.2.1 for how to use a notebook that you reopen in the same session.

Share subexpressions in memory. This tells *Mathematica* to use the same area of memory for identical parts of expressions in your notebook.

- Click in your *Mathematica* notebook beneath the last output, so that a horizontal line appears.
- Type Share[].
- Press [SHIFT] − [RET].

Save, quit, and start over. This is a more drastic solution. See Section 2.2.2 for how to use a notebook that you reopen in a different session. If you can give *Mathematica* more memory on your computer, it's a good idea to do it now.

If you frequently run short of memory, you can make sure *Mathematica* gets all the memory that has been allocated to it. Here are some ways to do that before starting *Joy* or *Mathematica*:

Close applications that are open before starting Joy. Applications that are open unnecessarily will restrict the amount of memory available to the operating system and may do the same for *Mathematica*.

Reduce the number of colors used by your monitor. Using many colors, especially on a large monitor, adds to the memory needed by the operating system. On the Macintosh, you can do this while *Mathematica* is running.

Memory gauges (Macintosh only). On the Macintosh, the front end and kernel have memory "thermometers" that show at a glance what fraction of the memory set aside for *Mathematica* is free.

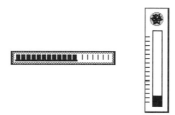

Front end (left) and kernel (right) memory thermometers, Macintosh only.

♡ Front end memory appears at the bottom of your notebook. To display kernel memory, choose *MathKernel* from the Macintosh **Applications** menu at the upper right of the screen. Then choose **Show Memory Usage** from the kernel's **File** menu.

When you see the thermometers filling up, try some of the techniques outlined above.

Give Mathematica more memory before it starts (Macintosh only). On the Macintosh, you can set the amount of memory assigned to an application before you open it.

Mathematica's front end requires more memory for extensive graphics and animation. The kernel requires more memory for large calculations such as exact integration and exact solutions of differential equations. You can adjust the memory for the kernel and front end separately according to your needs.

• In the Finder, find the *Mathematica* application.

   If the *Mathematica* name on the application icon is in *italics*, this is an "alias" for the application itself. Choose **Show Original** from the Finder's **File** menu (or press ⌘-R) to find the application.

• Select the *Mathematica* icon.

   This is *Mathematica*'s front end.

• Choose **Get Info** from the Finder's **File** menu (or press ⌘-I).
• If you wish, increase the **preferred size** of the front end.

   You may need to experiment to find a size that lets the front end work properly but leaves enough memory for the kernel and the operating system.

Setting the memory size, Macintosh only.

- Repeat for *MathKernel* if you wish to increase the kernel's memory size.

Double Buffering (Mathematica Version 4 only). Double Buffering is an option in *Mathematica* Version 4 that improves the screen display by reducing flicker but takes more memory. On the Macintosh this option sometimes seems to lead to the front end running out of memory unexpectedly, even though the memory thermometers are not full. In this case, the following message appears:

> There was not enough memory to display a cell. The cell has
> been closed. When you have freed some memory, you can open
> it by choosing Cell Open in the Cell Properties submenu.

To avoid this problem, *Joy* turns double buffering off for *Mathematica* Version 4 on the Macintosh. If you get the above message, here's how to make sure this setting has not been changed. This dialog appears for both Macintosh and Windows with Version 4.

- In **Joy ▷ Startup Prefs**, check **Double buffering off**.

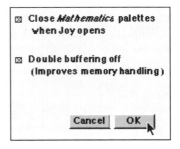

Startup Prefs dialog (Version 4 only).

Part II

How to Use *Joy* for …

Chapter 3
Graphing in Two Dimensions

This chapter includes *Joy* dialogs for graphing curves and plotting points in two dimensions, including parameterized curves, implicitly defined functions, and curves in polar coordinates. It also shows how to identify the coordinates of a point in a graph, combine several graphs, and create animations. Section 3.7 shows how to change various features of a graph, such as its size, where the axes cross, and how the axes should be labeled. Chapter 10 discusses graphing in three dimensions.

■ 3.1 How to Graph $y = f(x)$

For example, here's how to graph $y = 3\,x\,e^{-2x}$ for $-2 \le x \le 2, \ -3 \le y \le 3$.

- Choose **Graph ▷ Graph f(x)**.

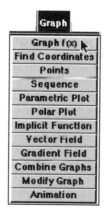

Joy's **Graph** menu.

- Fill in the function $3 \ \text{x}\_\text{E}^{-2\,x}$, where $\_$ means to leave a blank space to indicate multiplication.

43

If you are graphing more than one function, use commas to separate them. Section 1.2.1 illustrates graphing two functions together.

- Fill in the *x*-interval $-2 \le x \le 2$.
- Click the triangle ▼ for the *y*-interval to open the popup menu.
- Click **Specify**.
- Fill in $-3 \le y \le 3$.

Plot the function(s)

$$3 \, x \, E^{-2 \, x}$$

x – interval : $-2 \le x \le 2$

y – interval :   **Specify : ▼**

$$-3 \le y \le 3$$

☐ **Equal scales**
☐ **Axes at origin**
☐ **Legend**
☐ **Assign name**

`graph1`

Example **Cancel** **OK**

Graph f(x) dialog.

- Click **OK**.

Mathematica creates the output:

▷ Graph $3 \, x \, E^{-2 \, x}$ on $-2 \le x \le 2$ and $-3 \le y \le 3$.

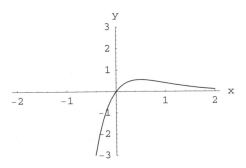

Out[2]= - Graphics -

■ 3.2 How to Graph Implicit Functions

Sometimes a curve, such as $x^2 + 3xy + 4y^2 = 4$, is not the graph of a function but instead defines two or more functions *implicitly*. Here is how to graph this kind of curve. We'll let x range between ±4.

- From the *Joy* menus, choose **Graph ▷ Implicit Function**.
- Fill in the equation $\mathbf{x}^2 + 3\ \mathbf{x\_y} + 4\ \mathbf{y}^2 = 4$

   The symbol ＿ means to leave a blank space to indicate multiplication.

You can choose two kinds of implicit plots in *Joy*. In this case, we choose the first kind, since *Mathematica* can solve for y in terms of x.

- Make sure the radio button for **solving for one variable in terms of the other** is selected.

   You can only fill in the input field for the button that is selected. The other field is "grayed out."

- Fill in the interval $-4 < \mathbf{x} \le 4$.

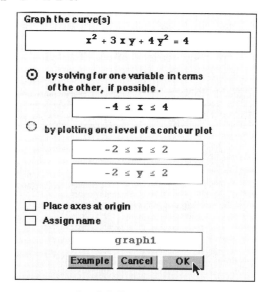

Implicit Function dialog.

- Click **OK**.

Mathematica graphs the curve:

▷ Graph of $x^2 + 3xy + 4y^2 = 4$ for $-4 \le x \le 4$.

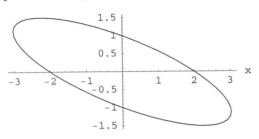

Out[2]= - Graphics -

⚠ *Joy* automatically clears any values that might earlier have been assigned to variables in an implicit equation. If you gave values to these variables, they will be lost when you graph the function.

The other kind of plot is good for equations where we cannot solve explicitly for all the values of *y*, such as $x^2 + \sin y = 1$. Here's how to use it.

- Click the **Last Dialog** button on the *Joy* palette to bring back the **Implict Function** dialog.
- Fill in the equation $x^2 + \text{Sin}[y] = 1$.
- Click the radio button for **plotting one level of a contour plot**.
- Fill in the intervals $-2 \le x \le 2$, $-4 \le y \le 8$.

Graph the curve(s)

$$x^2 + \text{Sin}[y] = 1$$

○ by solving for one variable in terms of the other, if possible.

$$-4 \le x \le 4$$

◉ by plotting one level of a contour plot

$$-2 \le x \le 2$$

$$-4 \le y \le 8$$

- Click **OK**.

The result is:

▷ Graph of $x^2 + \text{Sin}[y] = 1$ for $-2 \le x \le 2$, $-4 \le y \le 8$.

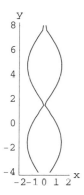

Out[3]= - ContourGraphics -

The first method would have given only part of this curve, without showing the repeating values of *y*.

The first method is the equivalent of graphing $y = \arcsin(1 - x^2)$, which only includes the values of *y* between $\pm\pi/2$.

3.2.1 Graphing Multiple Implicit Functions

You can graph more than one implicit function at a time by separating their equations with commas. Here is an example illustrating the graphs of $x^y + y = 2$, $x^y + y = 3$ for $0 \le x, y \le 4$.

- Click **Last Dialog** to reopen the **Implicit Function** dialog.
- Fill in $\mathbf{x^y + y = 2}$, $\mathbf{x^y + y = 3}$.
- Make sure **plotting one level of a contour plot** is selected.
- Fill in the intervals $0 \le \mathbf{x} \le 4$ and $0 \le \mathbf{y} \le 4$.

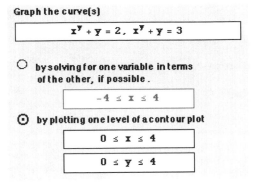

Graph the curve(s)

$$x^y + y = 2, \quad x^y + y = 3$$

○ **by solving for one variable in terms of the other, if possible.**

$$-4 \le x \le 4$$

⊙ **by plotting one level of a contour plot**

$$0 \le x \le 4$$

$$0 \le y \le 4$$

- Click **OK**.

▷ Graph of $\{x^y + y = 2,\ x^y + y = 3\}$ for $0 \le x \le 4$, $0 \le y \le 4$.

$x^y + y = 2$	Black	Solid
$x^y + y = 3$	Red	.....

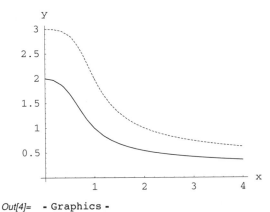

Out[4]=   - Graphics -

■ 3.3 How to Plot Points

The first example in this section shows how to plot points that are listed explicitly.

- From the *Joy* menus, choose **Graph ▷ Points**.
- The default example plots the points $\{3, 5\}$, $\{5, 3\}$, $\{2, 7\}$.

    Each point is in the form $\{x, y\}$, using curly brackets, and commas separate the points to be plotted.

Plot the points in the list

> {3, 5}, {5, 3}, {2, 7}

☐ **Position axes at the origin**

☐ **Assign name**

> graph1

[Example] [Cancel] [OK]

Plot Points dialog.

- Click **OK**.

Here is the plot:

▷ Plot of the points {{3, 5}, {5, 3}, {2, 7}}

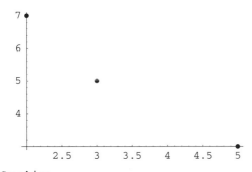

Out[2]= - Graphics -

Notice that *Mathematica*'s axes cross at $x = 2$, $y = 3$. *Mathematica* generally chooses the axes to emphasize the main features of the graph. To make the axes meet at $(0, 0)$, see Section 3.7.6.

3.3.1 How to Plot a List of Points

If you want to plot many points at once that are related by a formula, it is easier to do this by creating a list and using a name for the list in the **Graph ▷ Points** dialog. Here is how to plot $(n, n^2 \sin n)$, for $n = 0, 1, 2, \ldots, 15$.

- From the *Joy* menus, choose **Create ▷ List**.

Create dialog.

- Fill in n, n^2 Sin[n] for the entries in the list.
- Fill in $0 \leq n \leq 15$ in increments of 1.

    Be sure the popup menu is set to **x Equally Spaced**. The popup menu's current setting will depend on whether you have changed it during this session.

- Make sure **Display in decimal form** and **Display as a table** are checked.
- Check **Assign name**.
- Use the default name listA, or replace it with a name of your choice.

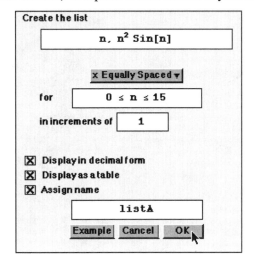

Create List dialog.

- Click **OK**.

Mathematica produces the list of points:

▷ Create the list with values {n, n^2 Sin[n]} for $0 \leq n \leq 15$.

listA =

Out[3]//TableForm=

0	0
1.	0.841471
2.	3.63719
3.	1.27008
4.	−12.1088
5.	−23.9731
6.	−10.059
7.	32.1923
8.	63.3189
9.	33.3816
10.	−54.4021
11.	−120.999
12.	−77.2665
13.	71.0082
14.	194.159
15.	146.315

♈ It's easier to read the result in table form, but if the list is long it will take up less space if you *uncheck* **Display as a table** when filling in the dialog.

Next, here is how to plot the points in the list.

- From the *Joy* menus, choose **Graph ▷ Points**.
- Fill in listA, or a name that you used instead.
- Click **Assign name** and fill in a name for the graph.

    Here we've changed the default name graph1 to graphList.

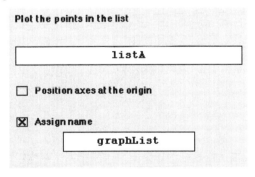

- Click **OK**.

Here is the result:

▷ Plot of the points listA

graphList =

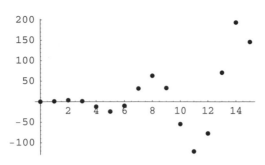

Out[4]=   **- Graphics -**

■ 3.4 How to Combine Graphs

Sometimes you may create two or more plots separately and later want to combine them in a single plot. Here's how to accomplish that. This example combines plots of the points $(n, n^2 \sin n)$, for $n = 0, 1, 2, \ldots, 15$ and of the curve $y = x^2 \sin x$, $0 \le x \le 15$.

- If you didn't plot the points $(n, n^2 \sin n)$ in 3.3.1, do that now.
- From the *Joy* menus, choose **Graph ▷ Graph f(x)**.
- Fill in x^2 **Sin[x]** and $0 \le x \le 15$.
- Click **Assign name** and fill in a name for the graph, such as **graphCurve**.

Plot the function(s)

$$x^2 \text{ Sin[x]}$$

x – interval : $0 \le x \le 15$

y – interval :   **Automatic ▾**

☐ **Equal scales**
☐ **Axes at origin**
☐ **Legend**
☒ **Assign name**

graphCurve

- Click **OK**.

The graph of this function is:

▷ Graph x^2 Sin[x] on $0 \le x \le 15$.

graphCurve =

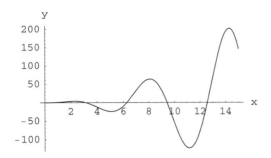

Out[5]=   - Graphics -

Now we will combine the two graphs. This example illustrates how to use output names to refer to previous output. If you didn't name the graphs, you can use their output numbers instead, e.g., `Out[4],Out[5]`. The output numbers appear just below the graphs.

- From the *Joy* menus, choose **Graph ▷ Combine Graphs**.
- Make sure the names `graphList, graphCurve` appear in the dialog.

When you open **Combine Graphs**, *Joy* automatically fills in the output names or numbers of the two most recent graphs. Using names for output rather than numbers makes it easier to reopen a notebook for later use. Section 2.2 gives more details about this.

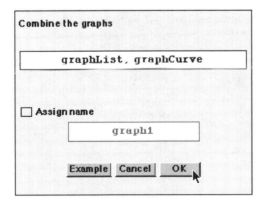

Combine Graphs dialog.

- Click **OK**.

Mathematica gives the result:

▷ Combining the graphs {graphList, graphCurve} gives

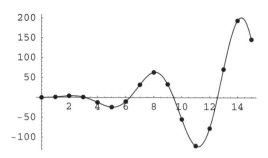

Out[6]= - Graphics -

♡ If the two graphs have different characteristics, such as axis labels, *Mathematica* uses the first graph you enter in the input field to determine the characteristics of the combined graph. If the results aren't what you want, try reversing the graphs in the **Combine Graphs** dialog. For example, entering `graphCurve, graphList` will label the axes.

■ 3.5 How to Graph 2D Parameterized Curves

Some curves are given in the form $x = f_1(t)$, $y = f_2(t)$, i.e., x and y are themselves functions of a third variable (a *parameter*) t. Here is how to graph a parameterized curve in two dimensions. We'll use the example $x = \sin t$, $y = \sin 2t$, $0 \leq t \leq 2\pi$.

- From the *Joy* menus, choose **Graph ▷ Parametric Plot**.
- Fill in `x = Sin[t]`, `y = Sin[2 t]`.
- Fill in $0 \leq t \leq 2\pi$.

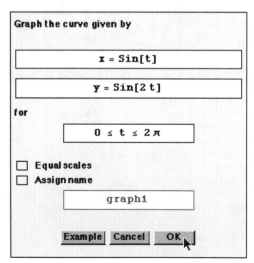

Parametric Plot dialog.

- Click **OK**.

Mathematica produces the result:

▷ Graph the curve (x, y) = (Sin[t], Sin[2 t]) for 0 ≤ t ≤ 2π.

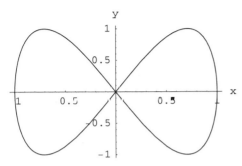

Out[2]=   - Graphics -

In this dialog, you can graph only one curve at a time. You can use **Graph ▷ Combine Graphs** (Section 3.4) to display several parametric curves together.

■ 3.6 How to Find Coordinates in a Graph

Here's how you can use *Mathematica* to estimate the coordinates of a point in a graph. Similar instructions appear when you choose **Graph ▷ Find Coordinates**.

- If you didn't graph $x^2 \sin x$ in Section 3.4, do it now.
- Click on the graph to select it.

    A rectangular frame appears around the graph.

- Hold down the ⌘ (Macintosh) or CTRL (Windows) key.
- Move the cursor around.

    As you move the cursor, its coordinates appear in the lower-left corner of your *Mathematica* notebook window.

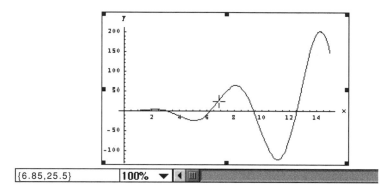

Cursor coordinates (lower left).

You can also copy and paste these coordinates into your *Mathematica* notebook:

- Click the mouse button *once*.
- Choose **Copy** from *Mathematica*'s **Edit** menu (Macintosh ⌘-C, Windows CTRL-C).
- Click where you would like to paste the coordinates into your notebook.
- Choose **Paste** (Macintosh ⌘-V, Windows CTRL-V).

Mathematica pastes the coordinates to more decimal places than it displays in the window:

    {6.85369, 25.4545}

⚠ *Mathematica* will give you the coordinates of the point where you clicked, to within the resolution of your monitor. But it is very hard to select a particular point on the graph and you should regard these values as estimates.

■ 3.7 How to Customize Graphs

3.7.1 How to Change the Size of a Graph

Sometimes you may want to make a graph appear larger or smaller without changing its relative proportions. Here's how to do that.

- If you didn't graph $x^2 \sin x$ in Section 3.4, do it now.
- Click on the graph to select it.

    A rectangular frame appears around the graph.

- Position the cursor over one of the small black "handles" at the corners and sides.

    The cursor changes to a two-headed arrow.

- Drag the frame in the direction of the arrow to make the graph larger or smaller.

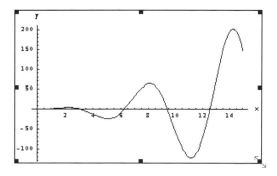

Dragging a graph by one of its corners.

3.7.2 How to Modify a Graph; Adding a Grid

After making a graph you can modify it in either of two ways:

- Redraw it by reopening the dialog and making different choices
- Choose **Graph ▷ Modify** from the *Joy* menus

The **Graph ▷ Modify** dialog includes several options that may not be present in the original graphing dialog. For example, here's how to add a grid to a graph to make it easier to estimate coordinates, slopes, etc. We'll illustrate this with the graph of $y = \sin(x^2)$.

- Choose **Graph ▷ Graph f(x)**.
- Fill in the function $\texttt{Sin}[\texttt{x}^2]$ and the interval $-2 \leq x \leq 2$.
- You may assign a name to the graph or not, as you wish.

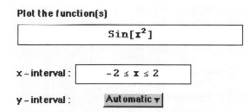

- Click **OK**.

▷ Graph $\texttt{Sin}[\texttt{x}^2]$ on $-2 \leq x \leq 2$.

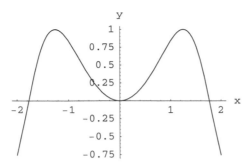

Out[2]=   - Graphics -

- Choose **Graph ▷ Modify Graph**.
- Make sure the output number of the preceding graph (in this case Out[2]), or its name if you assigned one, appears in the dialog.

    Mathematica automatically numbers its output. The number of the preceding graph or any name that you assigned to it appears by default when you choose **Modify Graph**.

- Check **Grid**.

  ☞ **Modify Graph** can be used with 2D and 3D graphs but some options, including **Grid**, apply only to graphs in two dimensions.

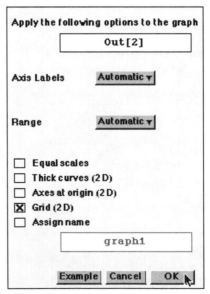

Modify Graph dialog.

- Click **OK**.

The result is (the paraphrase uses %, which is *Mathematica*'s shortcut for Out) :

▷ Redisplay %2.

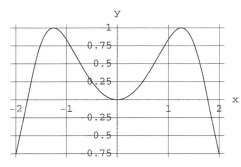

Out[3]= - Graphics -

3.7.3 How to Place Equal Scales on the Axes

Mathematica often draws graphs with different scales on the *x*- and *y*-axes. This is effective for many graphs, but it can cause some curves to appear distorted. Here is what it does to the circle $x = \cos t$, $y = \sin t$, $0 \leq t \leq 2\pi$.

- From the *Joy* menus, choose **Graph ▷ Parametric Plot**.
- Fill in the coordinates x = Cos[t], y = Sin[t], with 0 ≤ t ≤ 2 π.

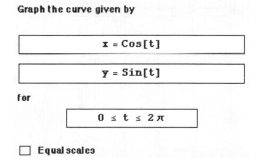

- Click **OK**.

The circle looks like an ellipse:

▷ Graph the curve (x, y) = (Cos[t], Sin[t]) for 0 ≤ t ≤ 2 π.

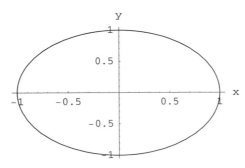

Out[4]=   - Graphics -

Here's how to set the the *Joy* dialog to make the curve look like a true circle.

- Click the **Last Dialog** button.
- Check **Equal scales**.

⊠ **Equal scales**

- Click **OK**.

Mathematica now draws a true circle.

▷ Graph the curve (x, y) = (Cos[t], Sin[t]) for 0 ≤ t ≤ 2π.

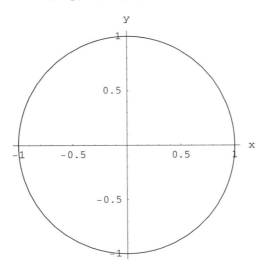

Out[5]=   - Graphics -

Another way. Several *Joy* dialogs include the **Equal scales** option, and you can choose the option when you create the graph initially if you want it to have this feature. For other dialogs, you must create the graph first and then use **Graph ▷ Modify Graph** to make the scales equal.

☿ If curves or lines that should be perpendicular to each other don't meet at right angles, try modifying the graph to make the scales equal.

3.7.4 How to Change the Range in a Graph

In some *Joy* graphing dialogs, *Mathematica* determines the range of a graph automatically. In others, you either may or must specify the range yourself. After you see a graph, you may decide to zoom in or out. For example, Section 1.2.2 (page 8) shows how to zoom in by decreasing the domain and range in the dialog **Graph ▷ Graph f(x)**.

Showing the entire range. When *Mathematica* determines the range automatically, it may omit some points in order to emphasize what may be more essential features. Here is how to display all points with x in the given domain. First we graph $y = x^3$ and $y = 3x - 2$ for $-5 \le x \le 5$.

- From the *Joy* menus, choose **Graph ▷ Graph f(x)**.
- Fill in the functions x^3, $3x - 2$ and the interval $-5 \le x \le 5$.

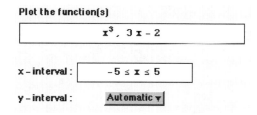

Plot the function(s)

x^3, $3x - 2$

x – interval : $-5 \le x \le 5$

y – interval : Automatic ▼

- Click **OK**.

The graph is:

```
▷ Graph

x³          Black       Solid
3 x - 2     Red         .....

on -5 ≤ x ≤ 5.
```

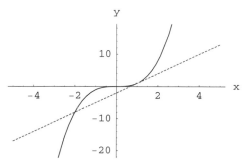

Out[6]= - Graphics -

Mathematica has cut off points on the cubic beyond about $y = \pm 20$. Now:

- Click the **Last Dialog** button to bring back the **Graph f(x)** dialog.
- Click the *y*-interval popup button and choose **Show All**.

x – interval : | $-5 \leq x \leq 5$ |

y – interval :   **Show All ▼**

- Click **OK**.

Mathematica now includes all the *y*-values it has computed for the given *x*-values but shows less detail near the *x*-axis:

▷ Graph

| x^3 | Black | Solid |
| $3\,x - 2$ | Red | . . . . . |

on $-5 \leq x \leq 5$.

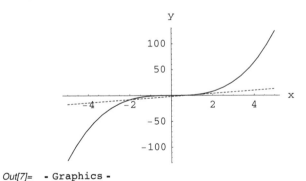

Out[7]= - Graphics -

When the dialog doesn't include an option for setting the range, you must create the graph first and then use **Graph ▷ Modify Graph** to choose **Show All**:

Apply the following options to the graph

Out[6]

Axis Labels   Automatic ▾

Range
Automatic
Specify :
Show All

Choosing **Show All** in **Modify Graph**.

Showing part of the range. Sometimes you may want to zoom in and show only part of the range. In dialogs where you cannot specify the range directly, you can use **Graph ▷ Modify Graph**. Here is an illustration using the parametrized curve $x = \sin t$, $y = \ln t$, $1 \le t \le 100$.

- From the *Joy* menus, choose **Graph ▷ Parametric Plot**.
- Fill in $x = \mathtt{Sin[t]}$, $y = \mathtt{Log[t]}$.
- Fill in $1 \le \mathtt{t} \le 100$.
- If necessary, *uncheck* **Equal scales**.

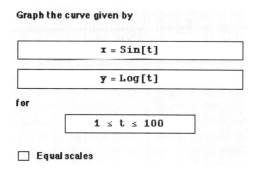

Graph the curve given by

x = Sin[t]

y = Log[t]

for

1 ≤ t ≤ 100

☐ **Equal scales**

- Click **OK**.

Mathematica draws the graph:

▷ Graph the curve $(\mathtt{x}, \mathtt{y}) = (\mathtt{Sin[t]}, \mathtt{Log[t]})$ for $1 \le \mathtt{t} \le 100$.

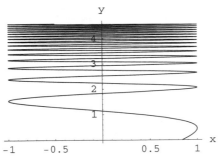

Out[8]= **- Graphics -**

Here's how to zoom in by showing only $4 \le y \le 5$.

- Choose **Graph ▷ Modify Graph**.
- Make sure the dialog shows the output name or number of the graph whose range you want to specify, in this case Out[8].
- Click the **Range** popup menu and choose **Specify**.
- Fill in $4 \le$ **height** ≤ 5.

    You can also fill in $4 \le$ y ≤ 5.

- If necessary, *uncheck* **Grid**.

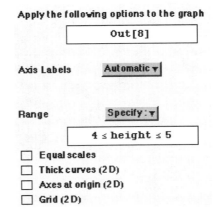

Specifying a particular range in **Modify Graph**.

The result is (**%** is *Mathematica*'s abbreviation for Out) :

▷ Redisplay **%**8 for 4 $\le$ height $\le$ 5.

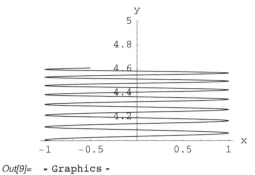

- Graphics -

⚠ Since you are only modifying how an existing graph is displayed, **Modify Graph** can only show points that *Mathematica* has already computed. Here only those points with $4 \le y \le 5$ where $1 \le t \le 100$ appear, not all points with $4 \le y \le 5$.

3.7.5 How to Specify Axis Labels

In most graphing dialogs, *Joy* automatically labels the axes according to the variables used in the equations being graphed. In some cases (e.g., plotting points and graphing in polar coordinates) the variables aren't apparent and *Joy* omits the labels. In other cases, the variables may stand for physical quantities and you may want to label the axes accordingly. Here's how to specify any labels you choose.

- From the *Joy* menus, choose **Graph ▷ Graph f(x)**.
- Fill in the function $-16 \, t^2 + 25 \, t + 150$ and the interval $0 \le t \le 4$.

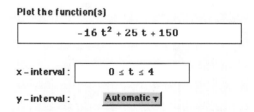

- Click **OK**.

The result is:

▷ Graph $-16\,t^2 + 25\,t + 150$ on $0 \le t \le 4$.

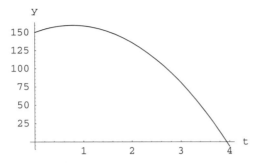

Out[10]=   ‑ Graphics ‑

- Choose **Graph ▷ Modify Graph**.
- Make sure the dialog contains the name or output number of the graph whose labels you want to change, in this case `Out[10]` (your output number may be different).
- Click on the **Axis Labels** popup menu and choose **Specify**.
- Fill in `"time","height"`.

  It's good practice to include quotation marks `" "` around each label. This guarantees that the text will appear in the label rather than any value you may have assigned to it.

- If necessary, change the **Range** to **Automatic**.

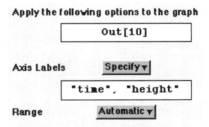

Specifying axis labels in **Modify Graph**.

- Click **OK**.

Mathematica redraws the graph with the new labels (`%` stands for the last output):

▷ Redisplay %10.

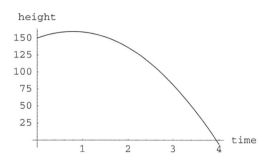

Out[11]= - Graphics -

3.7.6 How to Make the Axes Cross at the Origin

Depending on where the points in a plot are located, *Mathematica* may automatically set the axes to cross at a point other than the origin. Here we illustrate this and then show how to make them cross at (0,0). We use the example $y = 2 + \sin x$ with $-5 \le x \le 5$.

- From the *Joy* menus, choose **Graph ▷ Graph f(x)**.
- Fill in 2 + Sin[x] and $-5 \le x \le 5$.
- Make sure the popup menu for the *y*-interval is set to **Automatic**.

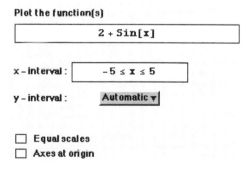

- Click **OK**.

In the resulting graph, the axes cross at (0, 1):

▷ Graph 2 + Sin[x] on −5 ≤ x ≤ 5.

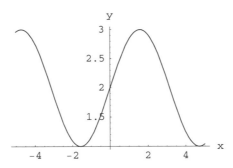

Out[12]= - Graphics -

- Click the **Last Dialog** button on the *Joy* palette to bring back the **Graph f(x)** dialog.
- Check **Axes at origin**.

⊠ **Axes at origin**

- Click **OK**.

Mathematica now draws the graph with the axes starting at the origin, although it leaves a gap in the *y*-axis.

▷ Graph 2 + Sin[x] on −5 ≤ x ≤ 5.

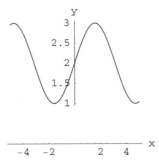

Out[13]= - Graphics -

⚠ **Axes at origin** can cause the graph to be compressed into a very small space if it contains points that are far from the origin. Depending on the curve, you may instead get better results by graphing *y* = 0 and your function simultaneously, perhaps setting the **Range** to **Show All**. This should remove the gap in the *y*-axis.

When the dialog doesn't include an **Axes at origin** option, create the graph first and then use **Graph** ▷ **Modify Graph** to put the axes' origin at (0, 0).

3.7.7 How to Add a Legend to a Graph

When you graph several functions simultaneously, *Joy* makes it easier to distinguish the curves in the output by displaying them in different colors and patterns. By default, these styles are displayed in the paraphrase that *Joy* includes with the graph. Instead, you can choose to have them appear in a legend alongside the graph. This can be especially useful if you choose to display *Mathematica*'s commands in place of the paraphrase.

Here we show how to add a legend to the graphs of $\sin x$, $\sin(x^2)$, and $\sin(x^3)$ for $-2 \le x \le 2$.

- From the *Joy* menus, choose **Graph ▷ Graph f(x)**.
- Fill in the functions `Sin[x]`, `Sin[x²]`, `Sin[x³]` and the interval $-2 \le x \le 2$.
- Check **Legend**.

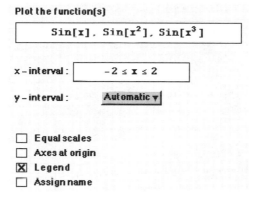

- Click **OK**.

The result is:

▷ Graph {Sin[x], Sin[x²], Sin[x³]} on $-2 \le x \le 2$.

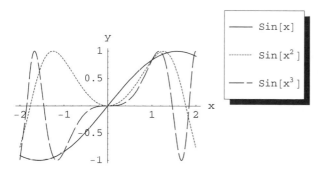

Out[14]= - Graphics -

⚠ Adding a legend to a graph reduces the amount of space available for the graph and may make it hard to read. Also, the expressions in the legend may extend beyond the legend box. You can remedy both of these problems when they occur by enlarging the graph and legend. Click anywhere within the output and drag it to a larger size by the one of the small black "handles" that appear around it. Section 3.7.1 illustrates this procedure.

3.7.8 How to Set Styles for Graphs with Several Curves

When a graph contains more than one curve, *Joy* automatically assigns different styles to the curves to make it easy to tell them apart. Here's how you can change the styles that *Joy* assigns. (This does *not* apply to curves that are combined in **Graph ▷ Combine Graphs**.)

- Choose *Joy* ▷ **Graph Styles**.

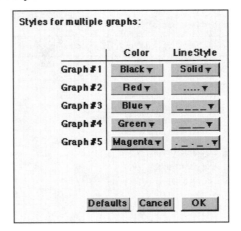

Graph Styles dialog.

- Use the popup menus to reassign the colors and styles as you wish.

  These changes will apply for the remainder of the session. To make them permanent, choose **Save** from *Mathematica*'s **File** menu.

- When you are done, click **OK**.

  ♡ To restore Joy's original settings, click **Defaults**.

■ 3.8 How to Graph in Polar Coordinates

Here is how to graph a curve whose equation is given in the form $r = f(\theta)$, where r and θ are polar coordinates. We use the example $r = \theta \cos \theta$, $0 \le \theta \le 4\pi$.

- From the *Joy* menus, choose **Graph ▷ Polar Plot**.
- Fill in the equation `r = θ_Cos[θ]`.

To enter Greek letters, use the **Greek** button on the *Joy* palette. The symbol ⎯ means to leave a space to indicate multiplication.

- Fill in the interval $0 \le \theta \le 4\,\pi$.

    You can also enter π using its own button on the *Joy* palette.

Plot the graph (in polar coordinates) of

r = θ Cos[θ]

for   0 ≤ θ ≤ 4π

☐ Assign name

graph1

[Example] [Cancel] [OK]

Polar Plot dialog.

- Click **OK**.

Here is the result:

▷ Graph of the function given in polar coordinates by $r = \theta \, Cos[\theta]$ for $0 \le \theta \le 4\,\pi$.

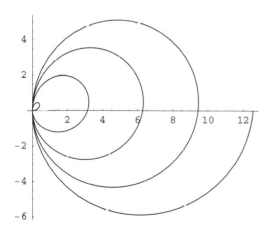

Out[2]= - Graphics -

When you are done with the **Greek** palette, here's how to close it:

- (Macintosh) Click the close box in the *upper-left corner of the palette*.

- (Windows) Click the close button X in the *upper-right corner of the palette*.

■ 3.9 How to Animate Graphs

Animation is an excellent way to understand change and motion. This section shows how to animate a sequence of graphs. The example illustrates the effect of the parameter k on the graph of $\sin kx$. We'll use *Joy* to graph $\sin kx$ for $k = 1, 2, 3, \ldots, 10$ and x between ±6. We'll then view the 10 graphs as successive frames, just as in a cartoon animation.

- From the *Joy* menus, choose **Graph ▷ Animation**.

   The popup menu shows **Graph f(x)**, which is the type of graph we will be animating.

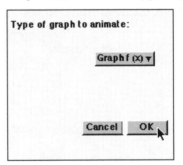

- Click **OK**.

   ♡ To animate a different type of graph, choose it from the popup menu. Click your choice to close the popup menu, even if you are not changing the type of graph to animate.

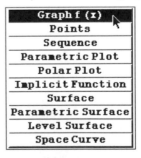

Animation popup menu

Two dialogs appear, one to specify the parameters of the animation and the other to create the graphs.

- In the **Graph f(x)** dialog (on the right), fill in `Sin[k__x]`.

    The symbol __ means to leave a space to indicate multiplication.

- Fill in the interval $-6 \le x \le 6$.
- If necessary, *uncheck* **Legend**.
- Click in the **Animation Info** dialog (on the left) to make it active.
- Fill in $1 \le k \le 10$.

    This specifies the 10 frames.

- For the common plot frame, fill in $-6 \le x \le 6$ and $-1 \le y \le 1$.

    All the graphs will be drawn in this frame. In this example, we happen to use the same *x*-values in both dialogs.

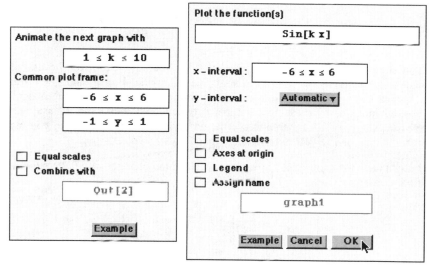

Animation Info (left) and **Graph f(x)** (right) dialogs.

- Click in the **Graph f(x)** dialog to make it active.
- Click **OK**.

Mathematica creates the 10 frames and collapses them, showing only the first, and sets the animation in motion.

▷ Graph `Sin[k x]` on $-6 \le x \le 6$. Animation for $k = 1$ to 10.

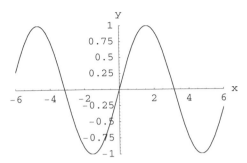

You can control the animation with the buttons that appear at the bottom of the *Mathematica* window when animation begins. Clicking anywhere in the notebook will stop the motion.

Animation control buttons:
Backward, Alternate directions, Forward, Pause, Slower, Faster.

♡ If you stop the animation, you can restart it by double-clicking on the graph. You can pause and restart by clicking **Pause** twice. Clicking **Pause** and then **Forward** or **Backward** makes the animation move one frame at a time. You can change the speed with the buttons or by pressing a number between 1 (slowest) and 9 (fastest). (On some platforms, you may not be able to use the numeric keypad for this and you may need to use the buttons before you can use the number keys.)

If you slow down the animation, you can see how each frame changes into the next. For example, here is the final frame:

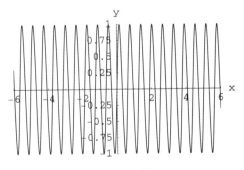

Graph of sin(10 *x*).

♡ The animation frames are grouped together and collapsed. Double-clicking on the cell bracket shown immediately below will expand the frames so you can see them all. Double-click again to collapse them. Notice the solid triangle at the bottom of the bracket. This shows that the bracket holds cells that are collapsed.

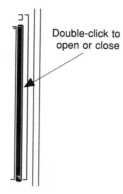

Double-click to
open or close

Sometimes, *Mathematica* doesn't collapse and animate the graphs automatically. In this case, you can collapse them by double-clicking on the bracket and then animate them by double-clicking on the graph that is visible.

Chapter 4
Manipulating Expressions

This chapter discusses various ways to work with algebraic, trigonometric, and other expressions. It includes substitution, making tables of values, factoring, and other kinds of algebraic manipulation. Chapters 6 and 11 show how to work with expressions that are defined as functions of one or more variables.

■ 4.1 How to Substitute into an Expression

Here's how to find the value of $3^x - x^3$ when $x = -5$.

- From the *Joy* menus, choose **Algebra ▷ Substitute**.

Algebra menu.

- Fill in the expression $3^x - x^3$.
- Make sure the radio button **replace** is selected, and change the fields to read "replace **x** by − 5."

    The default setting of the button is **replace**.

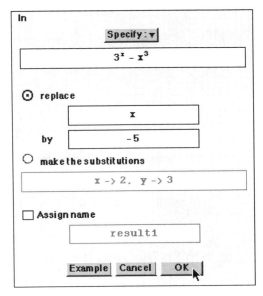

Substitute dialog.

- Click **OK**.

Here is the result:

▷ Substituting -5 for x in the expression $3^x - x^3$ gives

Out[2]= $\dfrac{30376}{243}$

     ♀ Substituting for *x* doesn't assign a permanent value to *x* or to $3^x - x^3$. You can still substitute a different value for *x* or use $3^x - x^3$ in algebraic expressions without its having any specific numerical value.

4.1.1 Substituting for More Than One Variable

You can use **Algebra** ▷ **Substitute** to substitute for more than one variable at a time. For example, here's how to evaluate $x^3 + \sin^3 y$ when $x = 7$, $y = \pi/4$.

- Click the **Last Dialog** button on the *Joy* palette to bring back the **Substitute** dialog.
- Fill in the expression $\mathtt{x^3\ +\ Sin[y]^3}$.
- Select **make the substitutions**, and fill in x -> 7, y -> π / 4 .

    You can type -> by typing - and then >.

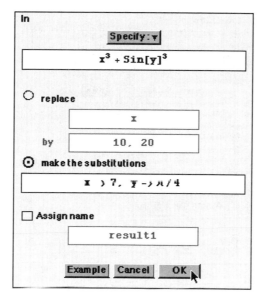

Substituting for *x* and *y* in one step.

- Click **OK**.

Mathematica makes the two substitutions at once:

▷ Making the substitutions {x -> 7, y -> π / 4} in the expression x³ + Sin[y]³ gives

$$Out[3]= \quad 343 + \frac{1}{2\sqrt{2}}$$

■ 4.2 How to Tabulate an Expression

Here's how to create a table of values for an expression such as

$$\frac{\ln(1+x)}{1+\ln x}, \text{ where } x = 1, \ 1001, \ 2001, \ \dots, \ 10001.$$

- Choose **Create ▷ List** from the *Joy* menus.
- Fill in **x**, $\frac{\text{Log}[1+x]}{1+\text{Log}[x]}$.
- Make sure the popup menu reads **x Equally Spaced**.
- Fill in 1 ≤ **x** ≤ 10001 in increments of 1000.
- Make sure the boxes **Display in decimal form** and **Display as a table** are checked.

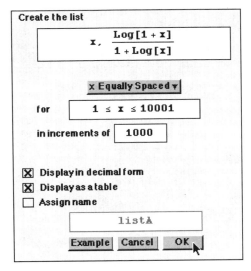

List dialog.

- Click **OK**.

Mathematica produces the result:

▷ Create the list with values $\left\{x, \frac{\text{Log}[1 + x]}{1 + \text{Log}[x]}\right\}$ for $1 \le x \le 10001$

    in steps of size 1000.

    Out[2]//TableForm=

1.	0.693147
1001.	0.873684
2001.	0.883798
3001.	0.889009
4001.	0.892434
5001.	0.89495
6001.	0.896921
7001.	0.898531
8001.	0.899886
9001.	0.901051
10001.	0.902071

4.2.1 A Table with *x* → 0

Here is an example of how to use *Joy* to create and plot a table of values for an expression as *x* approaches 0. In this table, the values of *x* will start with π and decrease by a factor of 2 with each step:

$\dfrac{\sin x}{x}$, where $x = \pi,\ \dfrac{\pi}{2},\ \dfrac{\pi}{2^2},\ ....,\ \dfrac{\pi}{2^{15}}$.

- Click the **Last Dialog** button on the *Joy* palette to bring back the **List** dialog.
- Fill in **x**, $\dfrac{\text{Sin}[x]}{x}$.
- From the popup menu, choose **x = f(n)**.

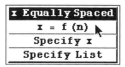

- Fill in $x = \pi\,/\,2^n$ and $1 \le n \le 15$.
- Make sure the default settings **Display in decimal form** and **Display as a table** are checked.

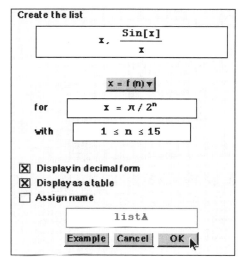

List dialog for **x = f(n)**.

- Click **OK**.

Mathematica produces the table:

▷ Create the list with values $\left\{ x,\ \dfrac{\text{Sin}[x]}{x} \right\}$ for $x = \pi\,/\,2^n$ with $1 \le n \le 15$.

Out[2]//TableForm=

1.5708	0.63662
0.785398	0.900316
0.392699	0.974495
0.19635	0.993587
0.0981748	0.998394
0.0490874	0.999598
0.0245437	0.9999
0.0122718	0.999975
0.00613592	0.999994
0.00306796	0.999998
0.00153398	1.
0.00076699	1.
0.000383495	1.
0.000191748	1.
0.0000958738	1.

The last five values are not identical but equal 1 to 6 significant digits.

Other ways. You can also use the other two choices in the popup menu to create lists. For example, if you just want $x = 1, \ 0.01, \ 0.0001$ in the preceding table, choose **Specify x**:

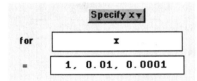

You can also create a list with specific entries by choosing **Specify List**:

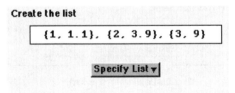

■ 4.3 How to Name an Expression

You can give an expression a name if you think you'll need to refer to it again. Here is how to do that with $x^2 - 10\,x$.

- Click in your *Mathematica* notebook beneath the last output, if any, so that a horizontal line appears.
- Type `quadratic = x`^{2}` - 10 x`.
- Press SHIFT − RET .

The result appears in your notebook:

In[2]:=   **quadratic = x² - 10 x**

Out[2]= $-10 \, x + x^2$

Mathematica automatically rearranges the terms so the lowest degree is first.

Now you can use the name *quadratic* whenever you need to instead of retyping $x^2 - 10\,x$. For example, here's how to evaluate this expression for $x = 0.99$.

- From the *Joy* menus, choose **Algebra ▷ Substitute**.
- Fill in the name `quadratic`.
- If necessary, click **replace** and change the fields to read "replace x by 0.99."

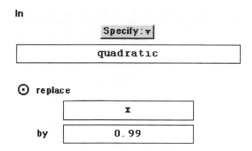

- Click **OK**.

Mathematica substitutes into the expression *quadratic*:

▷ Substituting 0.99 for x in the expression quadratic gives

Out[3]=   8.9199

You can accomplish the same thing by creating a function $h(x) = x^2 - 10\,x$ instead of an expression *quadratic* $= x^2 - 10\,x$. You might choose to create a function if you planned to evaluate $x^2 - 10\,x$ for several values of x. Chapter 6 discusses functions.

■ 4.4 How to Expand an Expression

Here's how to expand an expression such as $3\,(x - 1)\,(x + 1)^3$.

- From *Joy*'s menus, choose **Algebra ▷ Simplify**.
- Fill in 3 (x - 1) (x + 1)³.
- Click the radio button **Expand**.

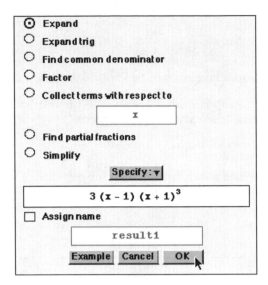

Simplify dialog.

- Click **OK**.

The result is:

▷ Expanding 3 (x − 1) (x + 1)³ gives

Out[2]=   − 3 − 6 x + 6 x³ + 3 x⁴

4.4.1 How to Expand a Trigonometric Expression

When an expression involves trigonometric functions, **Expand** in **Algebra** ▷ **Simplify** does not always expand the expression. In this case, you can use a special radio button, **Expand trig**. Here is how it works with $\sin(a^2 - b^2)$.

- Click **Last Dialog** to bring back the **Simplify** dialog.
- Fill in Sin[a² − b²].
- Click **Expand trig**.

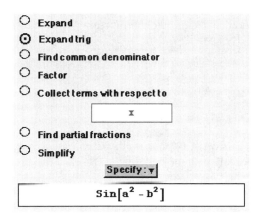

- Click **OK**.

The result is:

▷ Expand the trig expression Sin[$a^2 - b^2$] gives

Out[3]=   Cos[b^2] Sin[a^2] − Cos[a^2] Sin[b^2]

■ 4.5 How to Factor a Polynomial

Here's how to factor a polynomial such as $x^3 - x^2 + 5x - 5$.

- From *Joy*'s menus, choose **Algebra ▷ Simplify**.
- Fill in $x^3 - x^2 + 5 x - 5$.
- Click **Factor**.

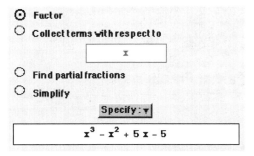

- Click **OK**.

Mathematica displays the factorization:

▷ Factoring $x^3 - x^2 + 5 x - 5$ gives

Out[2]=   (−1 + x) (5 + x^2)

Mathematica factors polynomials and rational functions (i.e., quotients of polynomials) into factors with real, integer coefficients.

> The quadratic $5 + x^2$ can't be reduced further using only integers and real numbers. You can see that it factors as $\left(\sqrt{5} + x\right)\left(\sqrt{5} - x\right)$.

■ 4.6 How to Collect Terms

Collecting terms is often a good way to simplify the way a polynomial is expressed. The following example illustrates this with the polynomial $(x^2 + x + 1)(x^2 - a^2 x + 2)$.

- From *Joy*'s menus, choose **Algebra ▷ Simplify**.
- Fill in $(x^2 + x + 1)(x^2 - a^2 x + 2)$.
- Click **Collect terms with respect to** and keep the default variable x.

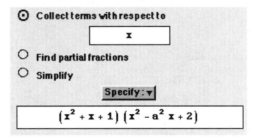

- Click **OK**.

Mathematica displays the result:

▷ Collecting terms in $(x^2 + x + 1)$ $(x^2 - a^2 x + 2)$ with respect to x gives

Out[2]= $2 + (2 - a^2) x + (3 - a^2) x^2 + (1 - a^2) x^3 + x^4$

Mathematica has expanded the product and then collected the terms according to powers of *x*.

■ 4.7 How to Find Partial Fractions

Expressions such as

$$\frac{x^2 + x - 1}{x^3 - 7x^2 + 8x + 16}$$

can be expanded into *partial fractions*, that is, into a sum of quotients $p(x)/q(x)$ where each denominator $q(x)$ is a factor of the denominator $x^3 - 7x^2 + 8x + 16$. Here's how to use *Joy* to find the partial fractions of this expression.

- From the *Joy* menus, choose **Algebra ▷ Simplify**.
- Fill in $\frac{x^2 + x - 1}{x^3 - 7x^2 + 8x + 16}$ and click **Find partial fractions**.

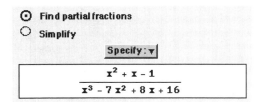

• Click **OK**.

Mathematica produces the partial fractions:

▷ The partial fraction decomposition of $\dfrac{x^2 + x - 1}{16 + 8\,x - 7\,x^2 + x^3}$ is

$Out[2]= \quad \dfrac{19}{5\,(-4+x)^2} + \dfrac{26}{25\,(-4+x)} - \dfrac{1}{25\,(1+x)}$

■ 4.8 How to Find a Common Denominator

Putting terms over a common denominator is a good way to simplify combinations of expressions with different denominators. Here's how to do that with the expression

$$\frac{1}{x+1} - \frac{x}{x-1}.$$

• From *Joy*'s menus, choose **Algebra ▷ Simplify**.
• Click **Find common denominator**.
• Fill in $\frac{1}{x+1} - \frac{x}{x-1}$.

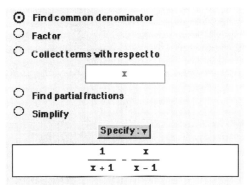

• Click **OK**.

The result is:

▷ Finding a common denominator for $\dfrac{1}{x+1} - \dfrac{x}{x-1}$ gives

Out[2]= $\dfrac{-1-x^2}{(-1+x)\ (1+x)}$

■ 4.9 How to Simplify Expressions in General

In addition to the specific methods for manipulating expressions, **Algebra ▷ Simplify** contains a generic **Simplify** command. Here we use this to carry the last result one step further.

- If you didn't find the common denominator in Section 4.8, do it now.
- From the main *Joy* menu, choose **Last Dialog**.
- Choose **Last Output** from the popup menu.

- Click **Simplify**.

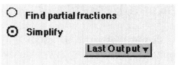

- Click **OK**.

Mathematica simplifies the expression:

▷ Simplifying $\dfrac{-1-x^2}{(-1+x)\ (1+x)}$ gives

Out[3]= $-\dfrac{1+x^2}{-1+x^2}$

In this example, applying **Simplify** directly to $\frac{1}{x+1}-\frac{x}{x-1}$ would not have combined the terms.

    ♡ **Simplify** is a catchall operation in which *Mathematica* tries several algebraic transformations and attempts to reduce the given expression to as few parts as possible. It can take a long time to execute when the expression is complicated and *Mathematica*'s idea of simplification may not match yours. If you have a particular method in mind for simplifying an expression, it is often better to try it first. You can combine it with **Simplify** if you wish, as in the example of $\frac{1}{x+1}-\frac{x}{x-1}$.

■ 4.10 How to Clear a Definition or Value

Sometimes you may give a value to a symbol and later wish to remove that value. For example, in one calculation you may want $x = 3$ and later want to find the expansion of $(x + y)^{10}$, where x is arbitrary. Here's what can happen if you don't clear the value of x in advance.

- Click in your *Mathematica* notebook beneath the last output, if any, so that a horizontal line appears.
- Type **x** = 3 and press SHIFT – RET.

This establishes a value for *x*:

In[2]:= **x = 3**

Out[2]= 3

- From the *Joy* menus, choose **Algebra ▷ Simplify**.
- If necessary, click **Example** so that the **Expand** button is selected and the expression is $(x + y)^{10}$.

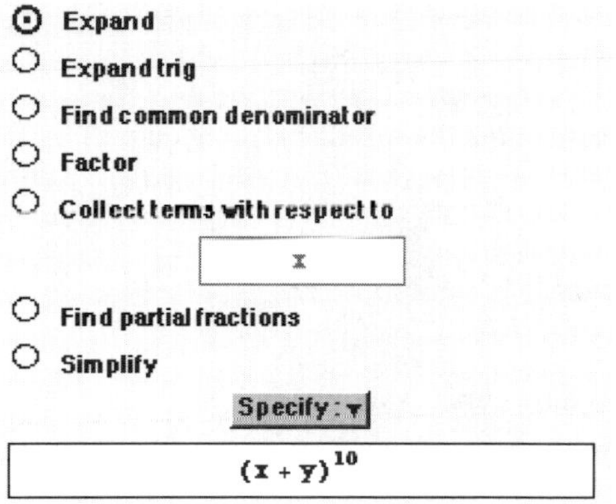

- ⊙ **Expand**
- ○ **Expand trig**
- ○ **Find common denominator**
- ○ **Factor**
- ○ **Collect terms with respect to**

  | x |

- ○ **Find partial fractions**
- ○ **Simplify**

  Specify: ▼

  $(x + y)^{10}$

- Click **OK**.

Mathematica expands $(3 + y)^{10}$:

▷ Expanding $(x + y)^{10}$ gives

Out[3]= $59049 + 196830\,y + 295245\,y^2 + 262440\,y^3 + 153090\,y^4 + 61236\,y^5 + 17010\,y^6 +$
$3240\,y^7 + 405\,y^8 + 30\,y^9 + y^{10}$

Here's how to remove the value from *x*.

- Choose **Create ▷ Clear**.
- Fill in **x**.

```
Clear any definitions or values for the following
symbols


┌──────────────────────────────────────────────┐
│                      x                         │
└──────────────────────────────────────────────┘




            [Example] [Cancel] [ OK ▖]
```

- Click **OK**.

Mathematica clears the value of *x* but doesn't display any output.

▷ Clear x.

- Choose **Recent** ▷ **Simplify**.
- Click **OK**.

Mathematica now treats *x* as an arbitrary variable:

▷ Expanding $(x + y)^{10}$ gives

Out[6]= $x^{10} + 10 \, x^9 \, y + 45 \, x^8 \, y^2 + 120 \, x^7 \, y^3 + 210 \, x^6 \, y^4 + 252 \, x^5 \, y^5 + 210 \, x^4 \, y^6 +$
$120 \, x^3 \, y^7 + 45 \, x^2 \, y^8 + 10 \, x \, y^9 + y^{10}$

⚠ In this example, the value of $(x + y)^{10}$ in Out[3] was correct but may not have been what you intended. If you assign a value to a symbol and later use that symbol in functions, derivatives, integrals, etc., the results may not make sense at all. In this case, clear the value and try again.

Chapter 5
Solving Equations

This chapter includes methods for solving equations algebraically and numerically, and for visualizing Newton's method for numerical solution.

■ 5.1 How to Solve Equations Algebraically

Here is how to use *Joy* to solve an equation such as

$$\frac{x-1}{\sqrt{x}} = 2.$$

- From the *Joy* menus, choose **Algebra ▷ Solve**.

Algebra menu.

- Fill in $\frac{x-1}{\sqrt{x}} = 2$.
- Fill in the variable **x**.
- Make sure **Solve algebraically** is selected.

> **For the following equation(s)**
>
> $$\frac{x - 1}{\sqrt{x}} = 2$$
>
> **for the variables**   x
>
> ⊙ **Solve algebraically**
> ○ **Solve numerically; initial estimate :**
>
>   x = 1, y = 2
>
> ○ **Polynomial : find all roots (numerically)**
> ☐ **Assign name to solutions**
>
>   solns1
>
>   Example Cancel   OK

Solve dialog.

- Click **OK**.

Mathematica displays the solution:

▷ Solve the equation $\dfrac{x - 1}{\sqrt{x}} = 2$ for x algebraically.

Out[2]= $\{\{x \rightarrow 3 + 2\sqrt{2}\,\}\}$

Mathematica shows that substituting $3 + 2\sqrt{2}$ for *x* will satisfy the equation, but doesn't assign a value to *x*. That is, *x* stands for a variable with no particular value.

⚠ If *x* was assigned a value earlier, then *Joy* will clear this definition before solving an equation containing *x*.

5.1.1 Solving Simultaneous Equations

You can use **Algebra** ▷ **Solve** to solve several equations simultaneously. For example, here is how to solve $x + 2y = 0$, $x^2 + 3y = 4$ simultaneously for *x* and *y*.

- Click the **Last Dialog** button on the *Joy* palette to bring back the **Solve** dialog.
- Click the **Example** button.

    These equations are the default example for the **Solve** dialog. Notice that the equations are separated by commas. So are the variables for which you want to solve.

- Click **Assign name**.

- If you wish, change the default name `solns1`.

> **For the following equation(s)**
>
> $$x + 2\,y = 0, \quad x^2 + 3\,y = 4$$
>
> **for the variables** $x, \; y$
>
> ⊙ Solve algebraically
> ○ Solve numerically; initial estimate :
>
> $$x = 1, \; y = 2$$
>
> ○ Polynomial : find all roots (numerically)
> ☒ Assign name to solutions
>
> `solns1`

- Click **OK**.

Here is the solution:

▷ Solve the equations $\{x + 2\,y = 0,\ x^2 + 3\,y = 4\}$ for $\{x, y\}$ algebraically.

`solns1 =`

Out[3]= $\left\{\left\{y \to \frac{1}{8}\,(-3 - \sqrt{73}\,),\ x \to \frac{1}{4}\,(3 + \sqrt{73}\,)\right\},\ \left\{y \to \frac{1}{8}\,(-3 + \sqrt{73}\,),\ x \to \frac{1}{4}\,(3 - \sqrt{73}\,)\right\}\right\}$

This notation means that there are two solutions,

$$(x, y) = \left(\tfrac{1}{4}\left(3 + \sqrt{73}\,\right), \tfrac{1}{8}\left(-3 - \sqrt{73}\,\right)\right),$$
$$(x, y) = \left(\tfrac{1}{4}\left(3 - \sqrt{73}\,\right), \tfrac{1}{8}\left(-3 + \sqrt{73}\,\right)\right).$$

5.1.2 How to Substitute a Solution into an Expression

You can use *Joy* to substitute the solutions to one or more equations into some expression. For example, the solutions to the equations in Section 5.1.1 are the points where the line $x + 2\,y = 0$ meets the parabola $x^2 + 3\,y = 4$. Here's how to find the distance from those points to the origin by substituting the solutions into the expression $\sqrt{x^2 + y^2}$.

- If you didn't solve the equations in 5.1.1, do it now.
- From the *Joy* menus, choose **Algebra ▷ Substitute**.
- Fill in $\sqrt{x^2 + y^2}$
- Click **make the substitutions** and fill in `solns1`, or the name that you assigned to the solutions.

    If you didn't assign a name, you can fill in the output number instead. In this example, the number is `Out[3]`.

- Click **Assign name** and fill in the name `distance`, or another name of your choice.

Substitute dialog.

- Click **OK**.

The result is:

▷ Making the substitutions solns1 in the expression $\sqrt{x^2 + y^2}$ gives

distance =

$$\textit{Out[4]}= \left\{ \sqrt{\frac{1}{64}\,(-3-\sqrt{73}\,)^2 + \frac{1}{16}\,(3+\sqrt{73}\,)^2}\,,\ \sqrt{\frac{1}{16}\,(3-\sqrt{73}\,)^2 + \frac{1}{64}\,(-3+\sqrt{73}\,)^2} \right\}$$

There are two values for $\sqrt{x^2 + y^2}$, one for each solution to the equations. Here's how to express them in decimal form.

- Click the **Numeric** button on the *Joy* palette.
- Fill in `distance`, or the name you chose instead.
- Press $\boxed{\text{SHIFT}}$ − $\boxed{\text{RET}}$.

Mathematica then expresses the value of $\sqrt{x^2 + y^2}$ in numeric form:

In[5]:= **N[distance]**

Out[5]= {3.22665, 1.5496}

■ 5.2 How to Solve Equations Numerically

Equations with transcendental functions, such as trigonometric, logarithmic, and exponential functions, generally cannot be solved exactly. Here's how to approximate the solutions to such an equation starting with an initial estimate, using the example $x^3 + 2 = e^x + x$. We begin by graphing the two functions together to make an initial approximation.

- From *Joy*'s menus, choose **Graph ▷ Graph f(x)**.
- Fill in $x^3 + 2$, $E^x + x$ and the interval $-5 \le x \le 5$.

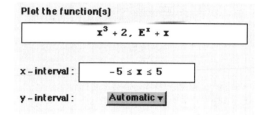

- Click **OK**.

Mathematica graphs the functions:

▷ Graph

$x^3 + 2$	Black	Solid
$E^x + x$	Red	.....

on $-5 \le x \le 5$.

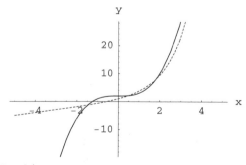

Out[2]= - Graphics -

There appear to be three solutions. Here's how to approximate the one near $x = 2$.

- From the *Joy* menus, choose **Algebra ▷ Solve**.
- Fill in the equation $x^3 + 2 = E^x + x$.
- Fill in the variable x.

- Click the radio button **Solve numerically** and fill in the initial estimate x = 2.

For the following equation(s)

$$x^3 + 2 = E^x + x$$

for the variables   x

○ Solve algebraically
◉ Solve numerically; initial estimate :

x = 2

- Click **OK**.

The solution is:

▷ Solve the equation $x^3 + 2 = E^x + x$ for x numerically
    with the initial estimates x = 2.

Out[3]=   {x → 1.80657}

The solution near $x = 2$ is approximately 1.80657. You can use this method to estimate the other two solutions as well.

> ♡ *Mathematica*'s primary technique for solving equations numerically is *Newton's method*. This method is explained in Section 5.4, which shows how to use *Joy* to illustrate it.

■ 5.3 How to Find Roots of Polynomials Numerically

Choosing **Algebra** ▷ **Solve** and **Solve algebraically** will find the exact roots of all polynomials of degree 4 or less, using the quadratic formula and analogous formulas for degrees 3 and 4. For degrees 3 and 4 the expressions for the roots can be quite messy, so it's often advantageous to estimate the roots numerically rather than find them algebraically. For degree 5 or more there is no general formula for finding the roots, and they usually need to be estimated numerically.

Solve numerically in **Algebra** ▷ **Solve** gives numerical estimates to the solutions one at a time, starting from initial estimates. For polynomial equations it is possible to estimate all the solutions simultaneously without any initial estimates. Here's how to do this with the roots of the polynomial $x^5 - 1331\,x + 11$.

- From the *Joy* menus, choose **Algebra** ▷ **Solve**.
- Fill in the equation $x^5 - 1331\,x + 11 = 0$.
- Fill in the variable x.
- Click **Polynomial**.

For the following equation(s)

$$x^5 - 1331\,x + 11 = 0$$

for the variables | x |

○ Solve algebraically

○ Solve numerically; initial estimate :

| $x = 1,\ y = 2$ |

⊙ Polynomial : find all roots (numerically)

- Click **OK**.

The roots are:

▷ Solve the polynomial equation $x^5 - 1331\,x + 11 = 0$ for x.

Out[2]= $\{\{x \to -6.04217\},\ \{x \to -0.00206611 - 6.04011\ I\},\ \{x \to -0.00206611 + 6.04011\ I\},$
$\{x \to 0.00826446\},\ \{x \to 6.03804\}\}$

Mathematica finds three real and two complex conjugate roots. Here I (or i in *Mathematica* Version 4) stands for $i = \sqrt{-1}$.

■ 5.4 How to Visualize Newton's Method

Section 5.2 shows how to estimate a solution to an equation that cannot be solved exactly. *Mathematica* usually uses Newton's method for this purpose. This section shows how to use *Joy* to give a graphical illustration of Newton's method using the example $2\sin(x^2) - \cos x = 0$, which *Mathematica* cannot solve algebraically.

Newton's method estimates a solution to an equation of the form $f(x) = 0$, i.e., a root of $f(x)$. The method starts with an initial estimate x_0 for the root. The unknown root is located where the curve in the accompanying sketch crosses the x-axis. The point x_1 where the tangent line at x_0 meets the x-axis is often a better estimate of the root than x_0.

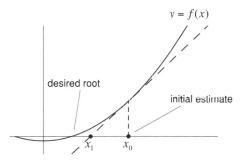

By applying this procedure to x_1 we get a second estimate x_2, and so on. If your initial guess is sufficiently close to a root of $f(x)$, then the estimates will converge to that root.

You can find an initial estimate for a root of $2 \sin(x^2) - \cos x$ by looking at its graph or by substituting a few values. Here we'll find the graph.

- From the *Joy* menus, choose **Graph** ▷ **Graph f(x)**.
- Fill in the function $2\,\text{Sin}\,[x^2] - \text{Cos}\,[x]$ and interval $-5 \le x \le 5$.

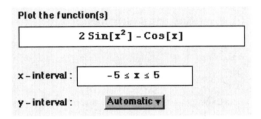

Plot the function(s)

$$2\,\text{Sin}[x^2] - \text{Cos}[x]$$

x – interval : $-5 \le x \le 5$

y – interval :   Automatic ▼

- Click **OK**.

Mathematica graphs the function:

▷ Graph $2\,\text{Sin}\,[x^2] - \text{Cos}\,[x]$ on $-5 \le x \le 5$.

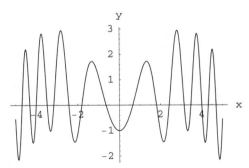

Out[2]= - Graphics -

This function has many roots, one of which is near $x = 2$. We'll take $x = 2$ as the starting value for Newton's method.

- From the *Joy* menus, choose **Calculus** ▷ **Newton's Method**.

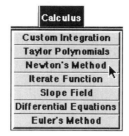

Calculus menu.

- Fill in $2 \operatorname{Sin}[x^2] - \operatorname{Cos}[x]$.

    Be sure not to type the right-hand side of the equation ($= 0$).

- Fill in the estimate $x = 2$.
- Keep the default number of iterations, 10.
- Check **Illustrate graphically**.

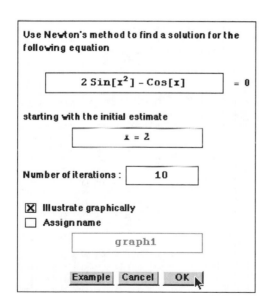

- Click **OK**.

The result is:

▷ An estimate for a solution of $2 \, \text{Sin} \, [x^2] - \text{Cos} \, [x] = 0$ (using
   Newton's Method with 10 iterations and the initial estimate x = 2):

Graphical illustration of Newton's Method

1.80496

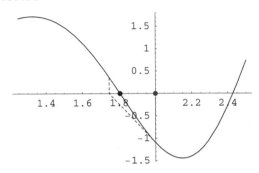

Out[3]=   - Graphics -

The graph shows two solutions near $x = 2$, and Newton's method finds the one with $x \approx 1.80496$. The output displays the successive tangents and approximations converging to the solution.

Chapter 6

Working with Functions of One Variable

This chapter shows how to create a function, evaluate it at individual points, make a table of values, and plot the points in the table. Chapter 11 discusses functions of more than one variable.

■ 6.1 How to Create a Function

For this example, we use the function $f(x) = x^3 - x^2 - 5x + 5$.

- From the *Joy* menus, choose **Create** ▷ **Function**.

Create menu.

- Fill in f [x] to name the function and the independent variable.
- Fill in the expression $x^3 - x^2 - 5x + 5$.

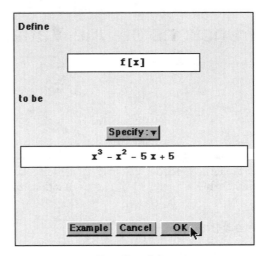

Function dialog.

- Click **OK**.

When you click **OK**, *Mathematica* creates the function but does not display any output:

▷ Define f[x] to be x³ – x² – 5 x + 5

⚠ If you have already defined the symbol *f*, then *Joy* clears your earlier definition before it creates the function.

■ 6.2 How to Evaluate a Function

You can find the value of $f(x)$ when x takes on any particular value, such as $x = -4$, directly in your *Mathematica* notebook.

- If you didn't create the function $f(x) = x^3 - x^2 - 5\,x + 5$ in Section 6.1, do it now.
- Click below the last cells, so that a horizontal line appears.
- Type f[-4].
- Press SHIFT – RET.

The result is:

In[3]:= f[-4]

Out[3]= -55

■ 6.3 How to Tabulate a Function

You can evaluate $f(x)$ for many values of x at once and create a table of values, using **Create ▷ List**. Here's how to create a table with $x = -5, -4, -3, ..., 3, 4, 5$.

- If you didn't create the function $f(x) = x^3 - x^2 - 5x + 5$ in Section 6.1, do it now.
- From the *Joy* menus, choose **Create ▷ List**.
- Fill in **x, f[x]** for the expression used to create the list.

   This will make the values of x and $f(x)$ appear in the table.

- Fill in $-5 < x < 5$ and increments of 1.

   Be sure the popup menu is set to **x Equally Spaced**. The popup menu's current setting will depend on whether you have changed it during this session.

- Check **Assign name** and use the default name **listA**, or choose a different name.

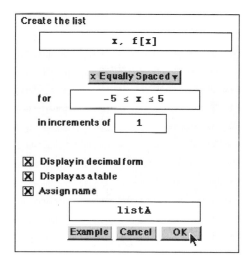

Create List dialog.

- Click **OK**.

Mathematica creates the list:

▷ Create the list with values {x, f[x]} for -5 ≤ x ≤ 5.

listA =

Out[4]//TableForm=

-5.	-120.
-4.	-55.
-3.	-16.
-2.	3.
-1.	8.
0	5.
1.	0
2.	-1.
3.	8.
4.	33.
5.	80.

Joy's default setting, **Display as table**, has *Mathematica* arrange the entries as a vertical table rather than a horizontal list. If there will be many entries, you can save space by not checking this option. Here's what horizontal output looks like.

- From the *Joy* menus, choose **Last Dialog**.
- Uncheck **Display as table**.

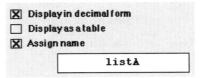

- Click **OK**.

Mathematica now displays the list horizontally.

▷ Create the list with values {x, f[x]} for -5 ≤ x ≤ 5.

listA =

Out[5]= {{-5., -120.}, {-4., -55.}, {-3., -16.}, {-2., 3.}, {-1., 8.},
       {0, 5.}, {1., 0}, {2., -1.}, {3., 8.}, {4., 33.}, {5., 80.}}

Mathematica treats the two lists identically, even though they are displayed differently.

  ⚲ In *Mathematica*, all lists are specified with curly brackets { } and the items in the list are separated by commas. Each pair of coordinates, starting with {-5., -120.}, is a list, and so is the entire list or table. *Joy* inserts brackets automatically in the *Mathematica* command it constructs for creating the list.

  Joy's default setting **Display in decimal form** is useful when you would rather see a decimal expression in the table instead of an exact value such as a fraction or sin(π/7). With this setting, decimal points in the table also appear when the values are integers.

Other ways. Section 4.2 shows how to make a table of values for an expression that you haven't already created as a function. It also shows how to use the popup menu in **Create ▷ List** to choose the values in the table in other ways.

■ 6.4 How to Plot a Table of Values

- If you didn't create a table of values in Section 6.3, do it now.
- From the *Joy* menus, choose **Graph ▷ Points**.
- Fill in `listA`, or use the name that you assigned if it is different from this.

```
Plot the points in the list

┌─────────────────────────────────┐
│             listA               │
└─────────────────────────────────┘

☐  Position axes at the origin

☐  Assign name
   ┌─────────────────────────────┐
   │          graph1             │
   └─────────────────────────────┘


     Example  Cancel    OK
```

Graph Points dialog.

- Click **OK**.

Mathematica plots the list:

▷ Plot of the points listA

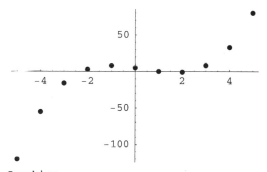

Out[6]= - Graphics -

■ 6.5 How to Iterate a Function

This sections shows how you can use *Joy* to calculate and illustrate the iterates of a function f starting from some initial value. This means beginning with some x_0 and finding $f(x_0)$, $f(f(x_0))$, $f(f(f(x_0)))$, ... We will illustrate this with $f(x) = \sin x - \cos x + 2$ and $x_0 = 0$.

- From the *Joy* menus, choose **Calculus ▷ Iterate Function**.

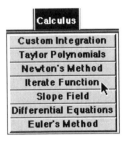

Calculus menu.

- Fill in 0 for the value of x_0.
- Fill in `Sin[x] - Cos[x] + 2` for the function.
- Keep the remaining settings, which include graphing the first 10 iterates.

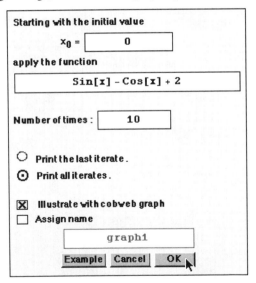

Iterate Function dialog.

- Click **OK**.

Mathematica calculates the first 10 iterates and displays a graph:

▷ The first 10 iterates of Sin[x] - Cos[x] + 2 applied to the starting
  value 0 and the associated cobweb graph:

{0, 1., 2.30117, 3.41207, 2.69645, 3.33314, 2.79134, 3.28242, 2.84973,
 3.24544, 2.89095}

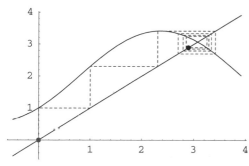

Out[2]= - Graphics -

The graph, called a *cobweb graph*, shows the graph of $f(x) = \sin x - \cos x + 2$ and the line $y = x$.
If we let

$$x_1 = f(x_0), \ x_2 = f(x_1), \ \dots$$

then the dotted line follows the path

$$(x_0, x_0) = (0, 0), \ (x_0, x_1) = (0, 1), \ (x_1, x_1) = (1, 1),$$
$$(x_1, x_2) = (1, 2.30117), \ \dots, \ (x_{10,} x_{10}) = (2.89095, 2.89095).$$

A large number of iterations takes less time if you don't graph the process:

- Click **Last Dialog** to bring back the **Iterate Function** dialog.
- Change **Number of times** to 100.
- *Uncheck* **Illustrate with cobweb graph**.

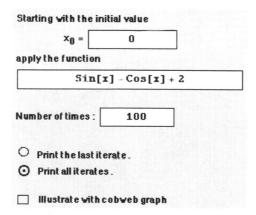

- Click **OK**.

The result is:

▷ The first 100 iterates of Sin[x] - Cos[x] + 2 applied to the starting value 0:

Out[3]=   {0, 1., 2.30117, 3.41207, 2.69645, 3.33314, 2.79134, 3.28242, 2.84973,
3.24544, 2.89095, 3.21678, 2.92206, 3.19378, 2.94648, 3.1749, 2.96614,
3.1592, 2.98224, 3.14601, 2.99557, 3.13486, 3.00671, 3.12539, 3.01607,
3.11733, 3.02397, 3.11044, 3.03066, 3.10456, 3.03634, 3.09952, 3.04117,
3.09521, 3.04529, 3.09152, 3.04879, 3.08836, 3.05179, 3.08565, 3.05435,
3.08333, 3.05653, 3.08134, 3.0584, 3.07964, 3.05999, 3.07818, 3.06136,
3.07693, 3.06253, 3.07586, 3.06353, 3.07494, 3.06438, 3.07415, 3.06511,
3.07348, 3.06574, 3.0729, 3.06628, 3.07241, 3.06674, 3.07199,
3.06713, 3.07162, 3.06746, 3.07131, 3.06775, 3.07105, 3.068,
3.07082, 3.06821, 3.07063, 3.06839, 3.07046, 3.06854, 3.07032,
3.06868, 3.07019, 3.06879, 3.07009, 3.06889, 3.07, 3.06897, 3.06992,
3.06904, 3.06986, 3.0691, 3.0698, 3.06915, 3.06975, 3.0692, 3.06971,
3.06924, 3.06968, 3.06927, 3.06965, 3.0693, 3.06962, 3.06932}

The iterates appear to be approaching $x \approx 3.0695$. In Lab 19.4, you can explore how iterates can converge or diverge.

♡ If you calculate many more iterates, you can avoid printing them all by choosing **Print the last iterate**.

Chapter 7
Differentiating Functions of One Variable

This chapter shows how to differentiate functions symbolically and numerically. It also includes implicit differentiation. For functions of several variables, see Chapter 12.

■ 7.1 How to Differentiate $y = f(x)$

Here's how to differentiate a function that you've already created in the form $f(x)$, such as $f(x) = x^4 \sin 10\,x$. For differentiating other expressions, see Section 7.5.

- From the *Joy* menus, choose **Create** ▷ **Function**.

Create menu.

- Fill in $f[x]$ and $x^4 \text{ Sin}[10\,x]$.

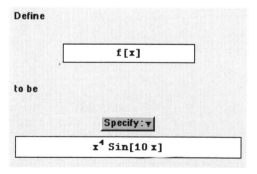

- Click **OK**.

Mathematica creates the function without any output:

▷ Define f[x] to be x⁴ Sin[10 x]

Next, here's how to find $f'(x)$.

- If necessary, click below the last output so that a horizontal line appears.
- Type f'[x].
- Press SHIFT — RET.

Mathematica calculates the derivative:

In[3]:= **f'[x]**

Out[3]= 10 x⁴ Cos[10 x] + 4 x³ Sin[10 x]

■ 7.2 How to Find *f'* (*a*)

You can find the derivative of $f(x)$ for some particular value of x, e.g., $x = \pi/15$, in the same way.

- If you didn't create $f(x) = x^4 \sin 10\,x$ in Section 7.1, do it now.
- Type f'[π/15].

    Type an apostrophe or single quotation mark ' ("prime") for a derivative. You can enter π using the π button on the *Joy* palette.

- Press SHIFT — RET.

Mathematica gives the result:

In[4]:= **f'[π / 15]**

Out[4]= $\dfrac{2\,\pi^3}{1125\,\sqrt{3}} - \dfrac{\pi^4}{10125}$

■ 7.3 How to Graph *f'* (*x*)

You can graph the derivative of $f(x)$ in the same way you would graph any function. For example, here's how to graph $f'(x)$ for $f(x) = x^4 \sin 10\,x$ with $-1 \le x \le 1$.

- If you didn't create $f(x) = x^4 \sin 10\,x$ in Section 7.1, do it now.
- From the *Joy* menus, choose **Graph ▷ Graph f(x)**.
- Fill in f'[x] and –1 ≤ x ≤ 1.

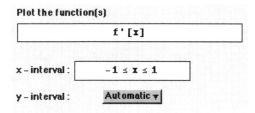

Plot the function(s)

f'[x]

x – interval : −1 ≤ x ≤ 1

y – interval : Automatic ▾

- Click **OK**.

Mathematica graphs the derivative:

▷ Graph f′[x] on −1 ≤ x ≤ 1.

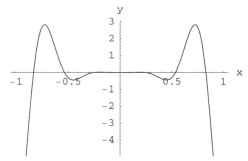

Out[5]= - Graphics -

■ 7.4 How to Find Higher-Order Derivatives of *f(x)*

You can calculate higher-order derivatives of $f(x)$ similarly. For example, here's how to find the fourth derivative of $f(x) = x^4 \sin 10\,x$.

- If you didn't create $f(x) = x^4 \sin 10\,x$ in Section 7.1, do it now.
- Type f''''[x].

  ⚠ Always combine single quotation marks ' to enter higher-order derivatives, *never* double quotations ".

- Press ⌨SHIFT⌨ — ⌨RET⌨.

The derivative is:

In[6]:- f''''[x]

Out[6]= 960 x Cos[10 x] − 16000 x³ Cos[10 x] + 24 Sin[10 x] − 7200 x² Sin[10 x] + 10000 x⁴ Sin[10 x]

■ 7.5 How to Differentiate an Expression in One Variable

You can differentiate functions without creating them explicitly in advance. Section 7.1 shows how to differentiate functions that have been defined explicitly. For example, here's how to find

$$\frac{d}{dx}\left(\frac{1 + e^x}{\cos(x^2)}\right).$$

- Click in your *Mathematica* notebook, beneath the last output if any, so that a horizontal line appears.
- Type

$$\frac{1 + \text{E}^x}{\text{Cos}[\text{x}^2]} \ .$$

   Type the numerator and select it before clicking the **Divide** button ÷ .

- Drag across the expression you want to differentiate.

$$\frac{1 + \text{E}^x}{\text{Cos}[\text{x}^2]} \ ]$$

   ᛞ Instead of dragging, you can click inside $\frac{1+\text{E}^x}{\text{Cos}[\text{x}^2]}$ and press the **Select** button on the *Joy* palette repeatedly until the whole expression is selected.

- On the *Joy* palette, click the **Derivative** button.

Derivative button.

An expression for the derivative appears, with the function filled in and the input field for the independent variable selected.

$$\frac{d}{d \ \square}\left(\frac{1 + \text{E}^x}{\text{Cos}[\text{x}^2]}\right)$$

Partially completed derivative.

- Type **x**.
- Press SHIFT − RET .

Mathematica computes the derivative:

$In[2]:=$ $\dfrac{d}{d\,x}\left(\dfrac{1 + E^x}{Cos[x^2]}\right)$

$Out[2]=$ $E^x \, Sec[x^2] + 2 \, (1 + E^x) \, x \, Sec[x^2] \, Tan[x^2]$

Another way. If you click the **Derivative** button without typing and selecting an expression first, the field for the independent variable will be selected.

$$\frac{d}{d\,\blacksquare}\ (\square)$$

In this case fill in the independent variable first, press TAB, and then fill in the expression to be differentiated.

■ 7.6 How to Find the Derivative of an Expression at a Point

Here's how to evaluate a derivative for some value of x by substitution. In this example, we evaluate

$$\frac{d}{dx}\left(\frac{1 + e^x}{\cos(x^2)}\right)$$

at $x = \sqrt{\pi}$.

- If you didn't find $\frac{d}{dx}\left(\frac{1+e^x}{\cos(x^2)}\right)$ in Section 7.5, do it now.
- From the *Joy* menus, choose **Algebra** ▷ **Substitute**.

Algebra menu.

- From the popup menu in the **Substitute** dialog shown below, choose **Last Output**.
- Make sure **replace** is selected and fill in the fields to read "replace **x** by $\sqrt{\pi}$."

Substitute dialog.

- Click **OK**.

Mathematica makes the substitution:

▷ Substituting $\sqrt{\pi}$ for x in the expression $E^x \text{Sec}[x^2] + 2 \ (1 + E^x) \ x \ \text{Sec}[x^2] \ \text{Tan}[x^2]$
  gives

Out[3]= $-E^{\sqrt{\pi}}$

To evaluate $f'(a)$ for a function $y = f(x)$ that you've already defined, see Section 7.2.

■ 7.7 How to Graph the Derivative of an Expression

For example, here's how to graph the derivative of

$$\frac{1 + e^x}{\cos(x^2)}$$

for $-2 \le x \le 2$.

- If you didn't calculate this in Section 7.5, do it now:

$$\frac{d}{dx}\left(\frac{1 + e^x}{\cos(x^2)}\right)$$

- From the *Joy* menus, choose **Graph ▷ Graph f(x)**.

- Drag across the derivative from the output in your *Mathematica* notebook, copy it (Macintosh ⌘-C, Windows [CTRL]-C), and paste it into the *Joy* dialog (Macintosh ⌘-V, Windows [CTRL]-V).

  Be careful to copy only the derivative and not any additional characters.

- Fill in $-2 \le x \le 2$.

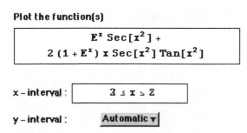

Plot the function(s)

$$E^x \, Sec[x^2] +$$
$$2 \, (1 + E^x) \, x \, Sec[x^2] \, Tan[x^2]$$

x – interval : $3 \le x \le 2$

y – interval :   **Automatic ▾**

- Click **OK**.

Mathematica graphs the derivative:

▷ Graph $E^x \, Sec[x^2] + 2 \, (1 + E^x) \, x \, Sec[x^2] \, Tan[x^2]$ on $-2 \le x \le 2$.

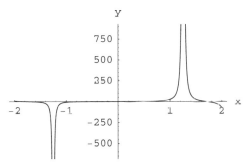

Out[4]= - Graphics -

⚠ You cannot graph the derivative of an expression by entering $\frac{d}{dx}$ (...) in the **Graph f(x)** dialog. You must fill in the expression for the derivative as shown here, use its output number or a name you have assigned, or else define a function $f(x)$ and fill in f'[x] as in Section 7.3.

■ 7.8 How to Find Higher-Order Derivatives of an Expression

Here's how to find the second derivative of an expression, for example,

$$\frac{d^2}{dx^2} \left(\ln \left(x + k \cos x \right) \right).$$

- Click in your *Mathematica* notebook, beneath the last output if any.
- Type `Log[x + k⎵ Cos[x]]`, where ⎵ means to type a blank space for multiplication.
- Select the expression you typed and click the **Derivative** button on the *Joy* palette.
- Type `x`.
- Select the resulting expression and click the **Derivative** button again.

$$\frac{d}{d\,\square}\left(\frac{d}{d\,x}\;(\text{Log}[x + k\,\text{Cos}[x]])\right)$$

Partially completed second derivative.

- Type `x` again.
- Press SHIFT — RET.

Mathematica calculates the second derivative, treating *k* as an unspecified constant:

$$In[2]:= \quad \frac{d}{d\,x}\left(\frac{d}{d\,x}\;(\text{Log}[x + k\,\text{Cos}[x]])\right)$$

$$Out[2]= \quad -\frac{k\,\text{Cos}[x]}{x + k\,\text{Cos}[x]} - \frac{(1 - k\,\text{Sin}[x])^2}{(x + k\,\text{Cos}[x])^2}$$

Now you can find the third derivative

$$\frac{d^3}{dx^3}\,(\ln\,(x + k\cos x))$$

by differentiating again.

- Drag across the last output to select it, being sure to select the initial – (minus sign).
- Click the **Derivative** button.

  An expression for the derivative appears immediately below with the input field for the independent variable selected.

- Type `x` and press SHIFT — RET.

The result is:

$$In[3]:= \quad \frac{d}{d\,x}\left(-\frac{k\,\text{Cos}[x]}{x + k\,\text{Cos}[x]} - \frac{(1 - k\,\text{Sin}[x])^2}{(x + k\,\text{Cos}[x])^2}\right)$$

$$Out[3]= \quad \frac{k\,\text{Sin}[x]}{x + k\,\text{Cos}[x]} + \frac{3\,k\,\text{Cos}[x]\,(1 - k\,\text{Sin}[x])}{(x + k\,\text{Cos}[x])^2} + \frac{2\,(1 - k\,\text{Sin}[x])^3}{(x + k\,\text{Cos}[x])^3}$$

■ 7.9 How to Approximate a Derivative Numerically

If you don't know how to compute a derivative algebraically, you can estimate it numerically. For any function g whose derivative exists,

$$g'(x) = \lim_{h \to 0} \frac{g(x+h) - g(x)}{h} \ .$$

So if h is near 0 then $\frac{g(x+h)-g(x)}{h}$ approximates $g'(x)$. Here's how to estimate the derivative of x^x when $x = 2$, using $h = 0.0001$.

- If necessary, click below the last output, if any, so that a horizontal line appears.
- Type

$$\frac{2.0001^{2.0001} - 2^2}{0.0001} \ .$$

- Press [SHIFT] – [RET].

The approximate derivative of x^x when $x = 2$ is:

In[2]:= $\dfrac{2.0001^{2.0001} - 2^2}{0.0001}$

Out[2]=   6.77326

Here's how you can use this idea to create a function $u(x)$ that is approximately equal to the derivative at each point.

- From the *Joy* menus, choose **Create ▷ Function**.
- Fill in u[x] and

$$\frac{(x + 0.0001)^{x+0.0001} - x^x}{0.0001} \ .$$

Define

u[x]

to be

Specify:▼

$\dfrac{(x + 0.0001)^{x+0.0001} - x^x}{0.0001}$

- Click **OK**.

Mathematica creates the function.

▷ Define u[x] to be $\dfrac{(x + 0.0001)^{x+0.0001} - x^x}{0.0001}$

Now you can work with this function as with any other, e.g., you can evaluate it for different values of *x*, graph it, etc.

You can find the derivative exactly by hand using logarithmic differentiation or by using *Joy*'s **Derivative** button (Section 7.5). The derivative is $x^x + x^x \ln x$, which equals $4(1 + \ln 2) = 6.77259$ when $x = 2$.

■ 7.10 How to Differentiate an Implicit Function

Sometimes a curve, such as $x^2 + 3\,xy + 4\,y^2 = 4$, is not the graph of a function but instead defines two or more functions *implicitly*. Here is how to differentiate these functions without solving for them explicitly.

- Click in your *Mathematica* notebook, beneath the last output if any.
- Type x² + 3 x⎵y[x] + 4 y[x]².

  The symbol ⎵ means to leave a blank space to indicate multiplication.

  ⚠ You need to type y[x] for *Mathematica* to treat *y* as a function of *x* rather than as a constant.

- Select the expression you typed and click the **Derivative** button.
- Fill in x for the independent variable.
- Press SHIFT – RET.

The derivative is:

In[2]:= $\dfrac{d}{d\,x}$ (x² + 3 x y[x] + 4 y[x]²)

Out[2]=   2 x + 3 y[x] + 3 x y'[x] + 8 y[x] y'[x]

Since $x^2 + 3\,xy + 4\,y^2$ is a constant (= 4) along the curve in question, its derivative is zero. We can then set the two derivatives equal to each other and solve for $y'(x)$.

- From the *Joy* menus, choose **Algebra** ▷ **Solve**.
- Fill in the equation Out[2] = 0, using your output number for the equation to be solved if it is different from this.
- Fill in y'[x] as the unknown quantity for which we want to solve.
- Click **Solve algebraically**.

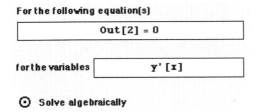

For the following equation(s)

Out[2] = 0

for the variables | **y' [x]** |

⊙ Solve algebraically

- Click **OK**.

Joy displays a message saying that $y'(x)$ doesn't appear explicitly in the equation you entered. Click **Continue** to ignore the message and solve anyway.

Mathematica gives the result:

▷ Solve the equation %2 = 0 for y′[x] algebraically.

Out[3]= $\{\{y'[x] \rightarrow -\dfrac{2\,x + 3\,y[x]}{3\,x + 8\,y[x]}\}\}$

This says that

$$y' = -\frac{2x+3y}{3x+8y}$$

is the derivative of the implicit functions defined by $x^2 + 3\,x\,y + 4\,y^2 = 4$.

♀ The paraphrase shows that %2 is another way to write Out[2] in *Mathematica*.

Chapter 8
Integrating Functions of One Variable

This chapter shows how to use *Joy* for calculating indefinite and definite integrals. It also contains numerical integration, including a special *Joy* dialog for numerical approximations using specific methods such as the midpoint approximation and Simpson's Rule.

■ 8.1 How to Find an Antiderivative (Indefinite Integral)

For example, here's how to calculate $\int (\ln x + e^{2x} \sin(x + a))\, dx$.

- Click in your *Mathematica* notebook below the last output, if any, so that a horizontal line appears.
- Type `Log[x] + E`^{2}`ˣ Sin[x + a]`.

    `Log[x]` is *Mathematica*'s notation for $\ln x$ and `E` stands for the constant e. You can use the **Functions** button on the *Joy* palette to show or paste in the correct syntax for a number of common functions.

- Drag across the expression you typed to select it.

    ♡ You can also click the **Select** button on the *Joy* palette enough times to select the whole expression.

$$\boxed{\text{Log[x] + E}^{2\,x}\ \text{Sin[x + a]}}$$

- In the *Joy* palette, click the **Antiderivative** button.

Antiderivative button.

An expression for the antiderivative appears, with the input field for the independent variable automatically selected.

$$\int \left(\text{Log} [x] + E^{2\,x}\,\text{Sin}[x+a] \right) d\square$$

Partially completed antiderivative.

- Type x.
- Press $\boxed{\text{SHIFT}} - \boxed{\text{RET}}$.

Mathematica calculates the antiderivative:

$In[2]:=$ $\int \left(\text{Log} [x] + E^{2\,x}\,\text{Sin}[x+a] \right) dx$

$Out[2]=$ $x\,\text{Log}[x] + \dfrac{1}{5}\,(-5\,x - E^{2\,x}\,\text{Cos}[a+x] + 2\,E^{2\,x}\,\text{Sin}[a+x])$

In Version 4, *Mathematica*'s output displays the constant e as $\mathbf{e}$.

⚠ *Mathematica* displays antiderivatives without an arbitrary constant of integration.

Another way. If you click the **Antiderivative** button without typing and selecting an expression, the field for the function to be integrated (the *integrand*) will be selected.

$$\int \blacksquare\, d\,\square$$

In this case fill in the function (it automatically goes in the selected field), press $\boxed{\text{TAB}}$, and then fill in the independent variable.

⚠ If the integrand contains more than one term, you will need to type parentheses around it. Notice that in the above example ($\text{In} [2]$) the parentheses appeared automatically.

8.1.1 When *Mathematica* Cannot Find an Antiderivative

When it calculates an antiderivative (indefinite integral) *Mathematica* uses symbolic techniques, i.e., integral tables and substitution methods. If its output contains an integral that is not evaluated, then *Mathematica* cannot integrate the function symbolically. For example:

$In[3]:=$ $\int \text{Log} [1 + \text{Sin} [x^2]]\,dx$

$Out[3]=$ $\int \text{Log} [1 + \text{Sin} [x^2]]\,dx$

For certain important functions that cannot be integrated symbolically, *Mathematica* gives a special name to the antiderivative. One example is the integral

$$\int e^{-x^2}\,dx\,,$$

which is related to the Gaussian, or normal, probability curve:

In[4]:= $\displaystyle\int \mathbf{E^{-x^2}}\ \mathbf{dx}$

Out[4]= $\dfrac{1}{2}\ \sqrt{\pi}\ \mathbf{Erf[x]}$

■ 8.2 How to Find a Definite Integral

For example, here's how to calculate $\int_1^4 \ln x\, dx$.

- Click in your *Mathematica* notebook below the last output, if any.
- Type Log[x],
- Drag across the expression you typed to select it.

- In the *Joy* palette, click the **Definite Integral** button.

Definite Integral button.

An expression for the integral appears, with the input field for the lower limit selected.

$$\int_{\square}^{\square} \mathbf{Log[x]}\ \mathbf{d}\square$$

- Type 1, press [TAB], type 4, and press [TAB] again.

  Pressing [TAB] moves the insertion point from the lower limit to the upper limit. Pressing it again moves it to the integrand.

- Type x.
- Press [SHIFT] − [RET].

Mathematica gives the exact value of the integral:

In[2]:= $\displaystyle\int_1^4 \mathbf{Log[x]}\ \mathbf{dx}$

Out[2]= $-3 + \mathbf{Log[256]}$

If you wish, you can use the **Numeric** button on the *Joy* palette to express the result in decimal form (Section 1.3.1).

The limits of integration can be symbols rather than specific numbers. For example, you can compute an integral such as $\int_1^x \ln t\, dt$:

In[3]:= $\int_1^x \text{Log}[t] \, dt$

Out[3]= $1 + x \, (-1 + \text{Log}[x])$

Another way. If you click the **Definite Integral** button without first typing and selecting a function, the field for the lower limit will be selected.

$$\int_\square^\square \square \, d\square$$

Fill in the lower limit, then the upper limits, function, and independent variable. Press [TAB] to move from one field to the next.

⚠ If the integrand contains more than one term, e.g., $\ln x + e^{2x} \sin(x + a)$, you will need to type parentheses around it. In this method, the parentheses will not appear automatically.

8.2.1 Editing Integrals

If you want to make changes in an integral that you have typed, click in the integral, select a part of it, and type the change you wish. It may be hard to select a small part of an integral, e.g., the upper or lower limit. This example, where we replace

$$\int_1^x \ln t \, dt \quad \text{by} \quad \int_a^x \ln t \, dt \, ,$$

shows how you to make the change. See below for how to keep both integrals in your notebook instead of replacing the old one.

• Click just to the right of the lower limit 1 in the integral $\int_1^x \text{Log}[t] \, dt$.

A short blinking cursor appears just to the right of 1. If you see a longer cursor, you need to click closer to 1.

In[3]:= $\int_{1|}^x \text{Log}[t] \, dt$

▲

Short blinking cursor.

• On the *Joy* palette, click the **Select** button.

Select

This highlights (selects) the lower limit, 1. You can extend a selection by pressing **Select** again.

♡ You can also select characters using the keyboard, by holding down the [SHIFT] key and pressing ← or →.

- Make sure only the lower limit 1 is selected, and type **a .**
- Press SHIFT — RET.

In[4]:= $\displaystyle\int_{a}^{x}$ **Log[t] d t**

Out[4]= **-a (-1 + Log[a]) + x (-1 + Log[x])**

> ♡ If you want to retain the old integral in your notebook, make a copy of it first. Drag across the integral or else click on the bracket for its cell (Section 1.4.2), copy the selection, click below the last output, and paste. Then edit the copy.

■ 8.3 How to Integrate Numerically

When no elementary antiderivative exists, *Mathematica* can estimate it numerically. In other cases there may exist an elementary antiderivative, but the function may be so complicated that it is more efficient to integrate numerically. Here's how to do that with

$$\int_{0}^{\pi} (x + \sin(\sin x))\, dx \ .$$

- If necessary, click below the last output, if any, so that a horizontal line appears.
- Type **x + Sin[Sin[x]]**.
- Drag across the expression you typed to select it.

- In the *Joy* palette, click the **Numeric Integral** button.

Numeric Integral button.

The letter **N** shows that the integral will be estimated numerically, using *Mathematica*'s built-in algorithm.

An expression for the numeric integral appears, with the input field for the lower limit selected.

- Type **0**, press TAB, type π, and press TAB again, to move to the input field for the variable (d□).

  You can enter π by clicking the π button on the *Joy* palette.

- Type **x** and press SHIFT — RET.

Mathematica calculates a numerical approximation to the integral:

In[2]:= $\text{N} \int_0^\pi (\text{x} + \text{Sin[Sin[x]]}) \, d\text{x}$

Out[2]= 6.72129

⚠ *Joy* defines expressions of the form $\text{N} \int_0^\pi \text{f[x]} \, d\text{x}$ for *Mathematica* and makes this equivalent to the *Mathematica* function `NIntegrate`. If you run *Mathematica* without *Joy* and type N before an integral, the expression will be undefined and will not produce a numeric integral.

Another way. If you click the **Numeric Integral** button before typing and selecting a function, the field for the lower limit will be selected.

$$\text{N} \int_{\square}^{\square} \square \, d \square$$

Fill in the lower limit, then the upper limits, function, and independent variable. Press [TAB] to move from one field to the next.

⚠ If the integrand contains more than one term, you will need to type parentheses around it. Notice that in the above example (`In[2]`) the parentheses appeared automatically.

If you try to evaluate a definite integral exactly and *Mathematica*'s output either contains an unevaluated integral or is expressed as a special function whose exact values are unknown, you can find a numerical approximation by using the **Numeric** button. For example:

- Enter the integral $\int_0^\pi (\text{x} + \text{Sin[Sin[x]]}) \, d\text{x}$ and press [SHIFT] − [RET].

  You can copy and paste it from the preceding calculation, being sure *not* to copy the symbol **N** (numeric).

In[3]:= $\int_0^\pi (\text{x} + \text{Sin[Sin[x]]}) \, d\text{x}$

Out[3]= $\dfrac{\pi^2}{2} + 2 \, \text{HypergeometricPFQ}\left[\{1\}, \left\{ \dfrac{3}{2}, \dfrac{3}{2} \right\}, -\dfrac{1}{4} \right]$

- Drag across the last output to select it.
- Click the **Numeric** button on the *Joy* palette.
- Press [SHIFT] − [RET].

The result is the same as before:

In[4]:= $\text{N}\left[\dfrac{\pi^2}{2} + 2 \, \text{HypergeometricPFQ}\left[\{1\}, \left\{ \dfrac{3}{2}, \dfrac{3}{2} \right\}, -\dfrac{1}{4} \right] \right]$

Out[4]= 6.72129

8.3.1 Simpson's Rule and Other Methods of Numerical Integration

Joy also includes several other methods for numerical integration. *Mathematica*'s numerical integration algorithm is adaptive in the sense that its precise operation and the number of steps it takes depend on the behavior of the function involved. *Joy* includes several other algorithms that are less efficient but are useful for understanding the principles of numerical integration. These are Simpson's Rule and the left and right endpoint, midpoint, and trapezoidal approximations.

For example, here is how to estimate $\int_0^\pi (x + \sin(\sin x))\,dx$ using Simpson's Rule with increasingly many subintervals.

- From the *Joy* menus, choose **Calculus ▷ Custom Integration**.

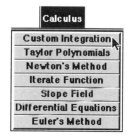

Calculus menu.

- Erase the default integral, which is automatically selected.
- Click the **Definite Integral** button on the *Joy* palette.
- Fill in the integral $\int_0^\pi (\texttt{x + Sin[Sin[x]]})\,d\texttt{x}$.

    Instead of filling in a new integral, you can edit the default integral as shown in Section 8.2.1.

- Keep the default settings, Simpson's Rule and $10, 20, 40, 80$ subintervals.

    For Simpson's Rule, the number n of subintervals must be *even*. You can find approximations for several values of n at once by separating them with commas.

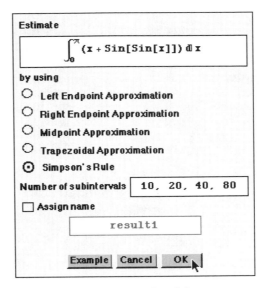

Custom Integration dialog.

- Click **OK**.

Mathematica's output includes all four approximations:

▷ The Simpson's Rule approximation for $\int_{0}^{\pi} (x + \mathrm{Sin}[\mathrm{Sin}[x]])\, dx$ with
   {10, 20, 40, 80} subintervals gives

Out[8]= {6.72152, 6.7213, 6.72129, 6.72129}

Chapter 9

Working with Sequences and Series

This chapter shows how to use *Joy* to work with sequences, including the sequence of partial sums of an infinite series. It contains making tables of values and graphing all or selected terms of the sequence. It also includes a special menu item for Taylor polynomials.

■ 9.1 How to Create a Sequence

Sections 9.1–9.4 illustrate working with sequences, using the example

$$a_k = \frac{(-1)^k}{\sqrt{k^2 + 1}}.$$

- From the *Joy* menus, choose **Create ▷ Sequence**.

Create menu.

- Replace the default sequence by

$$\frac{(-1)^k}{\sqrt{k^2 + 1}}.$$

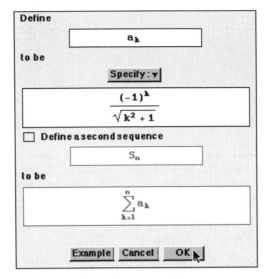

Create Sequence dialog.

- Click **OK**.

    When you click **OK**, *Mathematica* creates the sequence but does not display any output.

▷ Define a_k to be $\dfrac{(-1)^k}{\sqrt{k^2 + 1}}$

■ 9.2 How to Find a Value in a Sequence

Here's how to find the value of any particular term, such as a_7, in a sequence such as

$$a_k = \frac{(-1)^k}{\sqrt{k^2 + 1}} \; .$$

- If you didn't create the sequence a_k in Section 9.1, do it now.
- If necessary, click in your notebook below the last output so that a horizontal line appears.
- Type a_7.
- Press [SHIFT] − [RET].

Mathematica displays the result:

In[3]:= a_7

Out[3]= $-\dfrac{1}{5\sqrt{2}}$

Another way. If you haven't created the sequence in advance, you can choose **Algebra** ▷ **Substitute** and replace k by 7 in $\frac{(-1)^k}{\sqrt{k^2+1}}$.

■ 9.3 How to List the Values of a Sequence

Here's how to create a table of values of the elements of a sequence such as

$$a_k = \frac{(-1)^k}{\sqrt{k^2 + 1}},$$

for $k = 0, 1, 2, \ldots, 10$.

- If you didn't create the sequence a_k in Section 9.1, do it now.
- From the *Joy* menus, choose **Create** ▷ **List**.
- Replace the default expression by $\mathtt{k}$, a_k.
- Keep the default setting of the popup menu, **x Equally Spaced**.
- Fill in $0 \leq \mathtt{k} \leq 10$ in increments of 1.

    Be sure to change $\mathtt{x}$ to $\mathtt{k}$.

- Keep the default settings **Display in decimal form, Display as a table**.

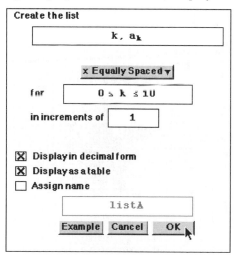

Create List dialog.

- Click **OK**.

Here is the result:

▷ Create the list with values $\{k, a_k\}$ for $0 \leq k \leq 10$.

Out[4]//TableForm=

0	1.
1.	−0.707107
2.	0.447214
3.	−0.316228
4.	0.242536
5.	−0.196116
6.	0.164399
7.	−0.141421
8.	0.124035
9.	−0.110432
10.	0.0995037

The popup menu gives you other choices for specifying how the variable *k* will change in the list. See Section 4.2 for other ways to make lists.

■ 9.4 How to Plot a Sequence

Joy gives two ways to graph a sequence: plotting the points $\{k, a_k\}$ and plotting the corresponding points $\{0, a_k\}$ on the *y*-axis. You can also combine the two in a single plot. Here's how to do this for the first 100 points of the sequence

$$a_k = \frac{(-1)^k}{\sqrt{k^2 + 1}} \, .$$

- If you didn't create the sequence a_k in Section 9.1, do it now.
- From the *Joy* menus, choose **Graph ▷ Sequence**.

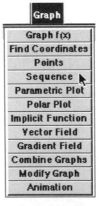

Graph menu.

- Replace the default sequence with a_k.
- Fill in the values $0 \le k \le 100$.
- Check the box **Plot values on y-axis**.

The box **Graph** is already checked by default.

For the sequence

$$a_k$$

for $0 \le k \le 100$

y-interval: Automatic ▾

[X] Graph
[X] Plot values on y‑axis
[ ] Position axes at the origin
[ ] Assign name

graph1

Example   Cancel   OK

Graph Sequence dialog.

- Click **OK**.

Mathematica displays the two plots:

▷ Plot of the sequence a_k for $0 \le k \le 100$.

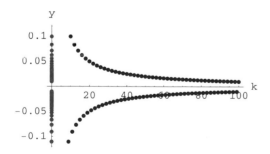

Out[5]=   - Graphics -

Another way. If you didn't create the sequence in advance, just fill in the expression $\frac{(-1)^k}{\sqrt{k^2+1}}$.

■ 9.5 How to Create a Series

Sections 9.5–9.8 show how to work with series, using the example

$$\sum_{k=1}^{\infty} \frac{1}{3^k} \, .$$

The method shown here simultaneously creates two sequences, the terms a_k and the partial sums S_n,

$$a_k = \frac{1}{3^k} \, ,$$
$$S_n = \sum_{k=1}^{n} a_k = \sum_{k=1}^{n} \frac{1}{3^k} \, .$$

- From the *Joy* menus, choose **Create ▷ Sequence**.
- Fill in the sequence a_k given by $\frac{1}{3^k}$.
- Check **Define a second sequence**.
- Keep the name S_n for the partial sums and the definition $\sum_{k=1}^{n} a_k$.

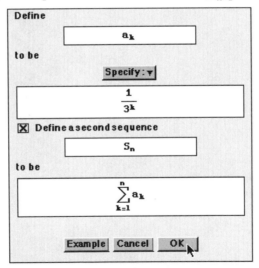

Create Sequence dialog.

- Click **OK**.

Mathematica defines the two sequences but displays no output:

▷ Define a_k to be $\frac{1}{3^k}$ and define S_n to be $\displaystyle\sum_{k=1}^{n} a_k$

■ 9.6 How to Find One Partial Sum of a Series

Here's how to find any particular partial sum of a series, such as S_{10}, for a series with partial sums S_n. We'll illustrate it with the example $S_n = \displaystyle\sum_{k=1}^{n} \frac{1}{3^k}$ of Section 9.5.

- If you didn't create the partial sums in 9.5, do it now.
- If necessary, click in your notebook below the last result so that a horizontal line appears.
- Type S_{10} and press SHIFT – RET.

    Be sure to type capital S, not lowercase s.

The result is:

In[4]:= $\mathbf{S_{10}}$

Out[4]= $\dfrac{29524}{59049}$

You can express it in decimal form by clicking the **Numeric** button **N[■]**, filling in S_{10}, and pressing SHIFT – RET:

In[5]:= $\mathbf{N[S_{10}]}$

Out[5]= 0.499992

■ 9.7 How to Tabulate the Partial Sums of a Series

Here's how to make a table of values for the partial sums of a series, for example $S_1, \ldots, S_{10}$, where $S_n = \displaystyle\sum_{k-1}^{n} \frac{1}{3^k}$ is the example in Section 9.5.

Since the partial sums form a sequence, this is the same method as in 9.3.

- If you didn't create the partial sums in 9.5, do it now.
- From the *Joy* menus, choose **Create ▷ List**.
- Fill in n, S_n for the expression to be listed.
- Keep the default setting of the popup menu, **x Equally Spaced**.
- Fill in the values $1 \le n \le 10$ in increments of 1.
- Keep the default settings **Display in decimal form, Display as a table**.

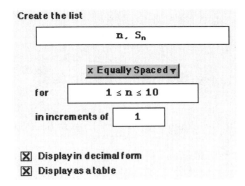

- Click **OK**.

Mathematica produces the list:

▷ Create the list with values {n, S$_n$} for 1 ≤ n ≤ 10.

Out[6]//TableForm=

1.	0.333333
2.	0.444444
3.	0.481481
4.	0.493827
5.	0.497942
6.	0.499314
7.	0.499771
8.	0.499924
9.	0.499975
10.	0.499992

■ 9.8 How to Plot the Partial Sums of a Series

Here's how to plot the partial sums of a series. Since the partial sums of a series form a sequence, this is the same method as in Section 9.4. We illustrate it with the first 10 partial sums of $\sum_{k=1}^{\infty} \frac{1}{3^k}$.

- If you didn't create the partial sums in 9.5, do it now.
- From the *Joy* menus, choose **Graph ▷ Sequence**.
- Fill in S$_n$.
- Fill in 1 ≤ n ≤ 10.

- Make sure **Graph** is checked and **Plot values on y-axis** is *unchecked*.

  In this example we won't graph the corresponding points on the *y*-axis.

For the sequence

$$S_n$$

for $1 \le n \le 10$

y-interval: Automatic ▾

☒ **Graph**
☐ **Plot values on y – axis**
☐ **Position axes at the origin**
☐ **Assign name**

graph1

Example | Cancel | OK ▸|

Graph Sequence dialog.

- Click **OK**.

Mathematica graphs the sequence:

▷ Plot of the sequence S_n for $1 \le n \le 10$.

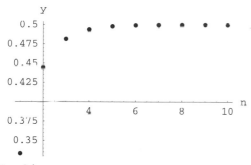

Out[7]= - Graphics -

🖋 *Mathematica* has set the axes to cross at $x = 2$, $y = 0.4$ because the points are clustered far from the origin. You can check the box **Position axes at the origin** in **Graph** ▷ **Sequence** to avoid this, but it may compress the graph into a very small space.

Another way. If you haven't created the sequence first, you can use the **Sum** button together with the **Graph** ▷ **Sequence** dialog. Here's how to do it.

- Choose **Last Dialog** to bring back the **Graph** ▷ **Sequence** dialog.
- Fill in the sequence $\frac{1}{3^k}$.
- Drag across the sequence to select it.

For the sequence

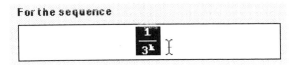

- Click the **Sum** button on the *Joy* palette.

Sum button.

The highlight field on the button shows that selecting the sequence first will place it inside the summation sign.

- Type k, press [TAB], and type 1.

$$\sum_{k=1}^{\square} \frac{1}{3^k}$$

Partially completed sum.

Pressing [TAB] moves the insertion point from one input field in the template to the next. The lower limit of the summation now reads k = 1.

- Press [TAB] again and type n.

This completes the summation.

- Fill in the graphing interval $1 \le n \le 10$.

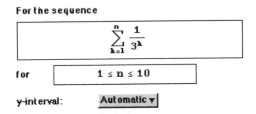

For the sequence

$$\sum_{k=1}^{n} \frac{1}{3^k}$$

for $1 \le n \le 10$

y-interval: Automatic ▾

- Click **OK**.

Mathematica creates the same graph as before:

▷ Plot of the sequence $\sum_{k=1}^{n} \frac{1}{3^k}$ for $1 \le n \le 10$.

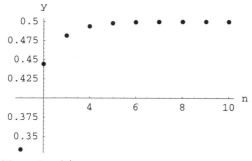

Out[8]=   - Graphics -

♡ You can make changes in a sum by clicking in it, selecting a part of it, and typing the change you wish. It may be hard to select a small part of a sum, e.g., part of the lower limit. You can click in the sum and then use the **Select** button on the *Joy* palette to help select the part that you want to change. Section 8.2.1 shows how to use the **Select** button in an analogous situation with integrals.

■ 9.9 How to Estimate a Sum Numerically

Calculating a partial sum S_n of a series exactly can be difficult when n is large. For example, if the terms of the series are all rational numbers, *Mathematica* will find a common denominator for the sum. When there are enough terms, the denominator will be huge, the calculation will take an inordinate amount of time and memory, and *Mathematica* may seem to hang up.

There are two shortcuts that you can use to tabulate or graph the partial sums S_n of a series when n is large:

> Approximate the sums numerically rather than calculate them exactly
> Let n increase in large steps rather than one at a time

We'll use the example

$$\sum_{k=0}^{\infty} \frac{(-1)^k}{\sqrt{k+1}} = 1 - \frac{1}{\sqrt{2}} + \frac{1}{\sqrt{3}} - \dots.$$

- From the *Joy* menus, choose **Create ▷ Sequence**.
- Fill in the sequence a_k given by

$$\frac{(-1)^k}{\sqrt{k+1}}.$$

- If necessary, check **Define a second sequence** and fill in S_n.
- Type N to the left of $\sum$ and change the starting value of k to 0, so that the expression becomes N $\sum_{k=0}^{n} a_k$.

You can edit the starting value by clicking just to its right and then clicking **Select** on the *Joy* palette. The letter N indicates that the partial sums are to be estimated numerically.

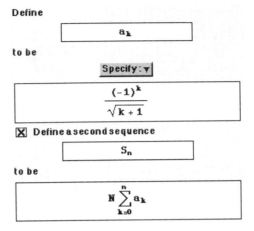

- Click **OK**.

Mathematica creates the two sequences:

▷ Define a_k to be $\dfrac{(-1)^k}{\sqrt{k+1}}$ and define S_n to be NSum[a_k, {k, 0, n}]

Another way. You can also create N $\sum_{k=0}^{n} a_k$ in a *Joy* dialog or in your *Mathematica* notebook by clicking the **Numeric sum** button on the *Joy* palette.

Numeric sum button.

Type the expression you want to sum up, select it, and click the button. Then fill in the lower and upper limits.

⚠ *Joy* defines expressions of the form N $\sum_{k=0}^{n} a_k$ for *Mathematica*. If you use *Mathematica* without *Joy* and type N before a summation, the expression will be undefined and will not produce the *Mathematica* function NSum shown in the paraphrase.

9.9.1 Tabulating and Graphing Numeric Sums

You can make tables and graphs for numeric sums just as for exact sums, as shown in Sections 9.7 and 9.8. Here is another method, in which we first make a table and then plot the points in the table. We'll take the partial sums of the terms up to $n = 1000$, in steps of size 25 to speed things up.

- Choose **Create ▷ List**.
- Fill in n, S_n, for $0 \le n \le 1000$ in increments of 25.
- *Uncheck* **Display as a table**.

  The list is too long to show as a table.

- Check **Assign name** and use the name listA or fill in a name of your own.

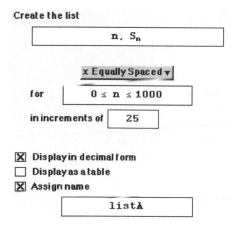

Create the list

n, S_n

x Equally Spaced ▼

for $0 \le n \le 1000$

in increments of   25

☒ Display in decimal form
☐ Display as a table
☒ Assign name

listA

- Click **OK**.

The result is:

▷ Create the list with values {n, S_n} for $0 \le n \le 1000$ in steps of size 25.

listA =

Out[3]= {{0, 1.}, {25., 0.507783}, {50., 0.674569},
      {75., 0.547733}, {100., 0.654527}, {125., 0.560443},
      {150., 0.645521}, {175., 0.567263}, {200., 0.640122}, {225., 0.571676},
      {250., 0.636427}, {275., 0.574829}, {300., 0.633694}, {325., 0.577227},
      {350., 0.631568}, {375., 0.57913}, {400., 0.629852}, {425., 0.580688},
      {450., 0.62843}, {475., 0.581993}, {500., 0.627226}, {525., 0.583108},
      {550., 0.62619}, {575., 0.584074}, {600., 0.625286}, {625., 0.584923},
      {650., 0.624488}, {675., 0.585675}, {700., 0.623777}, {725., 0.586348},
      {750., 0.623138}, {775., 0.586955}, {800., 0.62256}, {825., 0.587507},
      {850., 0.622033}, {875., 0.58801}, {900., 0.621551}, {925., 0.588472},
      {950., 0.621108}, {975., 0.588898}, {1000., 0.620698}}

Here's how to graph S_n.

- From the *Joy* menus, choose **Graph ▷ Points**.
- Fill in listA, or your name for the list if it is different from this.

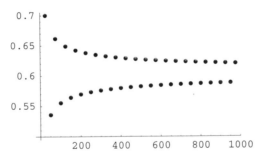

Plot Points dialog.

- Click **OK**.

Mathematica graphs the partial sums:

▷ Plot of the points listA

Out[6]= - Graphics -

■ 9.10 How to Create a Taylor Series

Joy provides a special menu item for computing and graphing the Taylor polynomials of a function, centered about a point of your choice. Here's how to create the Taylor polynomials $P_n(x)$ of $\ln x$ centered at $a = 1$. We'll list the polynomials of degrees 1 through 4 and graph them for $0 \le x \le 3$.

- From the *Joy* menus, choose **Calculus ▷ Taylor Polynomials**.

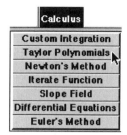

Calculus menu.

- Replace the default function by Log[x].

  You can type Log[x] for ln *x* or else use the **Functions** button on the *Joy* palette to enter it.

- Replace the center point by 1.
- Replace the degrees by 1,2,3,4.
- Replace the interval by 0 ≤ x ≤ 3.

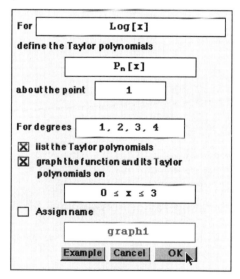

Taylor Polynomials dialog.

- Click **OK**.

Here is the result:

▷ Define $P_n[x]$ to be the Taylor polynomial for `Log[x]` about the point `1`.

For degrees n = {1, 2, 3, 4}, graph the function
  `Log[x]` and its Taylor polynomials on the interval 0 ≤ x ≤ 3.

`Log[x]`	Black	Solid
$P_1[x]$	Red	.....
$P_2[x]$	Blue	_ _ _ _
$P_3[x]$	Green	__ __
$P_4[x]$	Magenta	. _ . _ .

$$-1 + x$$
$$-1 - \frac{1}{2}(-1+x)^2 + x$$
$$-1 - \frac{1}{2}(-1+x)^2 + \frac{1}{3}(-1+x)^3 + x$$
$$-1 - \frac{1}{2}(-1+x)^2 + \frac{1}{3}(-1+x)^3 - \frac{1}{4}(-1+x)^4 + x$$

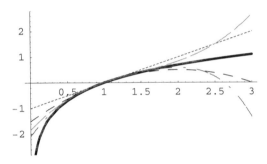

Out[2]=   - Graphics -

⚠ *Joy* automatically clears any value that might have been assigned to the variable *x* before defining Taylor polynomials in terms of *x*.

9.10.1 Evaluating and Tabulating Taylor Polynomials

Joy creates Taylor polynomials as functions that can be evaluated, tabulated, graphed, etc. For example, here's how to find $P_{10}(2)$.

- If you didn't create the Taylor polynomials in Section 9.10, do that now.
- Click in your *Mathematica* notebook, beneath the last output.
- Type $P_{10}[2]$.
- Press SHIFT − RET.

Mathematica calculates the result:

In[3]:= $P_{10}[2]$

Out[3]= $\dfrac{1627}{2520}$

You can express this in decimal form using the **Numeric** button on the *Joy* palette:

In[4]:=   **N[P$_{10}$[2]]**

Out[4]=   0.645635

Here's how to make a table of values for $P_n(2)$, $n = 2, 2^2, 2^3, \ldots, 2^{10}$.

- From the *Joy* menus, choose **Create ▷ List**.
- Fill in the values n, P$_n$[2].
- Click the popup menu and choose **x=f(n)**.

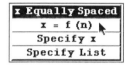

- Fill in the interval $n = 2^m$ and $1 \le m \le 10$.
- Check **Display in decimal form** and **Display as a table**.

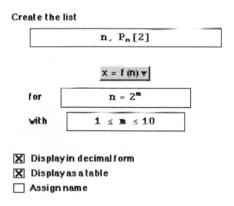

- Click **OK**.

Mathematica creates the table:

▷ Create the list with values $\{n, P_n[2]\}$ for $n = 2^m$ with $1 \le m \le 10$.

Out[5]//TableForm=

2.	0.5
4.	0.583333
8.	0.634524
16.	0.662872
32.	0.677766
64.	0.685396
128.	0.689256
256.	0.691198
512.	0.692172
1024.	0.692659

Chapter 10
Graphing in Three Dimensions

This chapter includes graphing surfaces, level sets, and space curves. For defining functions of several variables and using functional notation for surfaces, see Chapter 11. For parameterized curves in two dimensions, see Section 3.5.

■ 10.1 How to Graph a Surface

Here's how to graph the surface $z = \sin(y - x^2)$ for $-2 \le x \le 2$ and $-4 \le y \le 4$.

- From the *Joy* menus, choose **Graph 3D ▷ Surface**.

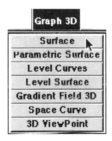

Graph 3D menu.

- In the dialog that appears, fill in the expression $\mathtt{Sin}\,[\,\mathtt{y} - \mathtt{x}^2\,]$.
- Fill in the intervals $-2 \le \mathtt{x} \le 2$ and $-4 \le \mathtt{y} \le 4$.

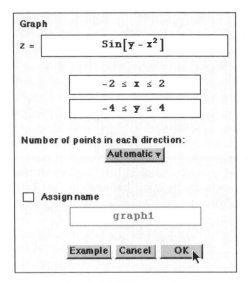

Graph 3D Surface dialog.

- Click **OK**.

Mathematica graphs the surface:

▷ Plot Sin[y − x²] for −2 ≤ x ≤ 2 and −4 ≤ y ≤ 4

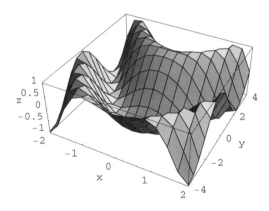

Out[2]= - SurfaceGraphics -

△ *Mathematica* draws a frame around the surface with coordinate scales along three sides. These sides are generally *not the x, y, z-axes but are parallel to them*. The three axes meet where $x = y = z = 0$, which for this example is in the center of the frame.

♡ If your graph does not have coordinate scales, it is an *interactive* graph (*Mathematica* Version 4). Section 10.2.1 discusses special issues in working with interactive graphs.

10.1.1 Making the Surface Smoother

The graph of a surface may sometimes appear jagged and be hard to visualize. You can make the surface look smoother by increasing the number of grid points (x, y) where *Mathematica* evaluates z.

> This is because *Mathematica* evaluates z at a fixed number of grid points regardless of how quickly z changes, and then fills in the surface between these points. When graphing curves in two dimensions, *Mathematica* uses an adaptive algorithm that evaluates y at more values of x when y is changing rapidly. So graphs of curves are usually smooth.

Here's how to do this with the graph of $z = \sin(y - x^2)$ for $-2 \le x \le 2$ and $-10 \le y \le 10$.

- Click **Last Dialog** on the *Joy* palette, or choose **Graph 3D ▷ Surface**.
- If you didn't graph the surface in Section 10.1, then fill in the dialog now.
- Change the y-interval to $-10 \le y \le 10$.

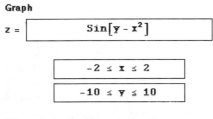

Graph

$z =$ $\boxed{\qquad \text{Sin}\big[\text{y} - \text{x}^2\big] \qquad}$

$\boxed{-2 \le \text{x} \le 2}$

$\boxed{-10 \le \text{y} \le 10}$

Number of points in each direction:
Automatic ▾

- Click **OK**.

Mathematica graphs the surface:

▷ Plot Sin[y – x²] for –2 ≤ x ≤ 2   and –10 ≤ y ≤ 10

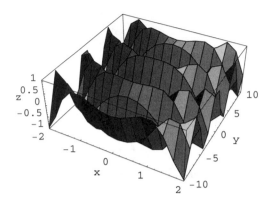

Out[3]= - SurfaceGraphics -

Now we'll make the graph smoother.

- Click **Last Dialog**.
- From the popup menu **Number of points in each direction**, choose **Specify**.
- Fill in 30.

  This doubles the default number of points.

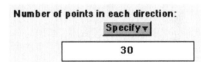

- Click **OK**.

The new graph is much smoother and shows more detail:

▷ Plot Sin[y − x²] for −2 ≤ x ≤ 2 and −10 ≤ y ≤ 10

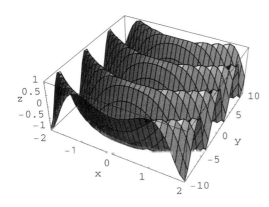

Out[4]=   - SurfaceGraphics -

⚠ Increasing the number of points to be evaluated makes graphing slower and takes more memory, so it's a good idea to use this sparingly.

■ 10.2 How to Change Your View of a Surface

Joy includes two methods for rotating a 3D graph to change your view of it, i.e., to change the point from which your eye sees the surface. Depending on your version of *Mathematica* and on the properties you want the graph to have, you may be able to perform the rotation interactively.

10.2.1 Interactive 3D Graphs (Requires *Mathematica* Version 4)

Mathematica Version 4 includes *interactive 3D graphs*. At this writing (version 4.0.1) the interactive mode is experimental, i.e., interactive 3D graphics may work differently in later versions of *Mathematica*.

Here's how to rotate a 3D graph interactively.

- If you didn't graph the surface $z = \sin(y - x^2)$ in Section 10.1, do it now.

In the interactive mode, the graph appears this way:

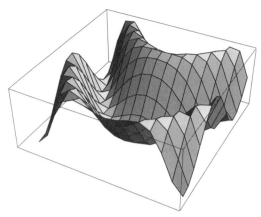

Out[2]= - SurfaceGraphics -

Notice that the graph has no coordinate scales or axis labels. This is characteristic of interactive graphs.

- Click on the graph to select it and drag it to a new orientation.

    A square frame appears around the graph when you select it.

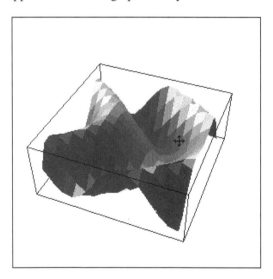

- Release the mouse button when you are satisfied with the new orientation.

The new graph is:

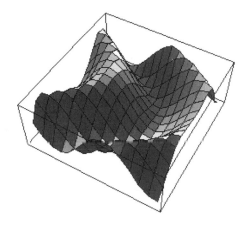

Out[2]=   - SurfaceGraphics -

You can now add coordinate scales and axis labels, which makes the resulting graph *noninteractive*. Here's how to do that.

- From the *Joy* menus, choose **Graph ▷ Modify Graph**.
- Make sure the name or output number of the graph appears, in this case Out[2].

    Modify Graph automatically fills in the output name or number of the most recent graph. Rotating an interactive graph doesn't change its name or number.

- In the **Axis Labels** popup menu, fill in "x","y","z".

    Adding labels to the graph also adds coordinate scales.

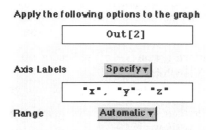

- Click **OK**.

Mathematica modifies the graph and makes it noninteractive:

▷ Redisplay %2 . (Noninteractive)

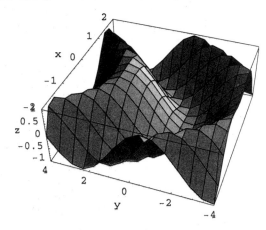

Out[5]= ▪ Graphics3D ▪

In *Joy*'s default setting, any 3D graphs that you create will initially be interactive. Section 2.3.3 shows how to change this setting if you wish.

     ◊ **Modify Graph** always produces a noninteractive graph. If you prefer all your 3D graphs to have scales and labels, you may want to choose the noninteractive setting rather than choose **Modify Graph** each time.

     ⚠ *Mathematica* will give an error message if you use Version 3 to open a notebook containing an interactive graph.

10.2.2 Noninteractive 3D Graphs

Mathematica Version 3 always creates noninteractive 3D graphs. In Version 4 you can choose whether 3D graphs should initially be interactive or not, and **Graph ▷ Modify Graph** always renders them noninteractive. Here's how to rotate a noninteractive graph.

- If you didn't graph the surface $z = \sin(y - x^2)$ in Section 10.1, do it now.
- Choose **Graph 3D ▷ 3D ViewPoint**.
- Fill in Out[2], or use the output number or name for your graph if it is different from this.

    If you are using *Mathematica* 4, make sure the graph is noninteractive. Otherwise, you can rotate it by dragging as shown in 10.2.1.

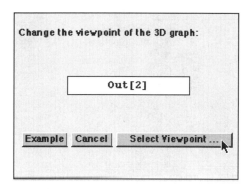

3D ViewPoint dialog.

- Click **Select Viewpoint …**.

Joy's instructions for changing the viewpoint appear:

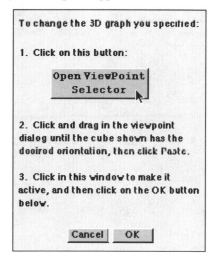

- Click **Open ViewPoint Selector**.

Mathematica's **ViewPoint Selector** opens and shows a cube whose sides are parallel to the x, y, z-axes. By changing the cube's orientation, you can specify how you want the surface to look.

- Click and drag the cube in the **ViewPoint Selector** until you are satisfied with its orientation.

    If you want to erase your changes and start again, click **Defaults**.

- When you are satisfied, click **Paste**.

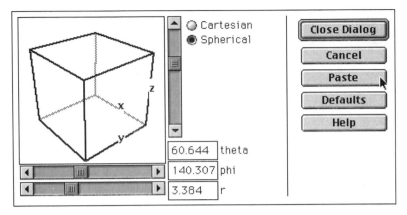

Mathematica's **ViewPoint Selector**.

- Click in the **Viewpoint Instructions** window to make it active.

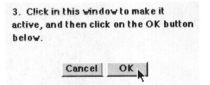

- Click **OK**.

Now *Mathematica* redraws the surface using the view you specified. The command that *Joy* creates shows the coordinates of the point from which you are viewing the surface.

∇ Changing the viewpoint of Out[2] gives

In[5]:= **Show[Out[2], ViewPoint -> {-2.269, 1.884, 1.659}]**

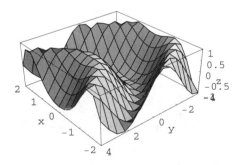

Out[5]=   - SurfaceGraphics -

To try another viewpoint:

- Click **Last Dialog** on the *Joy* palette.
- Repeat the preceding steps.

Mathematica's **ViewPoint Selector** will remain open. When you are finished, you may wish to clean up by clicking **Close Dialog** in the Selector window (it may be hidden behind other windows).

    ♡ The next time you open **ViewPoint Selector**, the cube will appear the way you left it. To restore it to its default position, click **Defaults**. Then drag it to whatever position you like.

■ 10.3 How to Plot the Level Curves of a Surface (Contours)

The *level curves* or *contours* of a surface are the curves in R^2 along which the height of the surface is constant. Here is how to plot the level curves of $z = \sin x \sin y$, for x and y between ±2.

- From the *Joy* menus, choose **Graph 3D ▷ Level Curves**.
- Fill in the expression `Sin[x]Sin[y]`.
- Fill in the intervals $-2 \leq x \leq 2$ and $-2 \leq y \leq 2$.

Graph the level curves

                            `Sin[x] Sin[y]`   = c

for

                      $-2 \leq x \leq 2$

                      $-2 \leq y \leq 2$

for values of c forming a contour plot with

                           10   **contour lines**.

Number of points in each direction   15

    ☐ **Contour Shading**
    ☐ **Assign name**

                    `graph1`

        Example **Cancel** **OK**

Level Curves dialog.

- Click **OK**.

Mathematica plots the result.

▷ Plot the level curves `Sin[x] Sin[y] = c` for $-2 \le x \le 2$ and $-2 \le y \le 2$

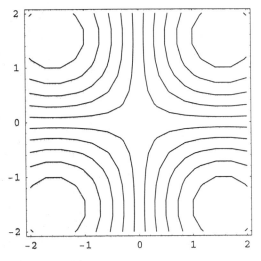

Out[2]=   ‐ ContourGraphics ‐

Some examples look better with contour shading that shows where the surface is higher and where it is lower. Here's how to shade the contours.

- On the *Joy* palette, click **Last Dialog**.
- Check **Contour Shading**.

☒ **Contour Shading**

- Click **OK**.

Mathematica redraws the curves with shading. Lighter regions correspond to larger values of $z = \sin x \sin y$, i.e., regions where the surface is higher.

▷ Plot the level curves Sin[x] Sin[y] = c for −2 ≤ x ≤ 2 and −2 ≤ y ≤ 2

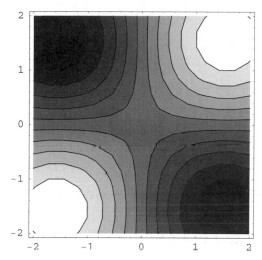

Out[3]= - ContourGraphics -

By default, *Mathematica* will draw 10 contour levels in the graph. For some examples, you may wish to show more levels to get a better sense of how rapidly the height of the surface is changing. To do this, just change the number of **contour lines** in the dialog.

✏ If the level curves or contours aren't smooth, you can have *Mathematica* calculate the height of the surface at more points. Just increase the **number of points in each direction** from the default value of 15.

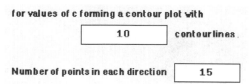

■ 10.4 How to Plot Level Surfaces (3D Contours)

The *level surfaces* or *contours* of a function $f : R^3 \to R$ are the surfaces in R^3 along which f is constant. Here is how to plot the level surface $x^2 + y^2 - z^2 = 1$, for x, y, z between ±3.

- From the *Joy* menus, choose **Graph 3D ▷ Level Surface**.
- Fill **x² + y² − z²**.
- Fill in the desired value **1**.
- Fill in the intervals −3 ≤ x ≤ 3, −3 ≤ y ≤ 3, −3 ≤ z ≤ 3.
- Check **Assign name** and fill in surface or a name of your choice.

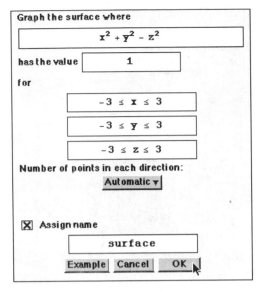

Level Surface dialog.

- Click **OK**.

Mathematica graphs the surface:

▷ Graph of the level surface $x^2 + y^2 - z^2 = 1$ for $-3 \le x \le 3$, $-3 \le y \le 3$, $-3 \le z \le 3$.

surface =

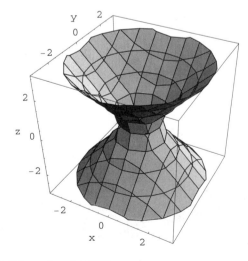

Out[2]= - Graphics3D -

⚠ *Joy* automatically clears any values that may have been given to the variables before graphing a level surface.

💡 If the level surface isn't smooth, you can have *Mathematica* calculate the value of the function at more points. Here's how to do it.

- Click **Last Dialog**.
- From the popup menu **Number of points in each direction**, choose **Specify**.
- Increase the number of points from its default value of 4.

Number of points in each direction:

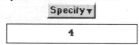

Specify ▾

4

⚠ Increasing the number of points to be evaluated makes graphing slower and takes more memory, so it's a good idea to use this sparingly.

■ 10.5 How to Combine 3D Graphs

Here's how to combine the graphs of $x^2 + y^2 - z^2 = 1$ and $x = 2$.

- If you didn't graph $x^2 + y^2 - z^2 = 1$ in Section 10.4, do it now.
- Click **Last Dialog** on the *Joy* palette to bring back the **Level Surface** dialog.

  In three variables, the equation $x = 2$ describes a level surface of $f(x, y, z) = x$.

- Fill in x for the function.
- Fill in 2 for the value.
- Keep the intervals $-3 \le x \le 3$, $-3 \le y \le 3$, $-3 \le z \le 3$.
- Change the name to `plane` or another name of your choice.

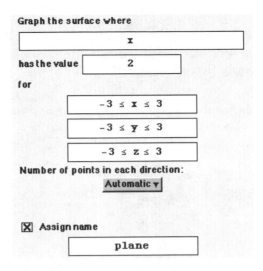

- Click **OK**.

Here is the graph of $x = 2$ in R^3:

```
▷ Graph of the level surface x = 2 for
 -3 ≤ x ≤ 3,  -3 ≤ y ≤ 3, -3 ≤ z ≤ 3.

plane =
```

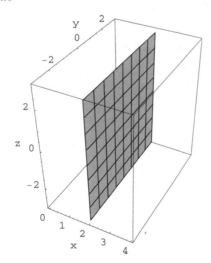

Out[3]= - Graphics3D -

- Choose **Graph** ▷ **Combine Graphs**.

- Make sure the names `surface,plane` or the ones you assigned instead appear in the dialog.

  Combine Graphs automatically displays the output names or numbers of the two most recent graphs. Using names for output rather than numbers makes it easier to reopen a notebook for later use. Section 2.2 gives more details about this.

Combine the graphs

> surface, plane

☐ **Assign name**

> graph1

[Example] [Cancel] [OK]

Combine Graphs dialog.

- Click **OK**.

The result is:

▷ Combining the graphs {surface, plane} gives

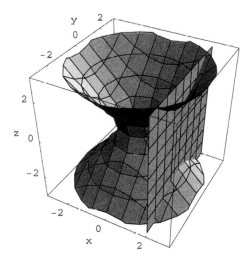

Out[4]=   - Graphics3D -

You can use **Graph ▷ Combine Graphs** to combine any 3D graphs, not just level surfaces.

■ 10.6 How to Graph a Curve in Space

Here's how to graph a parameterized curve in space. We will use the example $(x, y, z) = (t + 4 \sin t, t + \cos t, t)$, with $0 \le t \le 10$. For parameterized curves in the plane, see Section 3.5.

- From the *Joy* menus, choose **Graph 3D ▷ Space Curve**.
- Fill in the equations `x = t + 4 Sin[t]`, `y = t + Cos[t]`, `z = t`.
- Fill in the interval $0 \le$ `t` ≤ 10.

You may choose other symbols instead of *x*, *y*, *z*, *t* if you wish.

Graph the space curve given by

```
x = t + 4 Sin[t]
```

```
y = t + Cos[t]
```

```
z = t
```

for

```
0 ≤ t ≤ 10
```

☐ Thick curve
☐ Assign name

```
graph1
```

[Example] [Cancel] [**OK**]

Space Curve dialog.

- Click **OK**.

Here is the result:

▷ Graph the space curve (x,y,z) = (t + 4 Sin[t],Cos[t] + t,t) for 0 ≤ t ≤ 10

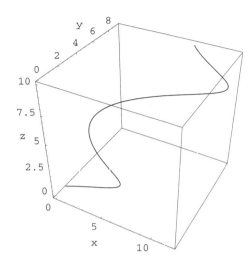

Out[2]= - Graphics3D -

■ 10.7 How to Graph a Parameterized Surface

Here is how to graph a surface expressed in terms of two parameters, such as $(x, y, z) = (u \cos v, \ u \sin v, \ u)$, where u, v are parameters and $-1 \le u \le 1$, $0 \le v \le 2\pi$.

- From the *Joy* menus, choose **Graph 3D ▷ Parametric Surface**.
- Fill in the equations x = u␣Cos[v], y = u␣Sin[v], z = u.

    The symbol ␣ means to leave a space to indicate multiplication.

- Fill in the intervals $-1 \le u \le 1$, $0 \le v \le 2\pi$.

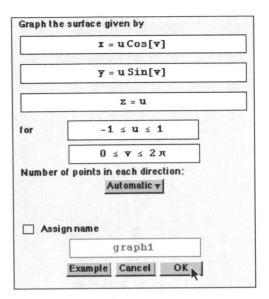

Parametric Surface dialog.

Mathematica graphs the surface:

▷ Parametric plot of the surface described by (x,y,z) = (Cos[v] u, u Sin[v],u) for −1 ≤ u ≤ 1 and 0 ≤ v ≤ 2π

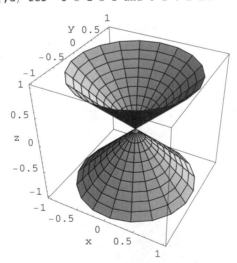

Out[2]= ‐ Graphics3D ‐

  ♀ If the surface isn't smooth, you can have *Mathematica* evaluate the coordinate functions at more values of the parameters. Here's how to do it.

- Click **Last Dialog**.
- From the popup menu **Number of points in each direction**, choose **Specify**.
- Increase the number of points from the default value of 20.

Number of points in each direction:

Specify ▾

20

■ 10.8 How to Change the Range in a 3D Graph

In some *Joy* dialogs for 3D graphing, *Mathematica* determines the range automatically. In others, you either may or must specify the range yourself. This section shows ways to modify the range.

Showing the entire range. *Mathematica* may cut off part of a surface when its height varies over a large range. Here's an example of this, and how to specify that all values should be shown. We will graph $z = e^{-x} \sin 2y$, for $-4 \le x \le 4$, $-\pi \le y \le \pi$.

- Choose **Graph 3D ▷ Surface**.
- Click **Example** to restore the default example.
- Change the *x*-interval to $-4 \le x \le 4$.

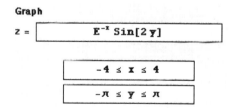

Graph

$z = $ $E^{-x} \, Sin[2\,y]$

$-4 \le x \le 4$

$-\pi \le y \le \pi$

- Click **OK**.

Mathematica graphs the surface but cuts off some values of z:

▷ Plot E^{-x} Sin[2 y] for $-4 \le x \le 4$ and $-\pi \le y \le \pi$

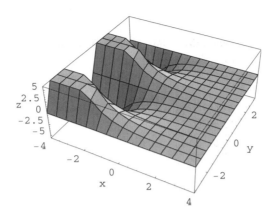

Out[2]= - SurfaceGraphics -

- From the *Joy* menus, choose **Graph ▷ Modify Graph**.
- Fill in Out[2], or your output number or name if it is different.

    Modify Graph automatically fills in the output number of the most recent graph.

- From the **Range** popup menu, choose **Show All**.

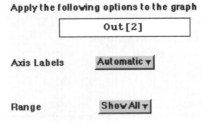

Choosing **Show All** in **Modify Graph**.

- Click **OK**.

Mathematica now shows all points for which *x, y* fall within the given domain (%2 is an abbreviation for Out[2]):

▷ Redisplay %2 showing all points.

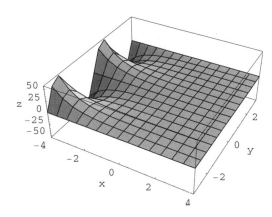

Out[3]= - SurfaceGraphics -

You can understand how the two graphs differ by comparing the scales on their *z*-axes.

Showing part of the range. Sometimes you may want to restrict your view to some of the height values *Mathematica* calculated rather than show all of them. In dialogs where you cannot specify the range directly, you can use **Graph ▷ Modify Graph**. In this case choose **Specify** from the **Range** popup menu and fill in the height values that you want.

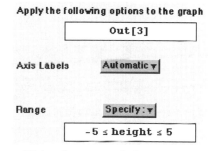

⚠ Since you are only modifying how an existing graph is displayed, **Modify Graph** can only show points that *Mathematica* has already computed.

■ 10.9 How to Animate 3D Graphs

Here's how to use animation to visualize the effect of the parameter *k* on the plane $z = x + ky$. We will use *Joy* to graph $x + ky$ for $k = -3, -2, -1, \ldots, 3$ and *x*, *y* between ±1. Then we'll view the seven graphs as successive frames, just as in a cartoon animation.

- From the *Joy* menus, choose **Graph ▷ Animation**.
- From the popup menu, choose **Surface**.

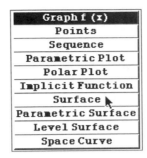

Animation popup menu.

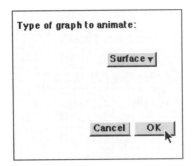

- Click **OK**.

Two dialogs appear, one to specify the parameters of the animation and the other to create the graphs.

- In the **Graph Surface** dialog (on the right), fill in $x + k\_y$.

  The symbol — means to leave a space to indicate multiplication.

- Fill in the intervals $-1 \le x \le 1$, $-1 \le y \le 1$.
- Click in the **Animation Info** dialog (on the left) to make it active.
- Fill in $-3 \le k \le 3$.

  This specifies the seven frames.

- For the common plot frame, fill in $-1 \le x \le 1$, $-1 \le y \le 1$, $-4 \le z \le 4$.

  All the graphs will be drawn in this frame. In this example, we happen to use the same *x*, *y*-values in both dialogs. We choose the *z*-values to show all the points on the planes as *x*, *y*, and *k* vary.

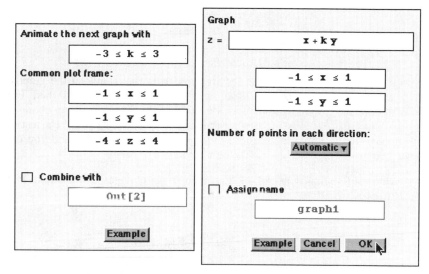

Animation Info (left) and **Graph Surface** (right) dialogs.

- Click in the **Graph Surface** dialog to make it active.
- Click **OK**.

Mathematica creates the seven frames and collapses them, showing only the first, and sets the animation in motion.

▷ Plot x + k y for −1 ≤ x ≤ 1 and −1 ≤ y < 1 Animation for k = −3 to 3.

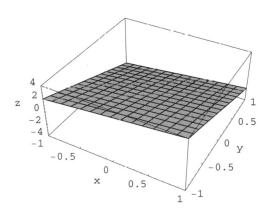

You can control the animation with the buttons that appear at the bottom of the *Mathematica* window. Clicking anywhere in the notebook will stop the motion.

Animation control buttons:
Backward, Alternate directions, Forward, Pause, Slower, Faster.

♀ If you stop the animation, you can restart it by double-clicking on the graphs. You can pause and restart by clicking **Pause** twice. Clicking **Pause** and then **Forward** or **Backward** makes the animation move one frame at a time. You can change the speed with the buttons or by pressing a number between 1 (slowest) and 9 (fastest). (On some platforms, you may not be able to use the numeric keypad for this and you may need to use the buttons before you can use the number keys.)

If you slow down the animation, you can see how each frame changes into the next. For example, here is the final frame:

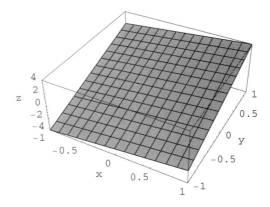

Graph of $z = x + 3y$.

Chapter 11
Working with Functions of Several Variables

This chapter shows how to create, evaluate, and graph functions of several variables. It includes both scalar- and vector-valued functions, i.e., functions $f : R^n \to R$ and $f : R^n \to R^m$.

■ 11.1 How to Create a Function of Several Variables

Here's how to create the function $f(x, y) = xy + e^{x-3y^2}$.

- From the *Joy* menus, choose **Create ▷ Function**.

Create menu.

- Fill in $f \lfloor \mathbf{x}, \mathbf{y} \rfloor$.

  You can create a function of any number of variables in this way.

- Fill in $\mathbf{x\_y} + \mathbf{E}^{x-3\,y^2}$.

  Here ⏜ indicates that you must type a space between $\mathbf{x}$ and $\mathbf{y}$ to denote multiplication.

173

```
Define


              ┌─────────────────────────┐
              │         f[x, y]         │
              └─────────────────────────┘

to be


                        ┌───────────┐
                        │ Specify: ▼│
              ┌─────────┴───────────┴───────────────┐
              │          x y + E^(x-3 y²)            │
              └─────────────────────────────────────┘




              ┌────────┐┌────────┐┌──────┐
              │Example ││Cancel  ││  OK  │
              └────────┘└────────┘└──────┘ ▶
```

Function dialog.

- Click **OK**.

  When you click OK, *Mathematica* creates the function but does not display any output.

▷ Define f[x, y] to be $x y + E^{x-3\,y^2}$

■ 11.2 How to Evaluate a Function of Several Variables

Here's how to use the notation of functions to evaluate $f(x, y) = xy + e^{x-3\,y^2}$ for $(x, y) = (3, -1.5)$.

- If you didn't create the function $f(x, y) = xy + e^{x-3\,y^2}$ in Section 11.1, do it now.
- Click below the last cells so that a horizontal line appears.
- Type f[3,-1.5].
- Press $\boxed{\text{SHIFT}}$ − $\boxed{\text{RET}}$.

Here is the result:

In[3]:= **f[3, -1.5]**

Out[3]= −4.47648

When the variables are integers, *Mathematica* will express the result in exact rather than decimal form. In Version 4, *Mathematica*'s output uses the symbol e instead of E.

In[4]:= **f[3, -2]**

Out[4]= $-6 + \dfrac{1}{E^9}$

If you wish, you can express the result numerically by using the **Numeric** button on the *Joy* palette (see Section 1.3.1):

In[5]:= $\mathbf{N}\left[-6 + \dfrac{1}{\mathbf{E}^9}\right]$

Out[5]= -5.99988

■ 11.3 How to Tabulate a Function of Several Variables

Here's how to create a table of values for $f(x, y) = xy + e^{x-3y^2}$, with $x = -3, -2, -1, 0, 1, 2, 3$ and $y = -4, -2, 0, 2, 4$. This will be a two-step process, where we vary y first and then x.

- If you didn't create the function $f(x, y) = xy + e^{x-3y^2}$ in Section 11.1, do it now.
- From the *Joy* menus, choose **Create ▷ List**.
- Fill in **f[x, y]**.
- Fill in $-4 \le \mathbf{y} \le 4$ with increments of 2.

    Be sure to change **x** to **y**.

- Uncheck **Display in decimal form** and **Display as a table**.
- Check **Assign name** and use the default name **listA** or a name of your choice.

List dialog.

- Click **OK**.

Mathematica's output shows how a typical row of the eventual table will appear, with y increasing from left to right.

▷Create the list with values f[x, y] for -4 ≤ y ≤ 4 in steps of size 2.

listA =

Out[6]= {E^{-48+x} − 4 x, E^{-12+x} − 2 x, E^x, E^{-12+x} + 2 x, E^{-48+x} + 4 x}

Now we use this typical row to create the whole table.

- On the *Joy* palette, click **Last Dialog** to bring back the **List** dialog.
- Fill in **listA**, using the name you assigned if it is different from this.
- Fill in − 3 ≤ x ≤ 3 with increments of 1.

  Be sure to change the variable back to **x**.

- Check **Display in decimal form** and **Display as a table**.
- Uncheck **Assign name** or else change the name for the new list.

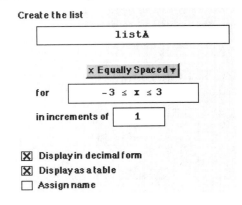

- Click **OK**.

Mathematica produces the table, with the rows corresponding to *x* and the columns to *y*.

▷Create the list with values listA for -3 ≤ x ≤ 3.

Out[7]//TableForm=

12.	6.	0.0497871	−6.	−12.
8.	4.	0.135335	−4.	−8.
4.	2.	0.367879	−2.	−4.
1.42516 × 10^{-21}	6.14421 × 10^{-6}	1.	6.14421 × 10^{-6}	1.42516 × 10^{-21}
−4.	−1.99998	2.71828	2.00002	4.
−8.	−3.99995	7.38906	4.00005	8.
−12.	−5.99988	20.0855	6.00012	12.

■ 11.4 How to Graph Functions of Several Variables

Chapter 10 shows how to use *Joy* to graph various kinds of surfaces and curves without creating a function. If you have created a function, you can use functional notation in the *Joy* dialogs. Here is an example.

We'll graph the function $f(x, y) = xy + e^{x-3y^2}$ for $-3 \le x \le 3$, $-4 \le y \le 4$. The graph is a surface in R^3.

- If you didn't create $f(x, y) = xy + e^{x-3y^2}$ in Section 11.1, do it now.
- From the *Joy* menus, choose **Graph 3D ▷ Surface**.

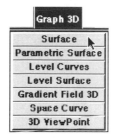

Graph 3D menu.

- Fill in the expression **f[x,y]**.
- Fill in the intervals $-3 \le x \le 3$ and $-4 \le y \le 4$.

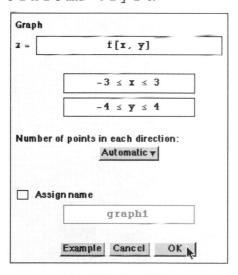

Graph Surface dialog.

- Click **OK**.

Mathematica graphs the surface:

▷ Plot f[x, y] for -3 ≤ x ≤ 3 and -4 ≤ y ≤ 4

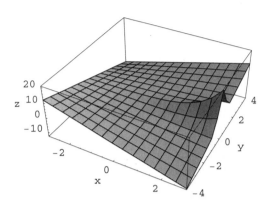

Out[8]= - SurfaceGraphics -

If you are using *Mathematica* Version 4 with interactive 3D graphs, the axis scales and labels will not appear. You can add them with **Graph ▷ Modify Graph** (Section 10.2.1).

■ 11.5 How to Create a Vector-Valued Function

Here's how to create a function of several variables whose values are vectors, such as $f(x, y, z) = (2^x - y^2, 2^y - z^2)$.

- From the *Joy* menus, choose **Create ▷ Function**.
- Fill in f[x,y,z].
- Fill in { $2^x - y^2$, $2^y - z^2$ }.

  Mathematica uses curly brackets { } to denote lists. Vectors, such as these points in two-dimensional space, are written as lists in *Mathematica*.

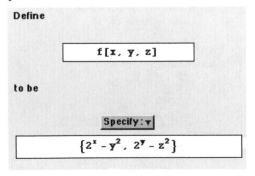

- Click **OK**.

Mathematica creates the function but does not display any output.

▷ Define f[x, y, z] to be $\{2^x - y^2, 2^y - z^2\}$

You can evaluate this function just as you would a scalar-valued function, e.g., in Section 11.2.

- Click below the last cells so that a horizontal line appears.
- Type f[5,-3,2].
- Press SHIFT − RET.

Mathematica calculates the result:

In[3]:= **f[5, -3, 2]**

Out[3]= $\left\{23, -\dfrac{31}{8}\right\}$

Chapter 12
Differentiating Functions of Several Variables

This chapter includes partial derivatives, gradient fields, and directional derivatives. Differentiating a function of one variable is discussed in Chapter 7.

■ 12.1 How to Find a Partial Derivative

Here's how to find

$$\frac{\partial}{\partial x}\left(e^{x-y}\sin\left(y^2+x\right)\right).$$

- If necessary, click in your *Mathematica* notebook beneath the last output so that a horizontal line appears.
- Type `E`$^{x-y}$ `Sin[y`^{2}` + x]`.
- Drag across the expression to select it.

$$\boxed{\text{E}^{x-y}\ \text{Sin}\left[y^2+x\right]}$$

    ♀ Instead of dragging, you can click inside the expression and then click the **Select** button on the *Joy* palette several times until all of it is selected.

- On the *Joy* palette, click the **Partial Derivative** button.

Partial Derivative button.

An expression for the partial derivative appears, with the function filled in and the input field for the independent variable selected.

$$\frac{\partial}{\partial \square} \left(E^{x-y} \, Sin\left[y^2 + x \right] \right)$$

Partly completed derivative.

- Type x for the independent variable.
- Press SHIFT — RET.

Mathematica computes the partial derivative:

In[2]:= $\dfrac{\partial}{\partial x} \left(E^{x-y} \, Sin\left[y^2 + x \right] \right)$

Out[2]= $E^{x-y} \, Cos\left[x + y^2 \right] + E^{x-y} \, Sin\left[x + y^2 \right]$

Other ways. If you click the **Partial Derivative** button without typing and selecting an expression first, the field for the independent variable will be selected.

$$\frac{\partial}{\partial \square} \left(\square \right)$$

In this case you fill in the independent variable first, press TAB, and then fill in the expression to be differentiated.

If you use **Create ▷ Function** to create $f(x, y) = e^{x-y} \sin(y^2 + x)$, then it is easier to differentiate this function by entering $\frac{\partial}{\partial x} \left(f[x, y] \right)$.

■ 12.2 How to Find a Partial Derivative at a Point

Here's how to evaluate a partial derivative at a particular point. We will evaluate $\frac{\partial}{\partial x} (e^{x-y} \sin(y^2 + x))$ at $(x, y) = (\pi, 1)$.

- If you didn't calculate $\frac{\partial}{\partial x} (e^{x-y} \sin(y^2 + x))$ in Section 12.1, do it now.
- From the *Joy* menus, choose **Algebra ▷ Substitute**.

Algebra menu.

- From the popup menu, choose **Last Output**.

    To use this method, your last output must be the value of $\frac{\partial}{\partial x} (e^{x-y} \sin(y^2 + x))$.

- Choose **make the substitutions** and fill in x->π, y->1.

You can type -> by typing – and then >.

Substitute dialog.

- Click **OK**.

Mathematica substitutes the values:

▷ Making the substitutions {x -> π, y -> 1} in the expression
  $E^{x-y} \cos[x + y^2] + E^{x-y} \sin[x + y^2]$ gives

Out[3]= $-E^{-1+π} \cos[1] - E^{-1+π} \sin[1]$

If you wish, you can express the result numerically by using the **Numeric** button on the *Joy* palette, as shown in Section 1.3.1:

In[4]:= $N[-E^{-1+π} \cos[1] - E^{-1+π} \sin[1]]$

Out[4]= -11.763

■ 12.3 How to Find Second- and Higher-Order Partial Derivatives

Here's how to find the partial derivative

$$\frac{\partial^2}{\partial x\, \partial z} (x^2 z^2 - y \sin(2x + z)) = \frac{\partial}{\partial x}\left(\frac{\partial}{\partial z}(x^2 z^2 - y \sin(2x + z))\right).$$

- If necessary, click in your *Mathematica* notebook beneath the last output so that a horizontal line appears.

- Type $x^2 \, z^2 - y \, \text{Sin}[2 \, x + z]$.
- Drag across this expression to select it, or click in it and then click the **Select** button until all of it is selected.

$$\boxed{x^2 \, z^2 - y \, \text{Sin}[2 \, x + z]} \; \text{I}$$

- Click the **Partial Derivative** button.

$$\frac{\partial}{\partial \, \square} \left(x^2 \, z^2 - y \, \text{Sin}[2 \, x + z] \right)$$

- Type z for the inner independent variable, select the resulting expression, and click **Partial Derivative** again.

$$\frac{\partial}{\partial \, \square} \left(\frac{\partial}{\partial \, z} \left(x^2 \, z^2 - y \, \text{Sin}[2 \, x + z] \right) \right)$$

- Type x for the outer variable.
- Press $\boxed{\text{SHIFT}} - \boxed{\text{RET}}$.

In[2]:= $\dfrac{\partial}{\partial \, x} \left(\dfrac{\partial}{\partial \, z} \left(x^2 \, z^2 - y \, \text{Sin}[2 \, x + z] \right) \right)$

Out[2]= $4 \, x \, z + 2 \, y \, \text{Sin}[2 \, x + z]$

You can calculate partial derivatives of any order in this way.

Another way. If you've already found a particular partial derivative, you can use it to find partials of the next order. For example, here's how to use the preceding result to find the third-order partial

$$\frac{\partial^3}{\partial x^2 \, \partial z} (x^2 \, z^2 - y \sin(2 \, x + z)) = \frac{\partial}{\partial x} \left(\frac{\partial^2}{\partial x \, \partial z} (x^2 \, z^2 - y \sin(2 \, x + z)) \right).$$

- Select the expression in the last output, $4 \, x \, z + 2 \, y \, \text{Sin}[2 \, x + z]$.
- Click the **Partial Derivative** button.

$$\frac{\partial}{\partial \, \square} \left(4 \, x \, z + 2 \, y \, \text{Sin}[2 \, x + z] \right)$$

- Type x and press $\boxed{\text{SHIFT}} - \boxed{\text{RET}}$.

The result is:

In[3]:= $\dfrac{\partial}{\partial \, x} \left(4 \, x \, z + 2 \, y \, \text{Sin}[2 \, x + z] \right)$

Out[3]= $4 \, z + 4 \, y \, \text{Cos}[2 \, x + z]$

■ 12.4 How to Find a Gradient

Here's how to use *Joy* to calculate the gradient of

$$z = \frac{x^2}{y} + e^{xy}$$

and evaluate it at $(x, y) = (1, 2)$. The gradient is the vector

$$\nabla z = \left(\frac{\partial z}{\partial x}, \frac{\partial z}{\partial y} \right).$$

First we'll define z.

- If necessary, click in your *Mathematica* notebook beneath the last output so that a horizontal line appears.

- Type $z = \dfrac{x^2}{y} + E^{x\_y}$.

   The symbol ⎯ means to leave a space to indicate multiplication.

- Press SHIFT − RET.

The result is

$$In[2]:= \quad z = \frac{x^2}{y} + E^{x\,y}$$

$$Out[2]= \quad E^{x\,y} + \frac{x^2}{y}$$

Now we form the gradient vector.

- If necessary, click in your *Mathematica* notebook beneath the last output so that a horizontal line appears.
- Type $\left\{ \frac{\partial}{\partial x} (z), \frac{\partial}{\partial y} (z) \right\}$.

   In *Mathematica*, vectors require curly brackets { }.

- Press SHIFT − RET.

The gradient at any (x, y) is:

$$In[3]:= \quad \left\{ \frac{\partial}{\partial x} (z), \frac{\partial}{\partial y} (z) \right\}$$

$$Out[3]= \quad \left\{ \frac{2x}{y} + E^{x\,y}\, y, \; E^{x\,y}\, x - \frac{x^2}{y^2} \right\}$$

Next we evaluate this result at (1, 2).

- From the *Joy* menus, choose **Algebra** ▷ **Substitute**.
- From the popup menu, choose **Last Output**.
- Click **make the substitutions** and fill in x -> 1, y -> 2.
- Check **Assign name** and fill in delz.

The gradient symbol ∇ is pronounced "del" so we call this *delz*.

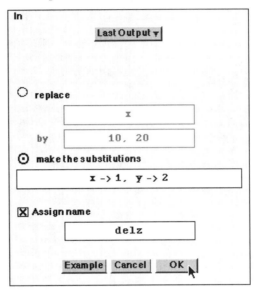

Substitute dialog.

- Click **OK**.

Mathematica evaluates the gradient:

▷ Making the substitutions

  {x -> 1, y -> 2} in the expression $\{\frac{2\,x}{y} + E^{x\,y}\,y,\ E^{x\,y}\,x - \frac{x^2}{y^2}\}$ gives

delz =

Out[4]= $\{1 + 2\,E^2,\ -\frac{1}{4} + E^2\}$

■ 12.5 How to Find a Directional Derivative

Here's how to use *Joy* to calculate the directional derivative of $z = x^2\,y + e^{x-y}$ at the point (1, 2) in the direction of the unit vector $u = (\frac{3}{5}, \frac{4}{5})$. The directional derivative is the dot product

$$D_u \, z = \nabla z \cdot u = \left(\frac{\partial z}{\partial x}, \ \frac{\partial z}{\partial y} \right) \cdot u,$$

where $x = 1$, $y = 2$.

- If you didn't find *delz* in Section 12.4, do it now.
- If necessary, click in your *Mathematica* notebook beneath the last output so that a horizontal line appears.
- Type `delz` . { $\frac{3}{5}$, $\frac{4}{5}$ }.

  In *Mathematica*, a period `.` denotes a dot or matrix product and vectors require curly brackets { }.

- Press SHIFT − RET.

Mathematica calculates the directional derivative:

$In[5]:=$ `delz` . $\left\{ \dfrac{3}{5}, \ \dfrac{4}{5} \right\}$

$Out[5]=$ $\dfrac{4}{5} \left(-\dfrac{1}{4} + E^2 \right) + \dfrac{3}{5} \left(1 + 2 \, E^2 \right)$

You can see that this can be written more simply as $\frac{2}{5} + 2 \, e^2$.

Chapter 13
Integrating with Several Variables

This chapter shows how to use *Joy* to compute multiple integrals as iterated integrals. It includes numeric integration and discusses how to interpret *Mathematica*'s output when it cannot find an integral exactly. Integrating a function of one variable is discussed in Chapter 8.

■ 13.1 How to Calculate a Multiple Integral

Mathematica carries out multiple integration as iterated integration, i.e., integrating with respect to one variable at a time. For example, here is how to calculate the integral

$$\int_{\pi/2}^{\pi} \int_{0}^{x} (\sin x + \sin y + xy \sin (x + y))\, dy\ dx .$$

You can calculate triple iterated integrals in a similar way.

- Type `Sin[x] + Sin[y] + x_y_Sin[x + y]`.

    The symbol __ stands for a blank space, to indicate multiplication.

- Drag across the expression to select it.

    `Sin[x] + Sin[y] + x y Sin[x + y]` ⌶

> ♥ Instead of dragging, you can click inside the expression and then click the **Select** button on the *Joy* palette several times until all of it is selected.

- In the *Joy* palette, click the **Definite Integral** button to create the inner integral.

Definite Integral button.

♡ The highlighted field on the **Definite Integral** button indicates that if you type an expression and select it before clicking, the integration command will apply to this expression.

$$\int_{\square}^{\square} (\mathrm{Sin[x]} + \mathrm{Sin[y]} + \mathrm{x\,y\,Sin[x+y]})\,d\square$$

Partly completed integral.

- Type 0, TAB, and **x** to enter the limits of the inner integral.
- Press TAB and type **y** to enter the variable for the inner integral.
- Click **Select** until the entire integral is selected.

$$\int_{0}^{x} (\mathrm{Sin[x]} + \mathrm{Sin[y]} + \mathrm{x\,y\,Sin[x+y]})\,dy$$

- Click the **Definite Integral** button again to create the outer integral.

$$\int_{\square}^{\square} \left(\int_{0}^{x} (\mathrm{Sin[x]} + \mathrm{Sin[y]} + \mathrm{x\,y\,Sin[x+y]})\,dy \right) d\square$$

- Type $\pi/2$, TAB, and π for the outer limits.
- Press TAB and type **x** for the outer variable.
- Press SHIFT − RET.

Mathematica calculates the result:

$$In[2]:= \int_{\pi/2}^{\pi} \left(\int_{0}^{x} (\mathrm{Sin[x]} + \mathrm{Sin[y]} + \mathrm{x\,y\,Sin[x+y]})\,dy \right) dx$$

$$Out[2]= 1 - \pi$$

Other ways. Clicking the **Definite Integral** button first, without typing and selecting an expression, creates the outer integral. Enter its limits and click the button again to create the inner integral. Then type the inner limits, the expression to be integrated, and the two independent variables.

⚠ In this alternate method, you will need to type any required parentheses yourself. Parentheses are required around sums and differences, and around the inner integral. In the example shown above, *Mathematica* inserts required parentheses automatically.

If you used **Create** ▷ **Function** to create $f(x, y) = \sin x + \sin y + xy \sin(x + y)$, it is easier to integrate this function by entering $\int_{\pi/2}^{\pi} \left(\int_{0}^{x} \mathrm{f[x, y]}\,dy \right) dx$.

Editing integrals. If you want to make changes in an integral that you have typed, click in the integral, select a part of it, and type the change you wish. If you make a mistake, click the **Undo** button on the *Joy* palette. It may be hard to select a small part of an integral, e.g., an upper or lower limit. You can click and use the **Select** button to accomplish this. For an illustration, see Section 8.2.1.

13.1.1 When *Mathematica* Cannot Calculate an Antiderivative

When *Mathematica* integrates, it uses symbolic techniques such as integral tables and substitution methods to find an antiderivative. If its output is the same as the input, or if the output is expressed in terms of another antiderivative, then *Mathematica* cannot integrate the function symbolically. For example:

$In[2]:=$ $\int_{1/2}^{1} \int_{1/2}^{t} \text{Log}[\text{Sin}[s^2] + \text{Sin}[t^2]] \, ds \, dt$

$Out[2]=$ $\int_{\frac{1}{2}}^{1} \left(t \, \text{Log}[2 \, \text{Sin}[t^2]] - \frac{1}{2} \, \text{Log}\left[\text{Sin}\left[\frac{1}{4}\right] + \text{Sin}[t^2]\right]\right) dt -$

$\qquad 2 \int_{\frac{1}{2}}^{1} \int_{\frac{1}{2}}^{t} \frac{s^2 \, \text{Cos}[s^2]}{\text{Sin}[s^2] + \text{Sin}[t^2]} \, ds \, dt$

For certain important functions that cannot be integrated symbolically, *Mathematica* gives a special name to the antiderivative. One example is the integral

$$\int e^{-x^2} \, dx,$$

which is related to the Gaussian or normal probability curve (E stands for the constant *e*):

$In[3]:=$ $\int E^{-x^2} \, dx$

$Out[3]=$ $\frac{1}{2} \sqrt{\pi} \, \text{Erf}[x]$

Such functions may then appear in multiple integrals (in Version 4, *Mathematica*'s output uses **e** rather than **E**):

$In[4]:=$ $\int_{-1}^{1} \int_{-1}^{1} E^{-x^2+y} \, dy \, dx$

$Out[4]=$ $\frac{(-1 + E^2) \sqrt{\pi} \, \text{Erf}[1]}{E}$

This result cannot be evaluated exactly, but you can use the **Numeric** button on the *Joy* palette to express it as a decimal (Section 1.3.1):

$In[5]:=$ $\text{N}\left[\frac{(-1 + E^2) \sqrt{\pi} \, \text{Erf}[1]}{E}\right]$

$Out[5]=$ 3.51067

■ 13.2 How to Calculate Multiple Integrals Numerically

When no elementary antiderivative exists, *Mathematica* can estimate an integral numerically. In other cases, an elementary antiderivative may exist but the function may be so complicated that it is more efficient to find a numerical approximation to it. Here's how to calculate

$$\int_{1/2}^{1} \int_{1/2}^{t} \ln\left(\sin\left(s^2\right) + \sin\left(t^2\right)\right) ds \ dt$$

numerically.

- Click in your *Mathematica* notebook below the last output, if any.
- Type `Log[Sin[s²] + Sin[t²]]`.

    You can type `Log[x]`, which is *Mathematica*'s notation for the natural logarithm function $\ln x$, or you can enter it with the **Functions** button on the *Joy* palette.

- Drag across the expression, or else click in it and then click the **Select** button until all of it is selected.

$$\boxed{\texttt{Log}\left[\texttt{Sin[s}^2\texttt{]} + \texttt{Sin[t}^2\texttt{]}\right]} \ \mathrm{I}$$

- Click the **Numeric Integral** button to create the inner integral.

$$\texttt{N} \int_{\square}^{\square} \blacksquare \, d\square$$

Numeric Integral button.

Mathematica displays

$$\texttt{N} \int_{\boxdot}^{\square} \texttt{Log}\left[\texttt{Sin[s}^2\texttt{]} + \texttt{Sin[t}^2\texttt{]}\right] d\square$$

Partly completed numerical integral.

- Type `1/2`, TAB, and `t` to enter the limits of the inner integral.
- Press TAB and type `s` to enter the inner variable.
- Type parentheses ( ) around the inner integral.

    With a *numeric* integral, *Mathematica* will not insert the required parentheses automatically.

- Drag across the expression or use the **Select** button.

$$\boxed{\left(\texttt{N} \int_{1/2}^{t} \texttt{Log}\left[\texttt{Sin[s}^2\texttt{]} + \texttt{Sin[t}^2\texttt{]}\right] d s\right)} \ \mathrm{I}$$

- Click the **Numeric Integral** button again to create the outer integral.

$$\mathbb{N} \int_{\square}^{\square} \left(\mathbb{N} \int_{1/2}^{t} \text{Log}[\text{Sin}[s^2] + \text{Sin}[t^2]] \, \text{d}s \right) \text{d}\square$$

- Type 1/2, TAB, and 1 to enter the outer limits.
- Press TAB and type t to enter the outer variable.
- Press SHIFT − RET.

Mathematica now calculates the integral:

In[2]:= $\mathbb{N} \int_{1/2}^{1} \left(\mathbb{N} \int_{1/2}^{t} \text{Log}[\text{Sin}[s^2] + \text{Sin}[t^2]] \, \text{d}s \right) \text{d}t$

Out[2]= 0.00548033

Other ways. Clicking the **Numeric Integral** button first, without typing and selecting an expression, creates the outer integral. Enter its limits and click the button again to create the inner integral. Then type the inner limits, the expression to be integrated, and the two independent variables. You will need to add the required parentheses.

If you used **Create ▷ Function** to create $f(s, t) = \ln(\sin(s^2) + \sin(t^2))$, it is easier to integrate this function by entering $\mathbb{N} \int_{1/2}^{1} \left(\mathbb{N} \int_{1/2}^{t} \text{f}[\text{s}, \text{t}] \, \text{d}s \right) \text{d}t$.

Chapter 14
Working with Vector Fields

This chapter shows how to create vector fields in R^2 and R^3. It also shows how to create gradient fields directly from *Joy* menus.

■ 14.1 How to Graph a 2D Vector Field

Here is how to graph a vector field in the plane. We'll take the example $(x^2 - y^2,\ x + y)$, with x, y between ±2.

- From the *Joy* menus, choose **Graph ▷ Vector Field**.

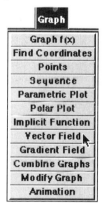

Graph menu.

- Fill in $\{x^2 - y^2,\ x + y\}$.

    Vectors in *Mathematica* are denoted by curly brackets { }.

- Keep the default intervals, $-2 \le x \le 2$, $-2 \le y \le 2$.
- Keep the default setting, **Normalize lengths**.

This equalizes the lengths of the vectors that *Mathematica* draws.

Graph the vector field

$$\{x^2 - y^2, \; x + y\}$$

for

$$-2 \le x \le 2$$

$$-2 \le y \le 2$$

☐ **3D**

$$-2 \le z \le 2$$

☒ **Normalize lengths**
☐ **Assign name**

    graph1

[Example] [Cancel] [OK]

Vector Field dialog.

- Click **OK**.

Mathematica plots the field:

▷ Graph of the vector field {x² - y², x + y} for -2 ≤ x ≤ 2, -2 ≤ y ≤ 2.

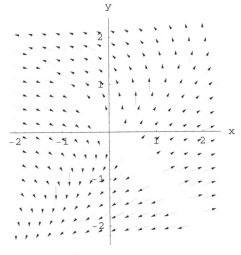

Out[2]= - Graphics -

Notice that the vectors all have the same length except along $y = -x$, where both coordinates are zero. In graphing a vector field, equalizing the lengths of the vectors helps to emphasize the direction of the field at each point. For example, it is useful for plotting the direction field of a system of differential equations.

Normalizing lengths can change the appearance of the field substantially when the lengths vary a lot within the x, y intervals shown in the graph. When the lengths are important, e.g., when they represent the magnitude of a force field, you may want to turn this option off. Here is how this affects the way $(x^2 - y^2, x + y)$ is displayed.

- If you graphed the field $(x^2 - y^2, x + y)$ in this section, click the **Last Dialog** button on the *Joy* palette. If you didn't graph it, fill in that dialog now.
- *Uncheck* **Normalize lengths**

☐ **Normalize lengths**

- Click **OK**.

Mathematica plots the field again:

▷ Graph of the vector field $\{x^2 - y^2, x + y\}$ for $-2 \le x \le 2$, $-2 \le y \le 2$.

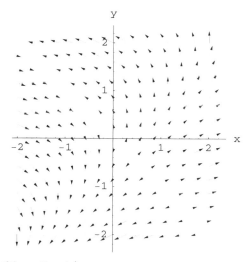

Out[3]= - Graphics -

Mathematica does not draw each vector with its actual length, but scales the vectors proportionally to fit the picture. So the lengths of the vectors drawn in the field are in the correct ratio to each other rather than being equal.

⚠ *Joy* automatically clears any values that may have been assigned to the variables before it plots a vector field.

■ 14.2 How to Graph a 3D Vector Field

For example, here is how to graph the vector field $(-x, -y, -z)$ for x, y, z between ± 1.

- From the *Joy* menus, choose **Graph ▷ Vector Field**.
- Fill in { -x, -y, -z }.

    Vectors in *Mathematica* are denoted by curly brackets { }.

- Fill in the intervals $-1 \le x \le 1$, $-1 \le y \le 1$.
- Check **3D** and fill in $-1 \le z \le 1$.
- If necessary, *uncheck* **Normalize lengths.**

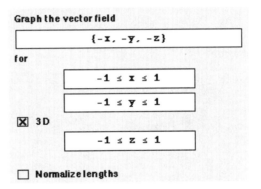

- Click **OK**.

Here is the result:

▷ Graph of the vector field {-x, -y, -z} for
  $-1 \le x \le 1$, $-1 \le y \le 1$, $-1 \le z \le 1$.

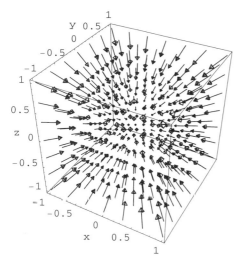

Out[2]= - Graphics3D -

If you are using *Mathematica* Version 4 with interactive 3D graphs, the axis scales and labels will not appear. You can add them with **Graph ▷ Modify Graph** (Section 10.2.1).

■ 14.3 How to Create a Function That Describes a Vector Field

For example, here is how to create the function $F(x, y, z) = (y + z, x + z, x + y)$, which describes a vector field in space. To create a function for a vector field in the plane, just omit the z-coordinate in the domain and range.

- From the *Joy* menus, choose **Create ▷ Function**.

Create menu.

- Fill in F[x, y, z].
- Fill in the expression {y + z, x + z, x + y}.

   Since the values of *F* are vectors, you must enter them with curly brackets { }.

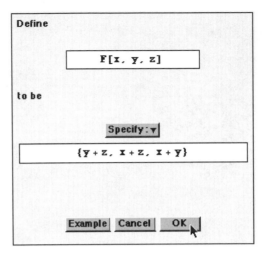

Create Function dialog.

- Click **OK**.

Mathematica creates the function but does not display any output.

▷ Define F[x, y, z] to be {y + z, x + z, x + y}

14.3.1 How to Plot the Vector Field for a Function Already Created

Here's how to plot the vector field for $F(x, y, z) = (y + z, x + z, x + y)$. We'll let x, y, z vary between ±1.

- If you haven't already done so, create the function $F(x, y, z) = (y + z, x + z, x + y)$ in Section 14.3.
- Choose **Graph ▷ Vector Field**.
- Fill in F[x,y,z].
- If necessary, fill in the intervals $-1 \leq x \leq 1$, $-1 \leq y \leq 1$.
- If necessary, check **3D** and fill in $-1 \leq z \leq 1$.

Graph the vector field

F[x, y, z]

for

−1 ≤ x ≤ 1

−1 ≤ y ≤ 1

☒ 3D

−1 ≤ z ≤ 1

☐ Normalize lengths

- Click **OK**.

The vector field is:

▷ Graph of the vector field F[x, y, z] for
  −1 ≤ x ≤ 1, −1 ≤ y ≤ 1, −1 ≤ z ≤ 1.

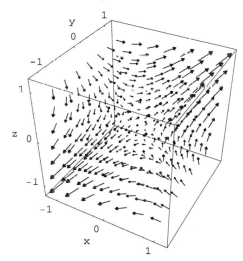

Out[4]= - Graphics3D -

■ 14.4 How to Plot a Gradient Field in 2D

Joy includes a menu item for gradient fields, which are particular examples of vector fields. Here is how to plot the gradient field of $z = 1 - x - y + xy$, for x, y between ±3.

- From the *Joy* menus, choose **Graph ▷ Gradient Field**.
- Fill in $1 - x - y + x\_y$.

  Here ⌣ indicates that you must type a space between **x** and **y** to denote multiplication.

- Fill in $-3 \leq x \leq 3$ and $-3 \leq y \leq 3$.

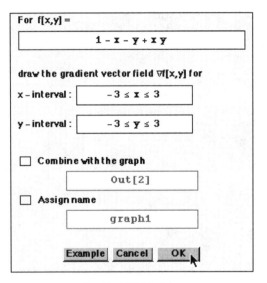

Gradient Field dialog.

- Click **OK**.

Mathematica plots the gradient field, adjusting the lengths of the gradient vectors to fit the picture.

```
▷ Graph the gradient field ∇f[x, y] for f[x, y] =
  1 - x - y + x y on - 3 ≤ x ≤ 3 and - 3 ≤ y ≤ 3
```

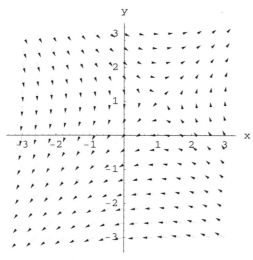

Out[2]= - Graphics -

You can click **Combine with the graph** to superimpose the gradient field on a graph you have already created, e.g., a plot of the level curves of the function. This is illustrated in Section 24.4.2.

■ 14.5 How to Plot a Gradient Field in 3D

Here's how to plot the gradient field in space of $w = xy + z$, for x, y, z between ± 1.

- From the *Joy* menus, choose **Graph 3D ▷ Gradient Field 3D**.
- Fill in **x ⎯y + z**.

    The symbol ⎯ means to leave a blank space to indicate multiplication.

- Fill in the intervals $-1 \le$ **x** $\le$ **1**, $-1 \le$ **y** $\le$ **1**, $-1 \le$ **z** $\le$ **1**.

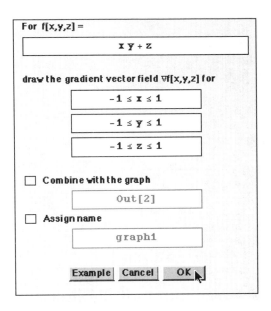

Gradient Field 3D dialog.

- Click **OK**.

Mathematica plots the field:

▷ Graph the gradient field ∇f[x,y,z] for f[x,y,z] = x y + z on
   -1 ≤ x ≤ 1, -1 ≤ y ≤ 1, -1 ≤ z ≤ 1

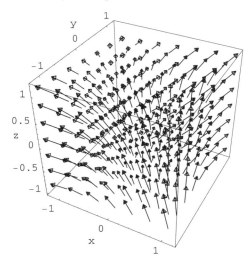

Out[2]= - Graphics3D -

By clicking **Combine with the graph**, you can superimpose the gradient field on an earlier graph.

If you are using *Mathematica* Version 4 with interactive 3D graphs, the axis scales and labels will not appear. You can add them with **Graph ▷ Modify Graph** (Section 10.2.1).

Chapter 15
Solving Differential Equations

This chapter shows how to solve differential equations and visualize their solutions. It includes symbolic and numeric solutions and systems of differential equations as well as individual equations.

■ 15.1 How to Graph the Slope Field of a First-Order DE

Here's how to graph the slope field of the differential equation

$$y'(t) = y - t^2,$$

for y and t between ±2.

- From the *Joy* menus, choose **Calculus ▷ Slope Field**.

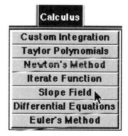

Calculus menu.

- Fill in the equation $\mathtt{y'[t] = y - t^2}$.

    You must enter derivatives in the form $\mathtt{y'[t]}$ rather than $\frac{dy}{dt}$.

- Fill in the intervals $-2 \le \mathtt{t} \le 2, -2 \le \mathtt{y} \le 2$.
- Check **Assign name** and fill in $\mathtt{slopes}$, or a name of your choice.

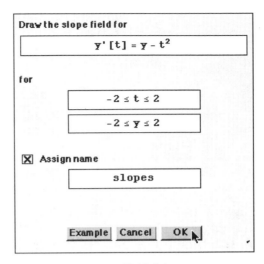

Slope Field dialog.

- Click **OK**.

Mathematica plots the slope field:

▷ The slope field for y′[t] = y − t² for −2 ≤ t ≤ 2 and −2 ≤ y ≤ 2

slopes =

Out[3]= - Graphics -

■ 15.2 How to Solve a DE Symbolically

This section shows how to solve a single differential equation in symbolic form, i.e., using algebraic techniques. Section 15.5 shows how to solve a system of two or more differential equations with the same *Joy* dialog.

Here's how to solve the differential equation

$$y'(t) = y - t^2.$$

We'll find the solution that satisfies the initial condition $y(-2) = 7/4$ and graph it over the interval $-2 \le t \le 2$.

- From the *Joy* menus, choose **Calculus ▷ Differential Equations**.

- Fill in the equation `y'[t] = y - t`^{2}.

    You must enter derivatives in the form `y'[t]` rather than $\frac{dy}{dt}$.

- Fill in the interval `- 2 ≤ t ≤ 2`.
- Fill in the initial condition `y = 7 / 4`.

    This sets the value of y at the left endpoint of the interval, $t = -2$. The box for **initial conditions** is checked by default.

- Check **Graph solution**.

```
Solve the differential equations

         y'[t] = y - t²

for          -2 ≤ t ≤ 2

[X] with the initial conditions

            y = 7 / 4

[ ] Solve numerically
[X] Graph solution
[ ] Combine with the graph

            Out[2]

[ ] Assign name

            result1

   Example   Cancel    OK
```

Differential Equations dialog.

- Click **OK**.

The result is:

```
▷ Solve  y'[t] = y - t²
   with  y = 7/4.
```

```
Define y[t] to be the solution; then y[t] =
```

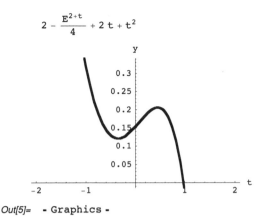

$$2 - \frac{E^{2+t}}{4} + 2\,t + t^2$$

Out[5]=   **- Graphics -**

In Version 4, *Mathematica* uses **e** rather than **E** to denote the constant *e*.

The solution is a function $y(t)$ that can be treated in the same way as any function in *Mathematica*, e.g., graphed, evaluated, or differentiated. *Mathematica* has changed the scale on the y-axis to bring out the features of the graph, which appears more curved than in the slope field.

      Ⅴ If you don't graph $y(t)$ when you solve the equation, you can always graph it later by choosing **Graph ▷ Graph f(x)**.

      ⚠ If you assigned values to *y* or *t* earlier, your old definitions will be lost and $y(t)$ will now be the new function.

If Mathematica cannot solve your equation. Many differential equations cannot be solved symbolically by *Mathematica* and can only be solved numerically. If *Joy* detects that *Mathematica* cannot solve an equation, it will display the message

```
Unable to solve this differential equation
```

and include *Mathematica*'s command `DSolve` in the output. If *Joy* does not detect this situation, *Mathematica* may display an error message while trying to solve the equation or take an inordinate amount of time with no apparent results. In this case, you can stop the calculation by pressing ⌘ – **.** (Macintosh) or ALT – **.** (Windows). You may need to do this repeatedly until *Mathematica* responds. Sections 15.3 and 15.4 show how to solve differential equations numerically.

15.2.1 How to Combine the Graphs of the Solution and Slope Field

- If you didn't graph the slope field in Section 15.1 and solve the differential equation in Section 15.2, do this now.
- Click **Last Dialog** to reopen **Differential Equations**.
- Check **Combine with the graph** and fill in `slopes`, or a different name that you chose in 15.1.

Solve the differential equations

$$y'[t] = y - t^2$$

for $-2 \le t \le 2$

☒ with the initial conditions

$$y = 7 / 4$$

☐ Solve numerically
☒ Graph solution
☒ Combine with the graph

slopes

- Click **OK**.

Mathematica combines the graphs:

▷ Solve $y'[t] = y - t^2$
   with $y = \frac{7}{4}$.

Define $y[t]$ to be the solution; then $y[t] =$

$$2 - \frac{E^{2+t}}{4} + 2t + t^2$$

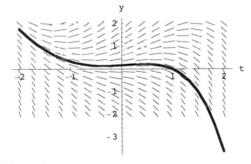

Out[6]= - Graphics -

Section 15.7 shows how to animate solutions to differential equations.

15.2.2 Initial Conditions and Constants of Integration

Here's how to solve the differential equation in Section 15.2 without an initial condition.

- If you filled in the **Differential Equations** dialog in Section 15.2, choose **Last Dialog** from the *Joy* palette to reopen it. If you didn't fill it out, do that now.

- *Uncheck* **with the initial conditions**.
- *Uncheck* **Graph solution**.
- *Uncheck* **Combine with the graph**.

☐ with the initial conditions

y = 7 / 4

☐ Solve numerically
☐ Graph solution
☐ Combine with the graph

- Click **OK**.

The result is:

▷ Solve y′[t] = y - t².

Define y[t] to be the solution; then y[t] =

Out[7]= 2 + 2 t + t² + Eᵗ C[1]

Here C[1] stands for a constant of integration, which is determined once we set an initial value for *y*.

⚠ You cannot graph a solution without specifying an initial condition, and trying to do this will result in error messages.

15.2.3 Higher-Order Equations

Here's how to solve higher-order differential equations using *Joy*. We'll use the example

$$t^2 x'' - 3 t x' + 4 x = t^2,$$

where $x = x(t)$ is the unknown.

- From the *Joy* menus, choose **Calculus ▷ Differential Equations**.
- Fill in t² x''[t] - 3 t␣x' + 4 x = t².

We write **x''[t]** to specify that the unknown function depends on *t*, but can omit this for *x'* and *x*. The symbol ␣ means to leave a space to indicate multiplication.

- If necessary, uncheck **with the initial conditions** and any other options.

Solve the differential equations

$$t^2 \, x''[t] - 3 \, t \, x' + 4 \, x = t^2$$

for $0 \le t \le 1$

☐ **with the initial conditions**

     $x = 1, \ y = 1$

☐ **Solve numerically**
☐ **Graph solution**
☐ **Combine with the graph**

- Click **OK**.

Mathematica solves the equation:

▷ Solve $t^2 \, x''[t] - 3 \, t \, x' + 4 \, x = t^2$.

Define x[t] to be the solution; then x[t] =

Out[8]= $t^2 \, C[1] + t^2 \, C[2] \, \text{Log}[t] + \dfrac{1}{2} \, t^2 \, \text{Log}[t]^2$

Here the solution contains two arbitrary constants, which can be determined by choosing initial conditions for x' and x.

■ 15.3 How to Solve a DE Numerically

This section shows how to use *Mathematica*'s built-in algorithm for approximating the solution to a differential equation numerically. The same *Joy* dialog can be used to solve systems of DE numerically.

Mathematica cannot solve

$$y'(x) = y^3 + x^3$$

in symbolic form, but it can find a numeric approximation to the solution over some interval, starting from a given initial value. We'll use the interval $-1 \le x \le 1$ and the condition $y(-1) = 0$.

- From the *Joy* menus, choose **Calculus ▷ Differential Equations**.
- Fill in the equation $y'[x] = y^3 + x^3$.
- Check **with the initial conditions**.

     You must check this box before you can fill in either the interval or the initial conditions.

- Fill in the interval $-1 \le x \le 1$ and the condition $y = 0$.

Be sure to use the same independent variable, *x*, for the interval as in the differential equation.

- Check **Solve numerically** and **Graph solution**.

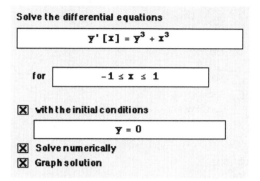

- Click **OK**.

The solution is:

▷ Solve $y'[x] = y^3 + x^3$ numerically
   with $y = 0$.

Define $y[x]$ to be the solution; then $y[x]$ =
       InterpolatingFunction[{{-1., 1.}}, <>][x]

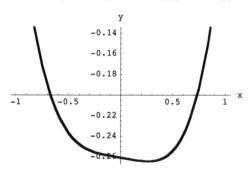

Out[2]= - Graphics -

Solve numerically creates a function, called an *interpolating function*, that approximates the solution over the interval. It is defined by a table of values rather than an equation and connects the points in the table by straight lines. You can evaluate an interpolating function as you would any other function. However, you cannot differentiate it (symbolically) because it is defined by a table of values.

For example, here's how to estimate the solution at $x = 0.735$.

- Click in your *Mathematica* notebook beneath the last output to make a horizontal line appear.
- Type y[0.735].
- Press SHIFT — RET.

Mathematica evaluates the function:

In[3]:= **y[0.735]**

Out[3]= −0.200014

> ♡ To solve a system of differential equations numerically, fill in the equations and initial conditions for each unknown, separated by commas.

> ⚠ You cannot solve an equation numerically without specifying an initial condition, and trying to do this will result in error messages. If you need to, you can stop the calculation by pressing ⌘ – . (Macintosh) or ALT – . (Windows). You may need to do this repeatedly until *Mathematica* responds.

■ 15.4 How to Solve a DE Using Euler's Method

Mathematica uses its own internal algorithms when you click **Solve numerically** in the **Differential Equations** dialog. *Joy* also includes Euler's method as a solution technique that is useful for understanding the principles of solving differential equations numerically, although it is less efficient than *Mathematica*'s built-in algorithm. You can use Euler's method to solve a single equation or a system of equations.

We'll apply Euler's method to the equation $y'(t) = \sin(t\,y)$, $y(-2) = 1$, on the interval $-2 \le t \le 2$.

- From the *Joy* menus, choose **Calculus ▷ Euler's Method**.
- Fill in the equation y'[t] = Sin[t⌣y].

  The symbol ⌣ means to insert a blank space to indicate multiplication.

- Fill in the interval $-2 \le t \le 2$ and the initial condition y = 1.
- Fill in 10 as the number of steps.
- Keep the setting of the output popup menu, **Graph**.

Solve with Euler's Method:

$$y'[t] = \text{Sin}[t\,y]$$

for $-2 \leq t \leq 2$

with initial conditions

$$y = 1$$

Number of Steps:

10

Output: Graph ▾

☐ Assign name

result1

Example Cancel OK

Euler's Method dialog.

- Click **OK**.

The result is:

```
  Equations:             y'[t] = Sin[t y]
  Initial Values:        y = 1
  Method of Solution:    Euler
▷ Name of Solutions:     y[t]
  Number of Steps:       10
  Output:                Graph
```

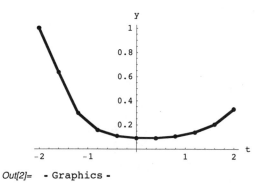

Out[2]=   - Graphics -

Joy creates and graphs a table of values, connecting the points by straight lines and defining the result as an interpolating function $y(t)$. This function approximates the solution to the differential equation over the interval in question. You can evaluate this function at the points in the table or in between them. For example, here is the value $y(1.5)$:

- Click in your *Mathematica* notebook beneath the last output to make a horizontal line appear.
- Type y[1.5].
- Press SHIFT – RET.

In[3]:= **y[1.5]**

Out[3]= 0.186477

△ *Joy* automatically clears any values that the variables y and t may have been assigned earlier.

You can apply Euler's method simultaneously for different numbers of steps in the same interval, to see how the step size affects the result. Here's how to take 10, 40, and 160 steps.

- Click the **Last Dialog** button on the *Joy* palette.
- Change the number of steps to 10,40,160.

Number of Steps.

10, 40, 160

- Click **OK**.

Mathematica plots the three interpolating functions:

```
Equations:          y'[t] = Sin[t y]
Initial Values:     y = 1
Method of Solution: Euler
Name of Solutions:  y[t]
Number of Steps:    {10, 40, 160}
Output:             Graph
```

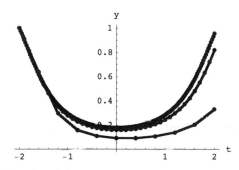

Out[4]= - Graphics -

When you apply Euler's method in this way, $y(t)$ refers to the solution for the last number of steps in the list, 160 in this case. Here we find the value of $y(1.5)$ for 160 steps.

- Click in your *Mathematica* notebook beneath the last output to make a horizontal line appear.
- Type y[1.5].
- Press SHIFT — RET.

In[5]:= **y[1.5]**

Out[5]= 0.509289

    ♀ To solve a system of differential equations by Euler's method, fill in the equations and initial conditions for each unknown, separated by commas.

15.4.1 How to Display the Values Calculated by Euler's Method

You can easily display the table of values that *Joy* constructs for Euler's method. Here's how to do it for the last example.

- If you filled out the **Euler's Method** dialog in Section 15.4, click the **Last Dialog** button on the *Joy* palette. Otherwise fill it in now.
- Change the number of steps to 10.
- Click on the popup menu.

    The popup menu determines how *Mathematica* will display its output. In every case, it creates the function $y(t)$.

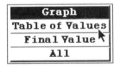

- Choose **Table of Values**.

Number of Steps:

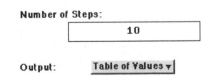

Output: Table of Values ▼

- Click **OK**.

Mathematica displays the table of values for Euler's method:

```
Equations:              y'[t] = Sin[t y]
Initial Values:         y = 1
Method of Solution:     Euler
Name of Solutions:      y[t]
Number of Steps:        10
Output:                 Table of Values
```

Out[6]//TableForm=

t	y
-2	1
-1.6	0.636281
-1.2	0.295847
-0.8	0.156805
-0.4	0.106759
-1.11022×10^{-16}	0.0896824
0.4	0.0896824
0.8	0.104029
1.2	0.137279
1.6	0.202876
2.	0.330448

Because of roundoff and similar numeric errors, a number that should be exactly zero is sometimes slightly different from zero. For example, this occurs in the table where we expect to see $t = 0$. It has no practical effect here.

When you want to take many steps, or several different numbers of steps, it is more convenient to display the final value instead of the whole table. Here's how to do this for 10, 40, and 160 steps.

- Click **Last Dialog**.
- Change the number of steps to 10,40,160.
- Change the output popup menu to **Final Value**.

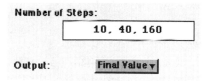

- Click **OK**.

The result is:

```
Equations:              y'[t] = Sin[t y]
Initial Values:         y = 1
Method of Solution:     Euler
Name of Solutions:      y[t]
Number of Steps:        {10, 40, 160}
Output:                 Final Value
```

Out[7]= {0.330448, 0.820815, 0.958103}

These are the estimates for $y(2)$ for the three different approximations.

Choosing **All** from the output popup menu displays the graph, table of values, and final value and creates the function $y(t)$.

■ 15.5 How to Solve a System of DE Symbolically

Here's how to find the solutions to the first-order system of two differential equations

$$x'(t) = x + t^2$$
$$y'(t) = x + y - t$$

in symbolic form, i.e., using algebraic techniques. We'll also graph the solution over the interval $-3 \le t \le 3$ with the initial conditions $x(-3) = -5$, $y(-3) = 1$.

- From the *Joy* menus, choose **Calculus ▷ Differential Equations**.
- Fill in the equations **x'[t] = x + t²**, **y'[t] = x + y - t**.

  ♡ Be sure to separate the equations by commas. You must enter derivatives in the form **y'[t]**. You can solve larger systems by including more differential equations.

- Fill in the interval **– 3 ≤ t ≤ 3**.
- Fill in the **initial conditions x = -5, y = 1**.

  These are the values of x, y at the left endpoint $t = -3$. If the box isn't checked, you will need to check it before filling in the interval or the initial conditions.

- If necessary, *uncheck* **Solve numerically**.
- If necessary, check **Graph solution**.

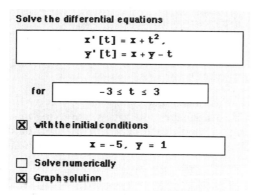

• Click **OK**.

Mathematica gives the solution:

▷ Solve   {x′[t] = x + t², y′[t] = x + y − t}
  with   {x = −5, y = 1}.

Graph styles:

x	Black	Solid
y	Red	.....

Define {x[t],y[t]} to be the solution; then {x[t],y[t]} =

$$\{-2 - 2t - t^2, \; 7 + 5t + t^2\}$$

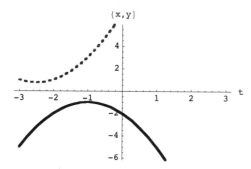

Out[2]= - Graphics -

You can treat these functions as you would any function, e.g., they can be graphed, differentiated, etc.

   ▽ If you don't want to graph the solution, just *uncheck* **Graph solution**. You can always graph the functions later by choosing **Graph ▷ Graph f(x)**.

   △ If you assigned values to *x*, *y*, or *t* earlier, your old definitions will be lost and *x(t)*, *y(t)* will now be the new functions.

If Mathematica cannot solve your system. Many differential equations cannot be solved symbolically by *Mathematica* and can only be solved numerically. If *Joy* detects that *Mathematica* cannot solve a system, it will display the message

```
Unable to solve this differential equation
```

and *Mathematica*'s output will contain the command `DSolve`. If *Joy* does not detect this situation, *Mathematica* may display an error message while trying to solve the system or take an inordinate amount of time with no apparent results. In this case, you can halt the calculation by pressing ⌘ – . (Macintosh) or ALT – . (Windows). Sections 15.3 and 15.4 show how to find numeric solutions.

15.5.1 How to Graph the Trajectory of a Solution (Phase Portraits)

Here's how to graph the trajectory $(x(t), y(t))$ of the solution to

$$x'(t) = x + t^2, \quad x(-3) = -5$$
$$y'(t) = x + y - t, \quad y(-3) = 1$$

- If you didn't specify the initial conditions and solve this system in Section 15.5, do that now.
- Choose **Graph ▷ Parametric Plot**.
- Fill in the functions `x = x[t]`, `y = y[t]`.

    These are the functions in Section 15.5 defined by the **Differential Equations** dialog.

- Fill in the interval $-3 \le t \le 3$.

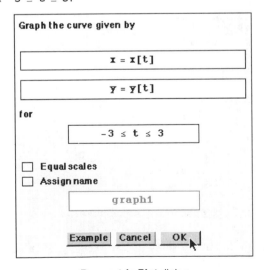

Parametric Plot dialog.

- Click **OK**.

▷ Graph the curve $(x,y) = (-2 - 2t - t^2, 7 + 5t + t^2)$ for $-3 \le t \le 3$.

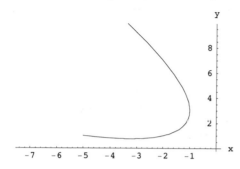

Out[3]=   - Graphics -

We also call the *xy*-plane the *phase plane* and the trajectory the *phase portrait* of the solution. Section 15.7 shows how to animate the motion of $(x(t), y(t))$ along the trajectory.

15.5.2 Initial Conditions and Constants of Integration for Systems of DE

Here's how to solve the system in Section 15.5 without specifying initial conditions.

- If you filled in the **Differential Equations** dialog in Section 15.5, reopen it from *Joy*'s **Recent** menu. If you didn't fill it out, do that now.
- *Uncheck* **with the initial conditions**.
- *Uncheck* **Graph solution**.

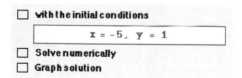

- Click **OK**.

The solution is:

▷ Solve $\{x'[t] = x + t^2, y'[t] = x + y - t\}$.

  Define $\{x[t], y[t]\}$ to be the solution; then $\{x[t], y[t]\} =$

  Out[4]= $\{-2 - 2t - t^2 + E^t C[1], 7 + 5t + t^2 + E^t t C[1] + E^t C[2]\}$

Here $C[1]$, $C[2]$ stand for two constants of integration, which are determined once we set initial values for x and y. The solution cannot be graphed without initial conditions.

■ 15.6 How to Plot the Direction Field of an Autonomous System

Here's how to plot the direction field for an *autonomous* first-order system of two differential equations, i.e., a system

$$x'(t) = f(x, y)$$
$$y'(t) = g(x, y)$$

where the right side doesn't depend on the independent variable t. The derivatives represent the velocities in the x, y directions at any point on a trajectory. For an autonomous system, these depend only on the location of the point. The accompanying sketch suggests that the tangent to the trajectory has the same direction of the vector $(x'(t), y'(t))$.

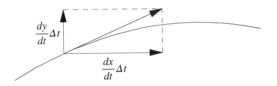

At a point (x, y) along a trajectory, during a short time interval $\Delta t > 0$ you move approximately a distance of $\frac{dx}{dt} \Delta t$ in the x-direction and $\frac{dy}{dt} \Delta t$ in the y-direction. Since $\frac{dy}{dt} = \frac{dy}{dx} \frac{dx}{dt}$ (the chain rule) and the slope of the tangent is $\frac{dy}{dx} = \frac{dy/dt}{dx/dt}$, the tangent lies along the diagonal of the rectangle.

Here's how to plot the direction field for

$$x'(t) = x - xy$$
$$y'(t) = x + xy$$

with x and y between ±4.

- From the *Joy* menus, choose **Graph ▷ Vector Field**.
- Fill in $\{x - x\_y, x + x\_y\}$.

    The symbol ⌣ means to insert a blank space to indicate multiplication. You must use curly brackets { } to enter vectors in *Mathematica*.

- Fill in $-4 \le x \le 4, -4 \le y \le 4$.
- If necessary, check **Normalize lengths**.

    This creates a field of vectors of equal length. It's easier to see the pattern of the trajectories if we scale the vectors so they all have the same length.

- Check **Assign name** and fill in `field` or a name of your choice.

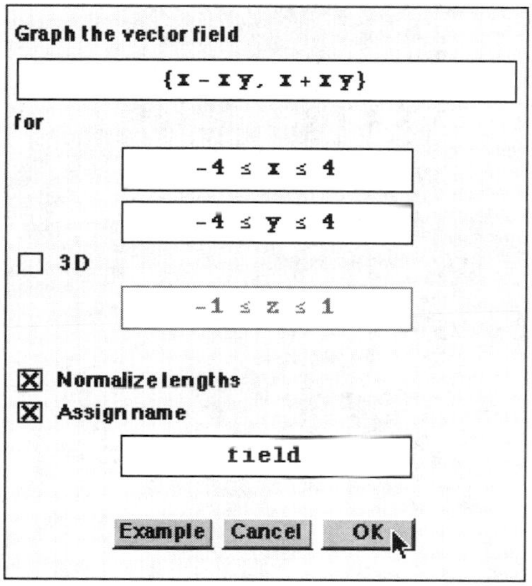

```
Graph the vector field

        {x - x y,  x + x y}

for

            -4 ≤ x ≤ 4

            -4 ≤ y ≤ 4

  ☐ 3D

            -1 ≤ z ≤ 1

  ☒ Normalize lengths
  ☒ Assign name

            field

      Example  Cancel   OK
```

Vector field dialog.

- Click **OK**.

⚠ Before graphing a vector field, *Joy* will clear any values you may have assigned to the variables earlier. So if you will be combining a phase portrait with a direction field, you should plot the field first and then solve the differential equations.

Mathematica plots the field:

▷ Graph of the vector field {x - x y, x + x y} for -4 ≤ x ≤ 4, -4 ≤ y ≤ 4.

field =

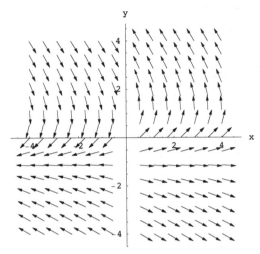

Out[2]=   - Graphics -

15.6.1 How to Combine the Phase Portrait and Direction Field

You can plot a phase portrait for the system

$$x'(t) = x - xy$$
$$y'(t) = x + xy$$

in Section 15.6 and combine it with the direction field. We'll choose the initial conditions $x(0) = 1$, $y(0) = -0.5$ and plot the solution for $0 \le t \le 3$. This system must be solved numerically.

- If you didn't plot the direction field in 15.6, do it now.
- Choose **Calculus** ▷ **Differential Equations**.
- Fill in **x'[t] = x - x⎵y, y'[t] = x + x⎵y**.

  Be sure to leave a space between *x* and *y* for multiplication, as indicated by ⎵.

- Check **with the initial conditions**, if necessary, and fill in **x = 1, y = -0.5**.
- Fill in the interval $0 \le t \le 3$.
- Check **Solve numerically**. You can *uncheck* **Graph solution**.

Solve the differential equations

$$\{x'[t] = x - x\,y,\ y'[t] = x + x\,y\}$$

for $0 \le t \le 3$

[X] **with the initial conditions**

$$x = 1,\ y = -0.5$$

[X] **Solve numerically**

[ ] **Graph solution**

[ ] **Assign name**

- Click **OK**.

Mathematica creates an interpolating function as the solution:

```
▷ Solve  {x'[t] = x - x y, y'[t] = x + x y} numerically
   with  {x = 1, y = -0.5}.
```

```
  Define {x[t],y[t]} to be the solution; then {x[t],y[t]} =

  Out[3]= {InterpolatingFunction[{{0., 3.}}, <>][t],
           InterpolatingFunction[{{0., 3.}}, <>][t]}
```

Now create the trajectory or phase portrait as in Section 15.5.1:

- Choose **Graph ▷ Parametric Plot**.
- Fill in **x = x[t], y = y[t]**.
- Fill in $0 \le t \le 3$ and click **OK**.

Next, make the curve thicker so it will be easier to see against the field:

- Choose **Graph ▷ Modify Graph**.
- Make sure the dialog contains the output name or number of the curve, e.g., Out[4] or your number if it is different.
- Check **Thick curves**.
- Check **Assign name** and fill in **trajectory** or a name of your choice.

Finally, combine the two graphs.

- Choose **Graph ▷ Combine Graphs**.
- Fill in the names **field,trajectory**, or the names you chose instead.

    If you didn't use names for these graphs, use their output numbers.

- Click **OK**.

The result is:

▷ Combining the graphs {field,trajectory} gives

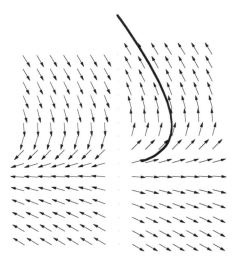

Out[6]=   - Graphics -

■ 15.7 How to Animate Solutions

You can use *Joy* to animate the solution to a differential equation or to a system of equations. Here we'll animate the trajectory of the system

$$x'(t) = x - y, \quad x(0) = 1$$
$$y'(t) = 2x - y, \quad y(0) = 1 \quad,$$

for $0 \le t \le 10$.

- From the *Joy* menus, choose **Calculus ▷ Differential Equations**.
- Fill in `x'[t] = x - y, y'[t] = 2 x - y`.
- Fill in the interval $0 \le t \le 10$.

   If the box for **initial conditions** isn't already checked, you will need to check it first.

- Fill in the initial conditions `x = 1, y = 1`.
- If necessary, *uncheck* **Solve numerically** and *check* **Graph solution**.

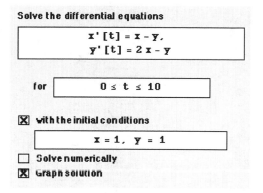

- Click **OK**.

Mathematica solves the system:

▷ Solve   {x′[t] = x − y, y′[t] = 2 x − y}
   with   {x = 1, y = 1}.

   Graph styles:

   x   Black   Solid
   y   Red   .....

   Define {x[t],y[t]} to be the solution; then {x[t],y[t]} =

$$\left\{ \text{Cos}[t], \, 2 \left(\frac{\text{Cos}[t]}{2} + \frac{\text{Sin}[t]}{2} \right) \right\}$$

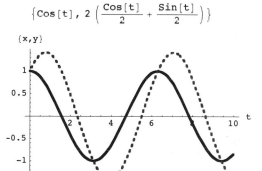

Out[2]= - Graphics -

We'll animate the trajectory $(x(t), y(t))$ in the phase plane by graphing it over five frames, first $0 \le t \le 2$, then $0 \le t \le 4$, ..., up to $0 \le t \le 10$. We can express this as $0 \le t \le 2k, 1 \le k \le 5$.

- From the *Joy* menus, choose **Graph ▷ Animation**.
- From the popup menu, choose **Parametric Plot**.

Learning Resources
Centre

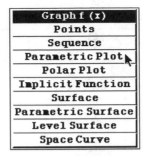

The popup menu shows various kinds of graphs that can be animated.

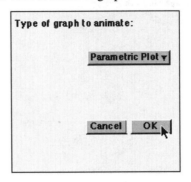

Choosing the type of graph for animation.

- Click **OK**.

Two animation dialogs appear, one that specifies the animation parameters and the other to create the graphs.

- In the **Parametric Plot** dialog (on the right), fill in $x = x[t]$ and $y = y[t]$.
- Fill in the interval $0 \le t \le 2k$.

    This specifies the *t*-values that each frame will cover.

- Click in the **Animation Info** dialog (on the left) to make it active.
- Keep the interval $1 \le k \le 5$ that specifies the five frames.
- For the plot frame, fill in $-1 \le x \le 1$ and $-2 \le y \le 2$.

    All the animations must be in the same size frame. In this case, we can read off the dimensions of a frame that will contain the trajectory from the graphs of $x(t)$ and $y(t)$ in the last output.

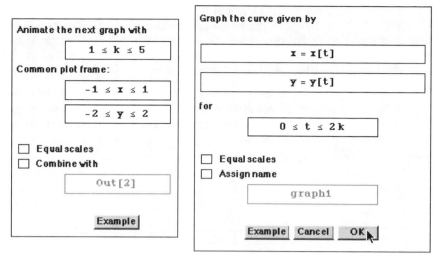

Animation Info (left) and **Parametric Plot** (right) dialogs.

- Click in the **Parametric Plot** dialog to make it active.
- Click **OK**.

Joy creates the five frames and collapses them, showing only the first, and sets the animation in motion.

▷ Graph the curve (x,y) – (Cos[t],2 ($\frac{1}{2}$ Cos[t] + $\frac{1}{2}$ Sin[t])) for 0 ≤ t ≤ 2 k.

  Animation for k = 1 to 5.

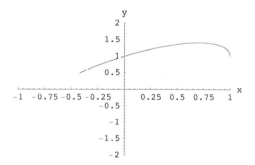

You can control the animation with the buttons that appear at the bottom of the *Mathematica* notebook window. Clicking anywhere in the notebook will stop the motion.

Animation control buttons:
Backward, Alternate directions, Forward, Pause, Slower, Faster.

♡ If you stop the animation, you can restart it by double-clicking on the graphs. You can pause and restart by clicking **Pause** twice. Clicking **Pause** and then **Forward** or **Backward** makes the animation move one frame at a time. You can change the speed with the buttons or by pressing a number between 1 (slowest) and 9 (fastest). (On some platforms, you may not be able to use the numeric keypad for this and you may need to use the buttons before you can use the number keys.)

By slowing down the animation you can see how the solution follows the trajectory. In this case the trajectory is an ellipse and the solution wraps around it counterclockwise, making one circuit and part of another for $0 \le t \le 10$.

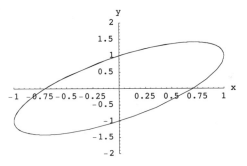

Last frame of the animation.

15.7.1 Animating a Solution on a Direction Field

If you have created a direction field for an autonomous system, you can animate its phase portrait using the field as a background. Here's how to do this for the system in Section 15.7,

$$x'(t) = x - y, \quad x(0) = 1$$
$$y'(t) = 2x - y, \quad y(0) = 1 \quad .$$

- From the *Joy* menus, choose **Graph ▷ Vector Field**.
- Fill in {x - y, 2 x - y}.
- Fill in − 1 ≤ x ≤ 1, − 2 ≤ y ≤ 2.

This will be the same frame size that you will later use in the animation.

- If necessary, check **Normalize lengths**.
- If necessary, check **Assign name** and fill in field or a name of your choice.

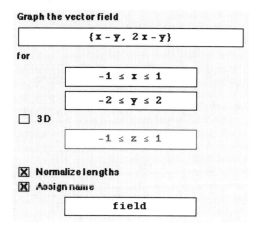

Graph the vector field

{x - y, 2 x - y}

for

-1 ≤ x ≤ 1

-2 ≤ y ≤ 2

☐ **3D**

-1 ≤ z ≤ 1

☒ **Normalize lengths**
☒ **Assign name**

field

- Click **OK**.

Mathematica plots the field:

▷ Graph of the vector field {x - y, 2 x - y} for -1 ≤ x ≤ 1, -2 ≤ y ≤ 2.

field =

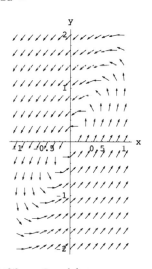

Out[4]= - Graphics -

Because **Vector Field** clears any values assigned to x and y, you'll need to solve the differential equations before animating the phase portrait:

- If you solved the differential equations in Section 15.7, choose **Recent ▷ Differential Equations** and click **OK**. Otherwise, solve the equations now.

Now follow the steps in 15.7 to animate the trajectory. When doing this, check **Combine with** in the **Animation Info** dialog and fill in the name `field` or the output number of the direction field.

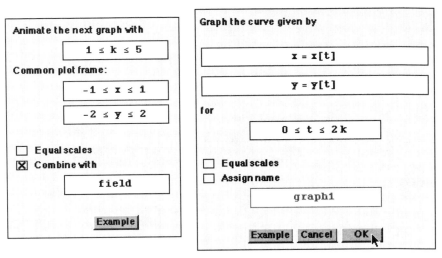

Animation Info (left) and **Parametric Plot** (right) dialogs.

The animation now includes the direction field:

▷ Graph the curve (x,y) = (x[t],y[t]) for 0 ≤ t ≤ 2 k. Animation for k = 1 to 5.

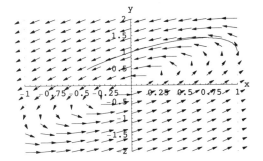

 ♀ The axis scales in the vector field were equal in Out[4], but are adjusted to fit the scales of the trajectory that is being animated. To prevent this adjustment, you can check **Equal scales** in either the **Animation Info** or the **Parametric Plot** dialog.

Chapter 16

Working with Vectors and Matrices

This chapter shows how to use *Joy* to work with vectors and matrices. It includes basic matrix algebra, row reduction in one step and via individual row operations, and how to calculate determinants, eigenvalues, and eigenvectors.

■ 16.1 How to Create a Vector

You can create vectors as rows or columns. This section illustrates one row format, in which a vector is a *list* of entries. You can also create vectors as matrices with one row or column. Section 16.4 shows how to create a matrix. Section 16.8.1 shows how to multiply matrices by vectors and considers all three formats.

Here's how to create the vector $v = (1, 2, 3)$.

- If necessary, click in your *Mathematica* notebook beneath the last output so that a horizontal line appears.
- Type v = {1, 2, 3}.

  Vectors in *Mathematica* require curly brackets { }. You can give a vector as many coordinates as you like.

- Press [SHIFT] − [RET].

Mathematica creates *v* as a list of numbers:

In[2]:= v = {1, 2, 3}

Out[2]= {1, 2, 3}

The coordinates of a vector don't need to have specific numerical values. For example, you can create $w = (-3, 11, a)$ where *a* is a scalar.

- Type w = {-3, 11, a}.
- Press [SHIFT] − [RET].

Mathematica creates *w*:

In[3]:= **w = {-3, 11, a}**

Out[3]= {-3, 11, a}

■ 16.2 How to Find Linear Combinations of Vectors

Here's how to calculate $3\,v - 5\,w$, where $v = (1, 2, 3)$ and $w = (-3, 11, a)$.

- If you didn't create the vectors *v* and *w* in Section 16.1, do it now.
- Type 3 v – 5 w.
- Press SHIFT – RET.

In[4]:= **3 v – 5 w**

Out[4]= {18, -49, 9 - 5 a}

Another way. You don't need to create and name *w* in advance in order to work with it. For example:

In[5]:= **3 v – 5 {-3, 11, a}**

Out[5]= {18, -49, 9 - 5 a}

■ 16.3 How to Find the Dot (Scalar) Product of Two Vectors

Here's how to compute $v \cdot w$, where $v = (1, 2, 3)$ and $w = (-3, 11, a)$.

- If you didn't create the vectors *v* and *w* in Section 16.1, do it now.
- Type v . w, i.e., type v, a period, and w.

    In *Mathematica*, dot products and matrix products are denoted by . (*period* or *dot*).

- Press SHIFT – RET.

The result is:

In[6]:= **v . w**

Out[6]= 19 + 3 a

■ 16.4 How to Create a Matrix

For example, here's how to create the matrix

$$A = \begin{pmatrix} 1 & 2 & 3 \\ 4 & 1 & 6 \\ 7 & 8 & 1 \end{pmatrix}.$$

By creating and naming the matrix in advance, you can refer to it by name in any *Joy* dialog or in your *Mathematica* notebook.

- If necessary, click in your *Mathematica* notebook beneath the last output so that a horizontal line appears.
- Type A = and click the **Matrix** button on the *Joy* palette.

Matrix button.

The **Paste Matrix** dialog appears for determining the size of the matrix.

- Keep the default setting of 3 rows and 3 columns.

Since the numbers on the diagonal are all the same, we can simplify entering the matrix by filling the diagonal all at once. Here's how to do it.

- Click **Fill diagonal with**.
- Keep the default setting, 1, for the entries that will go on the diagonal.

Number of Rows:

| 3 |

Number of Columns:

| 3 |

☐ **Fill with**

| 0 |

☒ **Fill diagonal with**

| 1 |

Example   Cancel   Paste

Paste Matrix dialog.

- Click **Paste**.

Joy pastes a 3×3 matrix template into the notebook, immediately to right of **A** =, with 1 along the diagonal. The first unfilled entry in row 1 is automatically selected (highlighted):

$$\mathbf{A} = \begin{pmatrix} 1 & \blacksquare & \square \\ \square & 1 & \square \\ \square & \square & 1 \end{pmatrix}$$

- Type 2 and press TAB.

  Pressing TAB moves the selection to the next matrix entry that is not yet filled in, from left to right and top to bottom.
- Fill in the remaining entries of the matrix.
- Press SHIFT − RET.

$$In[2]:= \quad \mathbf{A} = \begin{pmatrix} 1 & 2 & 3 \\ 4 & 1 & 6 \\ 7 & 8 & 1 \end{pmatrix}$$

$$Out[2]= \begin{pmatrix} 1 & 2 & 3 \\ 4 & 1 & 6 \\ 7 & 8 & 1 \end{pmatrix}$$

By using the checkboxes in the **Paste Matrix** dialog, you can make

> the entries all the same
> the diagonal have one value and rest of the matrix have a different value
> the diagonal have one value and not specify the rest of the matrix

You can create a vector in column format just by choosing the number of columns in the **Paste Matrix** dialog to be 1. For example:

$$In[3]:= \quad \mathbf{u} = \begin{pmatrix} 14 \\ -5 \\ c \end{pmatrix}$$

$$Out[3]= \begin{pmatrix} 14 \\ -5 \\ c \end{pmatrix}$$

Matrix form. *Mathematica* represents matrices internally as lists of lists, e.g.,

$$\begin{pmatrix} 1 & 2 & 3 \\ 4 & 1 & 6 \\ 7 & 8 & 1 \end{pmatrix} = \{\{1, 2, 3\}, \{4, 1, 6\}, \{7, 8, 1\}\}.$$

Section 2.3.4 shows how to change *Joy*'s settings so that matrices are displayed in list rather than rectangular format.

■ 16.5 How to Find Linear Combinations of Matrices

Here's how to calculate a linear combination such as $-5\,A + 3\,B$, where

$$A = \begin{pmatrix} 1 & 2 & 3 \\ 4 & 1 & 6 \\ 7 & 8 & 1 \end{pmatrix},\ B = \begin{pmatrix} 4 & 2 & 2 \\ 2 & 4 & 2 \\ 2 & 2 & 4 \end{pmatrix}.$$

- If you didn't create A in Section 16.4, do it now.
- If necessary, click in your *Mathematica* notebook beneath the last output so that a horizontal line appears.
- Type B = and click the **Matrix** button.
- Make sure that the **Paste Matrix** dialog is set for a 3×3 matrix.
- Check **Fill with** and fill in 2.
- If necessary, check **Fill diagonal with** and fill in 4.

   This puts 2 in each entry off the diagonal and 4 along the diagonal.

Number of Rows:

3

Number of Columns:

3

☒ **Fill with**

2

☒ **Fill diagonal with**

4

- Click **Paste**.
- Press [SHIFT] − [RET].

The matrix is:

$$In[4]:=\ \ B = \begin{pmatrix} 4 & 2 & 2 \\ 2 & 4 & 2 \\ 2 & 2 & 4 \end{pmatrix}$$

$$Out[4]= \begin{pmatrix} 4 & 2 & 2 \\ 2 & 4 & 2 \\ 2 & 2 & 4 \end{pmatrix}$$

- Type − 5 A + 3 B and press [SHIFT] − [RET].

Mathematica calculates the result:

In[5]:=   **−5 A + 3 B**

Out[5]= $\begin{pmatrix} 7 & -4 & -9 \\ -14 & 7 & -24 \\ -29 & -34 & 7 \end{pmatrix}$

Another way. You don't have to create and name *B* in advance in order to work with it. For example:

In[6]:=   **−5 A + 3** $\begin{pmatrix} 4 & 2 & 2 \\ 2 & 4 & 2 \\ 2 & 2 & 4 \end{pmatrix}$

Out[6]= $\begin{pmatrix} 7 & -4 & -9 \\ -14 & 7 & -24 \\ -29 & -34 & 7 \end{pmatrix}$

■ 16.6 How to Apply Elementary Row Operations

This section shows how to apply individual elementary row operations to a matrix. Section 16.7 shows how to reduce a matrix to row-echelon form in one step.

For example, here's how to apply a row operation that multiplies the first row of

$$\begin{pmatrix} 1 & 2 & 3 \\ 7 & -8 & 9 \\ 17 & -10 & 27 \end{pmatrix}$$

by −7 and adds it to the second row.

- From the *Joy* menus, choose **Matrices ▷ Row Operations**.

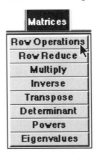

Matrices menu.

The **Row Operations** dialog opens with the default operation **Add Row Multiple**.

- Drag across the default matrix to select it.

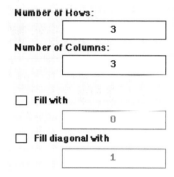

- On the *Joy* palette, click the **Matrix** button.
- Make sure that the **Paste Matrix** dialog is set for a 3×3 matrix.
- If necessary, *uncheck* the boxes for filling the entire matrix or the diagonal.

Number of Rows:

3

Number of Columns:

3

☐ Fill with

0

☐ Fill diagonal with

1

- Click **Paste**.

Joy pastes a 3×3 matrix template into the **Row Operations** dialog. The entry in the upper left corner is selected.

in the matrix   Specify : ▾

$$\begin{pmatrix} \square & \square & \square \\ \square & \square & \square \\ \square & \square & \square \end{pmatrix}$$

- Fill in

$$\begin{pmatrix} 1 & 2 & 3 \\ 7 & -8 & 9 \\ 17 & -10 & 27 \end{pmatrix}.$$

- Change the multiplier from -3 to -7, so the operation will add -7 times row 1 to row 2.

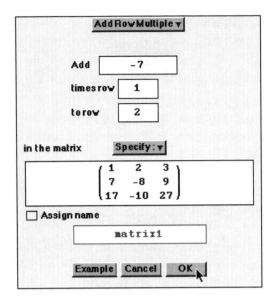

Row Operations dialog.

- Click **OK**.

Mathematica then displays the result of the operation:

▷ Adding $-7 \ast$ row 1 to row 2 of the matrix $\begin{pmatrix} 1 & 2 & 3 \\ 7 & -8 & 9 \\ 17 & -10 & 27 \end{pmatrix}$ gives

$Out[2]= \begin{pmatrix} 1 & 2 & 3 \\ 0 & -22 & -12 \\ 17 & -10 & 27 \end{pmatrix}$

16.6.1 Applying a Row Operation to the Last Output

Next, you can transform the output to get a new matrix whose third row also begins with 0.

- On the *Joy* palette, click **Last Dialog** to bring back the **Row Operations** dialog.
- Change the row operation to add -17 times row 1 to row 3.
- Click the popup menu for specifying the matrix.

Click the popup menu.

The current choice, **Specify**, is highlighted.

- Choose **Last Output**. The new choice is briefly highlighted as the popup menu closes.

    This makes the operation apply to the last output.

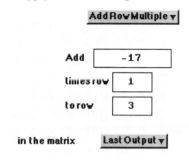

- Click **OK**.

Mathematica's result is:

▷ Adding −17∗row 1 to row 3 of the matrix $\begin{pmatrix} 1 & 2 & 3 \\ 0 & -22 & -12 \\ 17 & -10 & 27 \end{pmatrix}$ gives

$Out[3]=$ $\begin{pmatrix} 1 & 2 & 3 \\ 0 & -22 & -12 \\ 0 & -44 & -24 \end{pmatrix}$

16.6.2 Using Fractions in Row Operations

You can use fractions in elementary row operations and have the entries of the matrix appear as exact rational numbers rather than as numeric approximations.

- On the *Joy* palette, click **Last Dialog**.
- Click the popup menu to choose the type of row operation.

    The current choice, **Add Row Multiple**, is highlighted.

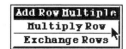

- Choose **Multiply Row**. The new choice is briefly highlighted as the popup menu closes.
- Multiply row 2 by − 1 / 22, and keep the setting **Last Output**.

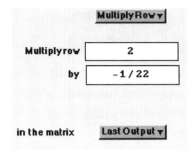

- Click **OK**.

Here is the result of the operation.

▷ Multipying row 2 of the matrix $\begin{pmatrix} 1 & 2 & 3 \\ 0 & -22 & -12 \\ 0 & -44 & -24 \end{pmatrix}$ by $-1/22$ gives

$Out[4]= \begin{pmatrix} 1 & 2 & 3 \\ 0 & 1 & \frac{6}{11} \\ 0 & -44 & -24 \end{pmatrix}$

16.6.3 Backtracking in Row Operations

If you change your mind, you can go back to an earlier matrix and choose a different operation. For example, here's how to go back to

$$\begin{pmatrix} 1 & 2 & 3 \\ 0 & -22 & -12 \\ 0 & -44 & -24 \end{pmatrix}$$

and apply a different operation to it. *Mathematica* gave this matrix an output number when creating it in Section 16.6.1. Here's how to use the output number to refer to the matrix.

- From the *Joy* palette, click **Last Dialog**.
- Using the popup menu for selecting operations, choose **Add Row Multiple**.
- Add -2 times row 2 to row 3.
- Using the popup menu for selecting the matrix, choose **Specify**.
- Replace the matrix by Out[3], using your output number if it is different from this.

    You can type the shorter form %3 in place of Out[3]. If you assigned a name to this matrix when you created it, you can type either the name or the output number.

Add Row Multiple ▼

Add	-2
times row	2
to row	3

in the matrix   Specify: ▼

Out[3]

- Click **OK**.

Here is the result:

▷ Adding -2*row 2 to row 3 of the matrix Out[3] gives

$$Out[5]= \begin{pmatrix} 1 & 2 & 3 \\ 0 & -22 & 12 \\ 0 & 0 & 0 \end{pmatrix}$$

Other ways. Instead of using an output number or name, you can copy the matrix

$$\begin{pmatrix} 1 & 2 & 3 \\ 0 & -22 & -12 \\ 0 & -44 & -24 \end{pmatrix}$$

from your *Mathematica* notebook and paste it into the **Row Operations** dialog. You can also drag across this matrix to select it and then choose **Selection** instead of **Specify** in the popup menu. In either case, you can apply whatever row operation you wish to this matrix.

■ 16.7 How to Row Reduce a Matrix in One Step

Here's how to find the reduced row-echelon form of a matrix in one step. Section 16.6 shows how to apply elementary row operations one at a time.

- From the *Joy* menus, choose **Matrices ▷ Row Reduce**.
- The default matrix is selected automatically.
- On the *Joy* palette, click the **Matrix** button.
- In the **Paste Matrix** dialog, change the number of rows and columns to 4.

Number of Rows:

4

Number of Columns:

4

- Click **Paste**.
- Fill in the matrix

$$\begin{pmatrix} 1 & 1 & 2 & 1 \\ 1 & 0 & 2 & 1 \\ 1 & 2 & 0 & 1 \\ 1 & 2 & 1 & 1 \end{pmatrix}.$$

- In the **Row Reduce** dialog, click **OK**.

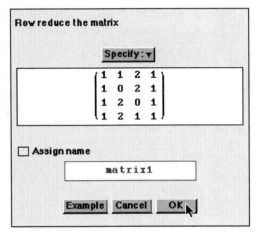

Row Reduce dialog.

Mathematica displays the result:

$$\triangleright\,\text{Row reduce the matrix}\ \begin{pmatrix} 1 & 1 & 2 & 1 \\ 1 & 0 & 2 & 1 \\ 1 & 2 & 0 & 1 \\ 1 & 2 & 1 & 1 \end{pmatrix}$$

$$\textit{Out[2]}=\ \begin{pmatrix} 1 & 0 & 0 & 1 \\ 0 & 1 & 0 & 0 \\ 0 & 0 & 1 & 0 \\ 0 & 0 & 0 & 0 \end{pmatrix}$$

⚠ If you apply **Matrices ▷ Row Reduce** to a matrix with symbolic entries, *Mathematica* assumes that the operations it chooses are well-defined. For example, if $u \neq 4$, then applying individual elementary row operations gives

$$\begin{pmatrix} 1 & 0 & u \\ 0 & 1 & -2 \\ 1 & 2 & 0 \end{pmatrix} \rightarrow \begin{pmatrix} 1 & 0 & u \\ 0 & 1 & -2 \\ 0 & 0 & 4-u \end{pmatrix} \rightarrow \begin{pmatrix} 1 & 0 & 0 \\ 0 & 1 & 0 \\ 0 & 0 & 1 \end{pmatrix}.$$

The last step includes dividing by $4 - u$, so if $u = 4$ the reduced row-echelon form is not the identity matrix, but rather

$$\begin{pmatrix} 1 & 0 & u \\ 0 & 1 & -2 \\ 0 & 0 & 4-u \end{pmatrix} = \begin{pmatrix} 1 & 0 & 4 \\ 0 & 1 & -2 \\ 0 & 0 & 0 \end{pmatrix}.$$

When you use **Row Reduce** with symbolic entries, *Mathematica* often gives the identity matrix without warning that this is not always valid. To get a result that is valid in all cases, you will need to choose **Matrices ▷ Row Operations** instead.

■ 16.8 How to Multiply Matrices

Here's how to find the product of two matrices. The same procedure is used to multiply several matrices together and to multiply a matrix by a vector.

We'll use the example

$$A \begin{pmatrix} a & b \\ c & d \\ e & f \end{pmatrix}, \text{ where } A = \begin{pmatrix} 1 & 2 & 3 \\ 4 & 1 & 6 \\ 7 & 8 & 1 \end{pmatrix}.$$

- If you didn't create A in Section 16.4, do it now.
- From the *Joy* menus, choose **Matrices ▷ Multiply**.
- Replace the default example by

$$A \cdot \begin{pmatrix} a & b \\ c & d \\ e & f \end{pmatrix},$$

i.e., type A, a period, and then use the **Matrix** button to enter the second matrix.

You can list as many matrices and vectors for multiplication in the **Multiply** dialog as you wish, separated by *periods (dots)*. *Mathematica* uses periods to indicate matrix products and dot (scalar) products of vectors. The shapes of the factors must be compatible to make the product well-defined.

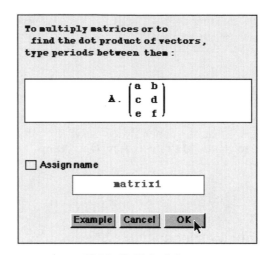

Matrix Multiply dialog.

- Click **OK**.

Mathematica displays the result:

▷ Evaluating A . $\begin{pmatrix} a & b \\ c & d \\ e & f \end{pmatrix}$ gives

$$Out[2]= \begin{pmatrix} a + 2\,c + 3\,e & b + 2\,d + 3\,f \\ 4\,a + c + 6\,e & 4\,b + d + 6\,f \\ 7\,a + 8\,c + e & 7\,b + 8\,d + f \end{pmatrix}$$

Another way. Instead of using the **Multiply** dialog, you can enter the product directly in your *Mathematica* notebook.

- Click in your *Mathematica* notebook below the last output, so that a horizontal line appears.
- Type A . $\begin{pmatrix} a & b \\ c & d \\ e & f \end{pmatrix}$, using the **Matrix** button to enter the second matrix.
- Press $\boxed{\text{SHIFT}} - \boxed{\text{RET}}$.

The result is the same as before:

$$In[3]:= \; A . \begin{pmatrix} a & b \\ c & d \\ e & f \end{pmatrix}$$

$$Out[3]= \begin{pmatrix} a + 2\,c + 3\,e & b + 2\,d + 3\,f \\ 4\,a + c + 6\,e & 4\,b + d + 6\,f \\ 7\,a + 8\,c + e & 7\,b + 8\,d + f \end{pmatrix}$$

16.8.1 Multiplying Matrices by Vectors

You can use the procedure in Section 16.8 to compute products such as

$$A \begin{pmatrix} a \\ c \\ e \end{pmatrix} \text{ and } (a \quad c \quad e) A.$$

For example, here are the products with $A = \begin{pmatrix} 1 & 2 & 3 \\ 4 & 1 & 6 \\ 7 & 8 & 1 \end{pmatrix}$:

In[4]:= **A .** $\begin{pmatrix} \mathbf{a} \\ \mathbf{c} \\ \mathbf{e} \end{pmatrix}$

Out[4]= $\begin{pmatrix} a + 2\,c + 3\,e \\ 4\,a + c + 6\,e \\ 7\,a + 8\,c + e \end{pmatrix}$

In[5]:= **(a   c   e) . A**

Out[5]= **(a + 4 c + 7 e   2 a + c + 8 e   3 a + 6 c + e)**

⚠ If you enter a vector with the **Matrix** button, you are treating it as a matrix with either one column or one row, and the factors in the product must have compatible sizes. For instance, **A . (a   c   e)** is undefined, and entering this will lead to an error message.

Another way. Section 16.1 shows how to create vectors in *list* form {a, c, e}, using curly brackets and commas. You can multiply matrices by vectors in this form also. *Mathematica* will correctly calculate both **A . {a, c, e}** and **{a, c, e} . A**, but will display the result as another list rather than as a column or row.

In[6]:= **A . {a, c, e}**

Out[6]= **{a + 2 c + 3 e, 4 a + c + 6 e, 7 a + 8 c + e}**

In[7]:= **{a, c, e} . A**

Out[7]= **{a + 4 c + 7 e, 2 a + c + 8 e, 3 a + 6 c + e}**

■ 16.9 How to Invert a Matrix

You can use *Joy* to find the inverse of a matrix in one step, rather than with elementary row operations. For example, here's how to invert

$$A = \begin{pmatrix} 1 & 2 & 3 \\ 4 & 1 & 6 \\ 7 & 8 & 1 \end{pmatrix}.$$

- If you didn't create the matrix A in Section 16.4, do it now.
- From the *Joy* menus, choose **Matrices ▷ Inverse**.
- Type A in place of the default matrix.

Inverse dialog.

- Click **OK**.

Mathematica computes the inverse:

▷ Invert the matrix A

$$Out[8]= \begin{pmatrix} -\dfrac{47}{104} & \dfrac{11}{52} & \dfrac{9}{104} \\ \dfrac{19}{52} & -\dfrac{5}{26} & \dfrac{3}{52} \\ \dfrac{25}{104} & \dfrac{3}{52} & -\dfrac{7}{104} \end{pmatrix}$$

⚠ If you later refer to A^{-1} in a *Joy* dialog or in your *Mathematica* notebook, *do not* type A^-1. This is *Mathematica*'s notation for the matrix whose entries are the *reciprocals* of those of A, i.e., $1/a_{ij}$. Instead, use the output number *Mathematica* assigned to A^{-1} or a name that you may have assigned in the **Inverse** dialog.

16.9.1 When a Matrix Is Not Invertible

If you enter a matrix that is not invertible in the **Inverse** dialog, such as

$$\begin{pmatrix} 1 & 2 & 3 \\ 7 & -8 & 9 \\ 17 & -10 & 27 \end{pmatrix},$$

Mathematica replies that the matrix is *singular*:

```
Inverse::sing : Matrix {{1, 2, 3}, {7, -8, 9}, {17, -10, 27}} is singular.
```

■ 16.10 How to Solve a Matrix Equation

You can solve an equation such as

$$A \begin{pmatrix} x \\ y \\ z \end{pmatrix} = \begin{pmatrix} 5 \\ -9 \\ -3 \end{pmatrix}, \text{ where } A = \begin{pmatrix} 1 & 2 & 3 \\ 4 & 1 & 6 \\ 7 & 8 & 1 \end{pmatrix},$$

in several ways.

Here's how to solve this equation using **Algebra ▷ Solve**.

- If you didn't create the matrix A in Section 16.4, do it now.
- From the *Joy* menus, choose **Algebra ▷ Solve**.

Algebra menu.

- Replace the default equations with $A \cdot \begin{pmatrix} \mathbf{x} \\ \mathbf{y} \\ \mathbf{z} \end{pmatrix} = \begin{pmatrix} 5 \\ -9 \\ -3 \end{pmatrix}$, using the **Matrix** button on the *Joy* palette.
- Replace the default variables with $\mathbf{x}$, $\mathbf{y}$, $\mathbf{z}$.
- Make sure the radio button for **Solve algebraically** is selected.

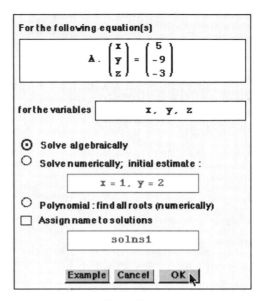

Solve dialog.

- Click **OK**.

Here is the solution:

▷ Solve the equation A.{{x}, {y}, {z}} = {{5}, {-9}, {-3}} for
  {x, y, z} algebraically.

Out[13]= $\left\{\left\{x \to -\dfrac{115}{26},\ y \to \dfrac{44}{13},\ z \to \dfrac{23}{26}\right\}\right\}$

Other ways. Because Section 16.9 showed that *A* is invertible, the solution to the equation is

$$\begin{pmatrix} x \\ y \\ z \end{pmatrix} = A^{-1} \begin{pmatrix} 5 \\ -9 \\ -3 \end{pmatrix} = \begin{pmatrix} -\frac{47}{104} & \frac{11}{52} & \frac{9}{104} \\ \frac{19}{52} & -\frac{5}{26} & \frac{3}{52} \\ \frac{25}{104} & \frac{3}{52} & -\frac{7}{104} \end{pmatrix} \begin{pmatrix} 5 \\ -9 \\ -3 \end{pmatrix} = \begin{pmatrix} -\frac{115}{26} \\ \frac{44}{13} \\ \frac{23}{26} \end{pmatrix},$$

which you can calculate using **Matrices ▷ Multiply** (Section 16.8).

⚠ If you refer to A^{-1} in a *Joy* dialog or in your *Mathematica* notebook, *do not* type A^{-1} but instead use an output number or name for this matrix. See the note in 16.9.

Another way to solve the equation is to apply **Matrices ▷ Row Reduce** (Section 16.7) to the augmented matrix. The process can be summarized as

$$\begin{pmatrix} 1 & 2 & 3 & 5 \\ 4 & 1 & 6 & -9 \\ 7 & 8 & 1 & -3 \end{pmatrix} \rightarrow \begin{pmatrix} 1 & 0 & 0 & -\frac{115}{26} \\ 0 & 1 & 0 & \frac{44}{13} \\ 0 & 0 & 1 & \frac{23}{26} \end{pmatrix},$$

from which you can read off the solutions $x = -\frac{115}{26}$, $y = \frac{44}{13}$, $z = \frac{23}{26}$.

16.10.1 When There Are Multiple Solutions

For example,

$$\begin{pmatrix} 1 & 2 & 3 \\ 7 & -8 & 9 \\ 17 & -10 & 27 \end{pmatrix} \begin{pmatrix} x \\ y \\ z \end{pmatrix} = \begin{pmatrix} 5 \\ -9 \\ -3 \end{pmatrix}$$

has multiple solutions. When you use **Algebra** ▷ **Solve**, *Mathematica* gives the result in this form:

▷ Solve the equation
   {{1, 2, 3}, {7, -8, 9}, {17, -10, 27}} . {{x}, {y}, {z}} = {{5}, {-9}, {-3}} for
   {x, y, z} algebraically.

       Solve::svars :
        Equations may not give solutions for all "solve" variables.

Out[14]= $\left\{\left\{x \rightarrow 1 - \frac{21\,z}{11}, \; y \rightarrow 2 - \frac{6\,z}{11}\right\}\right\}$

Mathematica's warning message means that it may not have solved for all the variables. In fact, z is a *free variable* whose value can be set arbitrarily. *Mathematica* tries to solve for the variables in the order in which you list them. For example, if you list them as y, z, x, then x will be the free variable.

16.10.2 When There Is No Solution

For example,

$$\begin{pmatrix} 1 & 2 & 3 \\ 7 & -8 & 9 \\ 17 & -10 & 27 \end{pmatrix} \begin{pmatrix} x \\ y \\ z \end{pmatrix} = \begin{pmatrix} 1 \\ 2 \\ 3 \end{pmatrix}$$

has no solution. *Mathematica* indicates this as follows:

▷ Solve the equation
   {{1, 2, 3}, {7, -8, 9}, {17, -10, 27}} . {{x}, {y}, {z}} = {{1}, {2}, {3}} for
   {x, y, z} algebraically.

Out[15]= {}

■ 16.11 How to Find a Determinant

Here's how to find the determinant of

$$A = \begin{pmatrix} 1 & 2 & 3 \\ 4 & 1 & 6 \\ 7 & 8 & 1 \end{pmatrix}.$$

- If you didn't create the matrix *A* in Section 16.4, do it now.
- From the *Joy* menus, choose **Matrices ▷ Determinant**.
- Replace the default matrix with **A**.

Determinant dialog.

- Click **OK**.

Mathematica displays the result:

▷ The determinant of the matrix A is

Out[16]= 104

■ 16.12 How to Transpose a Matrix

For example, here's how to create the transpose of

$$A = \begin{pmatrix} 1 & 2 & 3 \\ 4 & 1 & 6 \\ 7 & 8 & 1 \end{pmatrix}.$$

- If you didn't create the matrix *A* in Section 16.4, do it now.
- From the *Joy* menus, choose **Matrices ▷ Transpose**.

- Replace the default matrix by **A**.

Transpose dialog.

- Click **OK**.

Here is the result:

▷ Transpose the matrix A

$$Out[17]= \begin{pmatrix} 1 & 4 & 7 \\ 2 & 1 & 8 \\ 3 & 6 & 1 \end{pmatrix}$$

■ 16.13 How to Find Powers of a Matrix

For example, here's how to find the fifth power of

$$A = \begin{pmatrix} 1 & 2 & 3 \\ 4 & 1 & 6 \\ 7 & 8 & 1 \end{pmatrix}.$$

- If you didn't create the matrix A in Section 16.4, do it now.
- From the *Joy* menus, choose **Matrices ▷ Powers**.
- Replace the default matrix by **A**.
- Fill in **5** as the power.

Raise the matrix

Specify: ▼

A

to the power

5

☐ Assign name

matrix1

Example Cancel OK

Powers dialog.

- Click **OK**.

▷ Raising the matrix A to the power 5 gives

$$Out[18]= \begin{pmatrix} 32336 & 29008 & 29712 \\ 54566 & 47861 & 52524 \\ 71628 & 68882 & 55336 \end{pmatrix}$$

⚠ In *Mathematica*, entering A^5 raises each individual entry of A to the fifth power instead of giving the product $A\,A\,A\,A\,A$. To compute this product, you must use the *Joy* dialog or else enter the *Mathematica* command `MatrixPower[A,5]`.

■ 16.14 How to Find a Characteristic Polynomial

For example, here's how to find the characteristic polynomial of

$$B = \begin{pmatrix} 7 & 3 & -3 \\ 0 & 6 & 0 \\ -3 & -1 & 7 \end{pmatrix}.$$

This polynomial is the determinant of

$$B - \lambda I = \begin{pmatrix} 7-\lambda & 3 & -3 \\ 0 & 6-\lambda & 0 \\ -3 & -1 & 7-\lambda \end{pmatrix},$$

where I is the 3×3 identity matrix. Since Section 16.15 finds the eigenvalues and eigenvectors of B, the first step here will be to create B.

- Click in your *Mathematica* notebook, beneath the last output.
- Type B = and use the **Matrix** button on the *Joy* palette to fill in the matrix

$$\begin{pmatrix} 7 & 3 & -3 \\ 0 & 6 & 0 \\ -3 & -1 & 7 \end{pmatrix}.$$

- Press SHIFT − RET.

Mathematica creates the matrix:

$$In[19]:= \quad B = \begin{pmatrix} 7 & 3 & -3 \\ 0 & 6 & 0 \\ -3 & -1 & 7 \end{pmatrix}$$

$$Out[19]= \quad \begin{pmatrix} 7 & 3 & -3 \\ 0 & 6 & 0 \\ -3 & -1 & 7 \end{pmatrix}$$

- From the *Joy* menus, choose **Matrices** ▷ **Determinant**.
- Fill in B−, and then click the **Matrix** button.
- In the **Paste Matrix** dialog, check the boxes to fill the diagonal with λ and the rest of the matrix with 0.

  Use the **Greek** button on the *Joy* palette to enter λ.

$$\boxed{\alpha, \beta, \ldots}$$

Greek button.

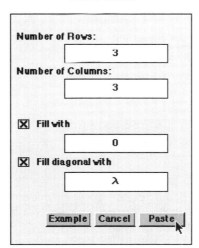

Paste Matrix dialog.

- Click **Paste**.

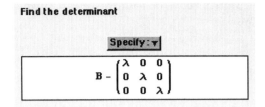

- In the **Determinant** dialog, click **OK**.

Mathematica calculates the result:

▷ The determinant of the matrix $B - \begin{pmatrix} \lambda & 0 & 0 \\ 0 & \lambda & 0 \\ 0 & 0 & \lambda \end{pmatrix}$ is

Out[20]= $240 - 124\,\lambda + 20\,\lambda^2 - \lambda^3$

⚠ You cannot use *I* in *Mathematica* as a name for an identity matrix. *Mathematica* reserves the symbol I for $\sqrt{-1}$. To create the $n \times n$ identity matrix you must either fill in the actual matrix or else refer to it as IdentityMatrix[n], which is the name *Mathematica* uses.

When you are done with the Greek palette, close it by clicking the close box in its upper-left corner (Macintosh) or the close button ⊠ in its upper-right corner (Windows).

■ 16.15 How to Find Eigenvalues and Eigenvectors

For example, here's how to find the eigenvalues and eigenvectors of the matrix

$$B = \begin{pmatrix} 7 & 3 & -3 \\ 0 & 6 & 0 \\ -3 & -1 & 7 \end{pmatrix}.$$

- If you didn't create the matrix *B* in Section 16.14, do it now.
- From the *Joy* menus, choose **Matrices ▷ Eigenvalues**.
- Fill in B.
- Click **Algebraically**.

    You have the choice of finding exact expressions for the eigenvalues and eigenvectors or of estimating them numerically. Numeric estimates are often faster to calculate and easier to understand, especially for larger matrices.

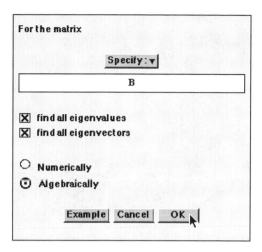

Eigenvalues dialog.

- Click **OK**.

Mathematica displays the result:

▷ The eigenvalues and eigenvectors of B are

Out[21]//MatrixForm=

$$\begin{pmatrix} 4 & \{1, 0, 1\} \\ 6 & \{0, 1, 1\} \\ 10 & \{-1, 0, 1\} \end{pmatrix}$$

The output shows that the matrix has three distinct eigenvalues, $\lambda = 4, 6, 10$. Each is followed in the listing by an eigenvector that spans all the eigenvectors for that eigenvalue.

  ☿ If you choose **Numerically**, you may get different eigenvectors than with **Algebraically**, but they will have the same span.

Here's an example that is less straightforward.

- Click the **Last Dialog** button on the *Joy* palette.

- Replace the matrix with

$$\begin{pmatrix} 0 & -2 & 0 & 0 \\ 1 & 0 & 0 & 0 \\ 0 & 0 & 1 & 1 \\ 0 & 0 & 0 & 1 \end{pmatrix} .$$

- Click **OK**.

Here is the result:

▷ The eigenvalues and eigenvectors of $\begin{pmatrix} 0 & -2 & 0 & 0 \\ 1 & 0 & 0 & 0 \\ 0 & 0 & 1 & 1 \\ 0 & 0 & 0 & 1 \end{pmatrix}$ are

Out[22]//MatrixForm=
$$\begin{pmatrix} 1 & \{0,\ 0,\ 1,\ 0\} \\ 1 & \{0,\ 0,\ 0,\ 0\} \\ -I\sqrt{2} & \{-I\sqrt{2},\ 1,\ 0,\ 0\} \\ I\sqrt{2} & \{I\sqrt{2},\ 1,\ 0,\ 0\} \end{pmatrix}$$

Here I (or ⅈ in *Mathematica* Version 4) stands for $i = \sqrt{-1}$, so two of the eigenvalues are imaginary. The eigenvalues are $\lambda = 1,\ -i\sqrt{2},\ i\sqrt{2}$. For $\lambda = 1$ *Mathematica* has listed two vectors, one of which is the zero vector. This means that while $\lambda = 1$ appears twice as a root of the characteristic polynomial of the matrix, its eigenvectors can be spanned by a single vector, $(0, 0, 1, 0)$. For each eigenvalue λ, *Mathematica* chooses the maximum number of linearly independent eigenvectors and shows in this way when that number is less than the multiplicity of λ as a root.

Another way. By checking the boxes in the **Eigenvalues** dialog, you can instruct *Mathematica* to display only the eigenvalues or only the eigenvectors.

Part III

Exercises and Labs

Chapter 17
Functions

Examples and Exercises

Prerequisites:
      Ch. 1   A Brief Tour of *Joy*

■ 17.1 Visualizing and Finding Roots

Additional Prerequisites:
      Sec. 4.2   How to Tabulate an Expression
      Secs. 5.1,2   Solving equations algebraically and numerically

Exercises

1. a) Factor the polynomial $2x^4 + x^3 + 4x^2 + 5x - 30$. (**Algebra** ▷ **Simplify**)

    b) From the factorization in a), what are the four roots of $2x^4 + x^3 + 4x^2 + 5x - 30$? Graph this polynomial. Why don't all the roots appear in the graph?

2. a) The graphs (i), (ii), (iii) below correspond to cubic polynomials with roots that are integers. For each graph, find the roots and then find the corresponding factors of the polynomial.

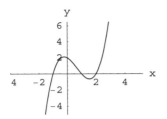 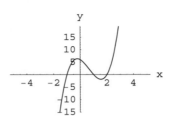 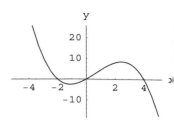

Cubic polynomials (i), (ii), (iii).

b) One point on graph (i) is given by $f(0) = 2$. Use this and your answer to a) to write an equation for the cubic in (i). Graph your answer for $-5 \le x \le 5$, compare it with the given graph, and modify your equation if necessary.

c) Repeat for graphs (ii) and (iii), given (ii) $f(0) = 6$ and (iii) $f(2) = 8$.

3. *Mathematica* cannot solve $e^x - \ln x = 4$ exactly. Here is a method for finding a solution numerically.

 a) Graph $y = e^x - \ln x$ and $y = 4$ together for $0 \le x \le 5$. (Enter $e^x - \ln x$ as $\mathtt{E^x - Log[x]}$.)

 > Recall that $\ln x$ is only defined for $x > 0$. *Mathematica* will display an error message for an interval containing negative numbers, but not for $x = 0$.

 b) Your graph in a) should show a solution to $e^x - \ln x = 4$ between $x = 1$ and $x = 2$. Make a table of values for $e^x - \ln x$ by choosing **Create ▷ List**. Fill in $\mathtt{x, E^x - Log[x]}$, with $1 \le \mathtt{x} \le 2$ in increments of 0.1.

 c) Find the two x-values in a) that bracket a solution to $e^x - \ln x = 4$. Create a table between those points, letting x increase in steps of 0.01. Make tables with smaller steps until you have found the solution rounded to three decimal places.

 d) Confirm your answer in c) by solving $e^x - \ln x = 4$ numerically, choosing **Algebra ▷ Solve**. You will need to enter an initial estimate for the solution, which you can take from the graph in a).

4. a) Graph $y = -x^4 + 5x^3 + 13x^2 + 5x + 14$ for $-5 \le x \le 5$.

 b) From the graph, what do you think are the limits $\lim_{x \to \infty} y$ and $\lim_{x \to -\infty} y$? Using the equation of the curve, would you predict the same limits?

 c) Factor $-x^4 + 5x^3 + 13x^2 + 5x + 14$. What are the roots of this polynomial? (**Algebra ▷ Simplify**)

 d) Use the factorization to graph the curve over an interval that contains all the real roots. Make the interval large enough to show what happens to y as $x \to \pm\infty$ and also show how y behaves near the roots. Now what do you think are the limits $\lim_{x \to \infty} y$ and $\lim_{x \to -\infty} y$?

5. a) Graph the polynomial $x^5 - x^4 - 51 x^3 + 49 x^2 + 98 x$ for $-5 \le x \le 5$. The graph should show three real roots. How can you predict from the equation that there must be exactly two more roots, and that they must be real? (Hint: What should happen to y as $x \to \pm\infty$?)

    b) Solve for the additional roots, and graph the polynomial over an interval that will show all the roots. Make the interval large enough to show what happens to y as $x \to \pm\infty$ and also show how y behaves near the two additional roots.

■ 17.2 Symmetry in Graphs

Exercises

Questions 1 and 2 relate symmetry to the concepts of *even* and *odd functions*. A function f is *even* if $f(-x) = f(x)$ and *odd* if $f(-x) = -f(x)$, for all x. In these questions, examine the graphs for symmetry about an axis and about the origin.

1. a) Graph x^2, x^4, x^6, x^8 for $-3 \le x \le 3$. What kind of symmetry do the various graphs have? Repeat with x, x^3, x^5, x^7.

    b) Graph $x^2 - x^4$, $x^4 - x^6$, $x^6 - x^8$ for $-2 \le x \le 2$. What kind of symmetry do the various graphs have? Repeat with $x - x^3$, $x^3 - x^5$, $x^5 - x^7$.

    c) Which functions in a) and b) are even and which are odd? What do you think is the connection between symmetry and the property of being even or odd?

    d) Make up a polynomial that is neither even nor odd. Graph the polynomial and see if the graph has the kind of symmetry you found in a).

2. a) Graph $\sin x$ and $\cos x$ for $-2\pi \le x \le 2\pi$. What kind of symmetry do these graphs have?

    b) Graph $\sin x \cos x$, $\sin^2 x$, $\cos^2 x$, $x + \sin x$, $x^2 + \cos x$, and $\sin x + \cos x$ for $-2\pi \le x \le 2\pi$. What kind of symmetry do these graphs have? (It may be easier to visualize the symmetry if you graph the curves separately.)

    c) Which functions in a) and b) are even and which are odd? What do you think is the connection between symmetry and the property of being even or odd?

    d) Suppose f and g are odd functions. Will the product fg and the sum $f + g$ be even, odd, or neither? Repeat for f and g both even. Repeat for f even and g odd. Then check your answers against the graphs in c).

■ 17.3 Transforming Graphs

17.3.1 Example: Shifting and Scaling

This example shows how making certain changes in the equation of a function leads to corresponding changes in its graph. For example, here is the effect of multiplying x by a constant, e.g., changing $y = \sin x$ to $y = \sin 2x$.

- Graph $\sin x$ and $\sin 2x$ for $-2\pi \le x \le 2\pi$.

The result is:

▷ Graph

```
Sin[x]        Black       Solid
Sin[2 x]      Red         .....
```

on $-2\pi \le x \le 2\pi$.

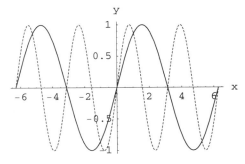

Out[2]= - Graphics -

Multiplying x by 2 compresses the sine curve horizontally by a factor of 2, since each x-coordinate only needs to be half of its earlier value for y to have the same value as before. This is called a *horizontal scaling* of the graph. It cuts the *period*, which is the length of one cycle, in half from 2π to π. It therefore doubles the *frequency*, which equals $\frac{1}{\text{period}}$, changing it from $\frac{1}{2\pi}$ (1 cycle per interval of length 2π) to $\frac{1}{\pi}$ (1 cycle per interval of length π).

Another transformation is to multiply the y-value by a constant, e.g., changing $y = \sin x$ to $y = 2 \sin x$:

- Graph $\sin x$ and $2 \sin x$ for $-2\pi \le x \le 2\pi$.

The result is:

▷ Graph

```
Sin[x]          Black      Solid
2 Sin[x]        Red        .....
```

on $-2\pi \le x \le 2\pi$.

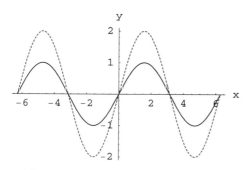

Out[3]= - Graphics -

Now we can see that multiplying y by 2 stretches the graph vertically by a factor of 2, i.e., it is a *vertical scaling*. It changes the *amplitude*, which is half the vertical distance between the minimum and maximum points on the graph, from 1 to 2.

Here is an example that shows the effect of adding a constant to x. We will change $y = x^3 - 3x + 1$ to $y = (x-2)^3 - 3(x-2) + 1$.

- Graph $y = x^3 - 3x + 1$ and $y = (x-2)^3 - 3(x-2) + 1$ for $-5 \le x \le 5$.

Mathematica graphs the two curves:

▷ Graph

```
x³ - 3 x + 1              Black      Solid
(x - 2)³ - 3 (x - 2) + 1  Red        .....
```

on $-5 \le x \le 5$.

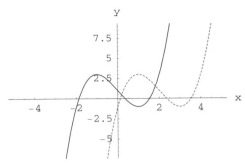

Out[4]= - Graphics -

In this case, we see the effect is a *horizontal shift* or *translation*, moving the curve to the right by two units.

A final operation is to add a constant to the *y*-value rather than to *x*. For example, here is the effect of changing $y = x^3 - 3x + 1$ to $y = x^3 - 3x + 3$, i.e., of adding 2 to *y*.

• Graph $y = x^3 - 3x + 1$ and $y = (x - 2)^3 - 3(x - 2) + 3$ for $-5 \le x \le 5$.

The result is:

▷ Graph

$x^3 - 3x + 1$	Black	Solid
$x^3 - 3x + 3$	Red	.....

on $-5 \le x \le 5$.

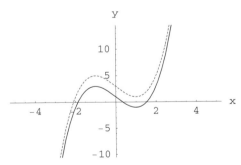

Out[5]= - Graphics -

Here the effect is a *vertical shift* or *translation*, moving the curve up by two units at each point. (This is harder to see where the curve is steeper, but it is true at those points, too.)

We can summarize the four operations on $y = f(x)$ as:

$$y = f(k\,x) \qquad \text{horizontal scaling}$$

$$y = k\,f(x) \qquad \text{vertical scaling}$$

$$y = f(x + k) \qquad \text{horizontal shift}$$

$$y = f(x) + k \qquad \text{vertical shift}$$

Exercises

1. Compare the effect of adding positive and negative constants to y by graphing $y = \sin x + \cos x$, $y = \sin x + \cos x + 2$, and $y = \sin x + \cos x - 2$ for $-2\pi \le x \le 2\pi$. What is the effect of adding -2 to y instead of 2?

2. Compare the effect of multiplying x by factors larger than 1 and smaller than 1, by graphing $y = \cos x$, $y = \cos 2x$, and $y = \cos\left(\frac{x}{2}\right)$ for $-2\pi \le x \le 2\pi$. What is the effect of multiplying x by $\frac{1}{2}$ instead of 2?

3. a) Compare the effect of multiplying x by positive and negative constants, by graphing $y = x^2 - x^3$, $y = (2x)^2 - (2x)^3$, and $y = (-2x)^2 - (-2x)^3$, for $-1 \le x \le 1$. What is the effect of multiplying x by -2 instead of 2?

   b) Compare the effect of multiplying y by positive and negative numbers, by graphing $y = x^2 - x^3$, $y = 2(x^2 - x^3)$, $y = -2(x^2 - x^3)$ for $-1 \le x \le 1$. What is the effect of multiplying y by -2 instead of 2?

4. a) Compare the effect of adding positive and negative constants to x by graphing $y = \sin x$, $y = \sin(x - \frac{\pi}{2})$, and $y = \sin(x + \frac{\pi}{2})$, for $-2\pi \le x \le 2\pi$. What is the effect of adding $-\frac{\pi}{2}$ to x instead of $\frac{\pi}{2}$?

   b) Compare the effect of adding positive and negative constants to x by graphing $y = x^3 - 3x + 1$, $y = (x - 2)^3 - 3(x - 2) + 1$, and $y = (x + 2)^3 - 3(x + 2) + 1$ for $-5 \le x \le 5$. What is the effect of adding -2 to x instead of 2?

5. What is the equation of the curve you obtain by first scaling the graph of $y = x^2 - x - 1$ vertically by a factor of 2 and then shifting it to the left by one unit? Graph the old and new curves and see if your equation is correct.

6. a) By hand, apply the transformations in Question 5 in the reverse order, i.e., shifting left by 1 and then scaling vertically by a factor of 2. Does the order matter?

   b) Scale the graph of $y = x^2 - x - 1$ horizontally by a factor of 2 and then shift left by 3 Graph the old and new curves to see if your equation is correct. Then reverse the order and see if it makes a difference.

7. a) By hand, complete the square for the quadratic $2x^2 - 12x + 13$. Use this to explain how to get the graph of this quadratic from the graph of $y = x^2$ by scaling and shifting. Reminder: the process for completing the square is:

$$ax^2 + bx + c = a\left(x^2 + \tfrac{b}{a}x + \tfrac{c}{a}\right) = a\left(x^2 + \tfrac{b}{a}x + \left(\tfrac{b}{2a}\right)^2 - \left(\tfrac{b}{2a}\right)^2 + \tfrac{c}{a}\right)$$
$$= a\left(x + \tfrac{b}{2a}\right)^2 + \left(c - \tfrac{b^2}{4a}\right).$$

b) Graph $y = x^2$ and $y = 2x^2 - 12x + 13$ simultaneously and confirm that the scaling and shifting operations you chose in a) do transform the first graph into the second.

8. In this graph, the solid curve is the graph of $y = \sin x$ and the dashed curve is that of an unknown function $y = f(x)$.

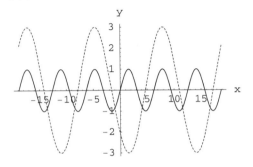

a) What is the amplitude of f? What is the period of f (in terms of π)? (The example in this section, Example 17.3.1, defines amplitude and period.)

b) By how much can you stretch and shift the graph of $\sin x$, horizontally and/or vertically, to produce the graph of f?

c) Use your answers to a) and b) to find a possible equation for $y = f(x)$. Graph $\sin x$ together with your answer for $-6\pi \le x \le 6\pi$. Compare your result with the given graph to see if you are correct.

9. a) Section 3.9 shows how to animate graphs. Animate the graphs of $\sin(x + k)$ with $-6 \le x \le 6, -1 \le y \le 1$, for $-3 \le k \le 3$. (**Graph ▷ Animation**)

b) Write a short paragraph describing the effect of adding k to x, and how the value and sign of k affect the result.

■ 17.4 Rates of Growth

In Exercises 1–3, graph the given functions to help decide which grows faster as $x \to \infty$. In order to answer these questions definitively you need to use tools such as inequalities or derivatives, but graphing gives evidence that you can use to help build your intuition.

Since the functions may behave one way when x is small and another way when x is large, you may have to try different x-intervals in order to see the eventual growth pattern.

1. a) Which grows faster as $x \to \infty$, 2^x or x^2? Graph these functions for $0 \le x \le 8$.

b) Repeat for 10^x and x^{10}. If you choose $0 \le x \le 8$, you will not see all the points where the graphs cross. What interval should you choose? To see all the y-values, you may need to choose **Show All** in the popup menu. (**Graph** ▷ **Graph f(x)**)

2. a) Which grows faster as $x \to \infty$, e^x or $\ln x$? (Enter e as **E**; enter $\ln x$ as **Log[x]**, or use the **Functions** button on the *Joy* palette.)

   b) Repeat for x^2 and $\ln x$.

3. a) Which grows faster as $x \to \infty$, $\sqrt{x}$ or $\ln x$?

   b) Show that $x^{1/3}$ grows faster than $\ln x$. You will need a very large interval to demonstrate this.

■ 17.5 Inverse Functions

In this section, you'll verify the principle that the graphs of a function $f(x)$ and its inverse $f^{-1}(x)$ are symmetric about the line $y = x$. When *Mathematica* graphs these functions, the symmetry appears only if the two axes have the same scale.

In Exercises 1–3:

- Find an equation (by hand) for the inverse $f^{-1}(x)$ of the given function $f(x)$.
- Check your equation by graphing the original function, its inverse, and the line $y = x$ simultaneously. Be sure to check **Equal scales** in **Graph** ▷ **Graph f(x)**.

   ⚠ You cannot enter $f^{-1}(x)$ in the **Graph f(x)** dialog. Instead, use the equation you found for the inverse in each exercise.

If $f^{-1}(x)$ is undefined for some of the values of x that you are asked to include in your graph, *Mathematica* may give you a warning message that you can ignore. You need to include these values in the graph to bring out the symmetry.

1. a) By hand, find an equation for the inverse of $f(x) = 3x + 5$. Check for x and y between -3 and 3.

   b) Repeat for $f(x) = -3x + 5$.

2. a) By hand, find an equation for the inverse of $f(x) = e^{2x}$. Check for x and y between -3 and 3.

   b) Repeat for $f(x) = \ln 2x$.

3. a) By hand, complete the square for the quadratic $f(x) = x^2 - 4x + 7$. (Exercise 17.3.7 shows how to complete the square.) Use this to find a possible equation for the inverse of $f(x)$. Check for x and y between 0 and 10.

   b) The graph of f is a parabola and is symmetric about the vertical line $x = 2$. The graph of f^{-1} should be its mirror image around the line $y = x$. Why isn't it symmetric about the horizontal line $y = 2$?

■ 17.6 Polar Coordinates

> Additional Prerequisite:
>
>   Sec. 3.8   How to Graph in Polar Coordinates

In these exercises, use **Graph** ▷ **Polar Plot**. You can enter θ by using the **Functions** button on the *Joy* palette.

1. a) Graph $r = \cos\theta$ in polar coordinates for $0 \le \theta \le \pi$. Start with $\theta = 0$ and trace the path of (r, θ) on the graphs, paying attention to the value of θ and the sign of r.

    b) Where is the point on the graph with $\theta = \frac{3\pi}{4}$? Explain why (r, θ) appears in the fourth quadrant when $\frac{\pi}{2} \le \theta \le \pi$.

2. a) Graph $r = \cos 3\theta$ in polar coordinates for $0 \le \theta \le \pi$. Starting with $\theta = 0$, trace the path of (r, θ) on the graphs.

    b) Where is the point on the graph with $\theta = \frac{\pi}{4}$? Explain why (r, θ) appears in the third quadrant when $\frac{\pi}{6} \le \theta \le \frac{\pi}{2}$.

    c) For which values of θ does the curve pass through the origin? For which values of θ is (r, θ) in the second quadrant?

3. a) Predict the shape of the graph of $r = \cos 5\theta$ for $0 \le \theta \le \pi$. Then graph the curve and see if your prediction was correct. (If not, try graphing over smaller intervals until you understand the path that (r, θ) follows.)

    b) Predict the shape of the graph of $r = \cos n\theta$ for $0 \le \theta \le \pi$, when n is *odd*. Check your prediction by graphing an example.

4. a) Graph $r = \cos 2\theta$ for $0 \le \theta \le \pi$. Start with $\theta = 0$ and trace the path of (r, θ) on the graphs, paying attention to the value of θ and the sign of r.

    b) Where is the point on the graph with $\theta = \frac{\pi}{2}$? Explain why (r, θ) appears in the third quadrant when $\frac{\pi}{4} \le \theta \le \frac{\pi}{2}$.

    c) Repeat a) for $\pi \le \theta \le 2\pi$. For which values of θ is (r, θ) in the first quadrant? in the second quadrant?

    d) Predict the shape of the graph for $0 \le \theta \le 2\pi$. Then graph the curve and see if your prediction was correct.

5. a) Section 3.9 shows how to animate graphs. Animate the graph of $r = \cos 4\theta$ for $0 \le \theta \le 2\pi\frac{k}{16}$, with x, y between ± 1 and $1 \le k \le 16$. This shows how (r, θ) traces out the graph. (**Graph** ▷ **Animation, Polar Plot**)

b) Describe the path that (r, θ) takes for $0 \le \theta \le 2\pi$. How many petals are there? At which values of θ do the petals meet the origin?

c) Predict the shape of the graph of $r = \cos n\theta$ for $0 \le \theta \le 2\pi$, when n is *even*. Check your prediction by graphing an example.

6. a) (For Valentine's Day.) Predict the shape of the graph of $r = -1 - \sin\theta$ for $0 \le \theta \le 2\pi$, by tracing the path of (r, θ) beginning with $\theta = 0$.

   b) Check your prediction by graphing the curve. The graph is called a *cardioid*.

7. a) Graph $r = 1 - 2\cos\theta$ for $0 \le \theta \le 2\pi$. Start with $\theta = 0$ and trace the path of (r, θ) on the graph, paying attention to the value of θ and the sign of r.

   b) Predict the shape of the graph of $r = 1 - 2\cos 2\theta$ for $0 \le \theta \le 2\pi$. Check your prediction by graphing the curve. (If you were not correct, try animating the graph or graphing over smaller intervals until you understand the path that (r, θ) follows.)

   c) Repeat for $r = 1 - 2\cos 3\theta$.

8. Circles, ellipses, parabolas, and hyperbolas are called *conic sections* (because they can be formed by intersecting a plane with a cone). In polar coordinates, noncircular conic sections have equations of the form

$$r = \text{constant} \times \frac{e}{1 + e\cos\theta}, \quad e > 0. \tag{*}$$

The number e is called the *eccentricity* of the conic section and determines the shape of the curve (it does not stand for the constant 2.718…). In this exercise, you'll use animation to see the role that e plays. Section 3.9 discusses animating graphs.

a) Follow these steps to animate a typical conic section as e varies. We'll use 13 frames and take the constant in equation (*) equal to 1.

- Choose **Graph ▷ Animation** and then **Polar Plot**. Click **OK**.
- Fill in

$$r = \frac{k\,/\,6}{1 + \frac{k}{6}\_\text{Cos}[\theta]}, \quad 0 \le \theta \le 2\pi \,.$$

  The symbol ⎯ means to leave a space to indicate multiplication.

- Let $0 \le k \le 12$, $-3 \le x \le 5$, $-3 \le y \le 3$.
- If necessary, *uncheck* **Combine with**.
- Click **OK** in the **Polar Plot** window.

You can control the animation with the icons that appear at the bottom of the *Mathematica* window. Clicking anywhere in the notebook will stop the motion. If you stop the animation, you can restart it by double-clicking on the graphs.

b) What values does the eccentricity e take in this animation? Open the animation frames by double-clicking on the cell bracket that contains them and match each frame to a value of e.

c) *Mathematica* graphs a hyperbola with its asymptotes, since it tries to connect the two branches of the hyperbola. Only the part of the hyperbola that fits into the plot frame will appear. Which values of e seem to correspond to a hyperbola?

d) Only one value of e determines a parabola, and it is included in the animation. Which value does that seem to be? The remaining nonzero values of e determine ellipses.

Lab 17.1 Fitting Cosines to Periodic Data (Hours of Daylight)

Prerequisites:
Ch. 1	A Brief Tour of *Joy*
Sec. 3.3,4	Plotting points and combining graphs

This lab shows how the number of hours of daylight at a fixed location changes during the year. In the course of the lab, you'll see how to fit a trigonometric function to given astronomical data.

■ Before the Lab (Do by Hand)

The following table shows the number of hours between sunrise and sunset during 1998 in Boston, Massachusetts, in 60-day intervals (data taken from *The Old Farmer's Almanac*, Yankee Publishing Inc., Dublin, New Hampshire, 1997).

Day	Date	Daylight	
1	Jan 1	9 hrs 8 min	
61	Mar 2	11 hrs 17 min	
121	May 1	14 hrs 4 min	
172	Jun 21	15 hrs 18 min	(longest day)
181	Jun 30	15 hrs 14 min	
241	Aug 29	13 hrs 17 min	
301	Oct 28	10 hrs 31 min	
355	Dec 21	9 hrs 5 min	(shortest day)
361	Dec 27	9 hrs 6 min	

Because the earth revolves around the sun once a year, you can expect a graph of the number of hours of daylight to repeat its values each year. The number of hours doesn't repeat exactly every 365 days, but this is a good approximation.

Let $y = f(t)$ be the amount of sunlight during the day, measured in hours. We will let t be the number of days since January 1, 1998. In the lab, you will see that the data points appear to follow a sine or cosine curve $y = f(t)$.

To fit a cosine function to data that repeat periodically, you need to identify the amplitude and period of the data. These determine the vertical and horizontal multipliers that will stretch or compress the graph of $y = \cos t$ to fit the data. You must also determine the vertical and horizontal shifts that slide the cosine curve from its original position near the origin to the data points.

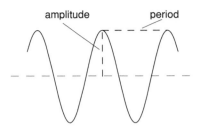

1. Example 17.3.1 shows how changing the equation of a function $y = f(t)$ affects its graph. If you haven't read this example, do it now.

2. What is the amplitude of the number of hours of daylight? Find an equation that corresponds to scaling $y = \cos t$ vertically to have the same amplitude as the daylight data. Round your final answer off to 1 decimal place.

3. Find an equation that shifts your last result vertically so that the greatest value of y is 15 hours and 18 minutes, the length of the longest day, and the amplitude is the number you found in Question 2.

4. Transform your last result so that its period is 365 days, the length of the year.

5. Find an equation that shifts your last result horizontally so that a maximum of the graph occurs on June 21, the longest day. This should be the function that fits the daylight data.

■ In the Lab

Using *Joy*, you'll plot the sunlight data and the equation of the cosine function you derived before the lab. You can then combine the two graphs to see how successful you were.

1. From *Joy*'s menus, choose **Graph ▷ Points**. Plot the points corresponding to the sunlight data in the table on page 275, assigning the name hours.

    For example, type $\{1, 9 + \frac{8}{60}\}$ for 9 hours and 8 minutes of daylight on Day 1 (January 1, 1998). After you enter all the data, you will probably need to scroll down to find the **Assign names** input field and the **OK** button.

2. Graph the function you found in Question 5 (Before the Lab) for $0 \le t \le 365$, and assign the name curve.

3. a) Combine *hours* and *curve*. (**Graph 2D ▷ Combine Graphs**).

    b) Does the curve appear to fit the almanac data fairly well? If not, reexamine the way in which you found the equation and try to improve your result.

4. Print your work.

■ After the Lab

1. According to the graph, how many days are there during the year when day and night have the same number of hours? Which days are those?

2. When during the year is the number of hours of daylight increasing? When is it decreasing?

3. When during the year does the number of hours of daylight change most rapidly? When does it change least rapidly?

4. According to your function, how many hours of daylight should there be today in Boston? If you wish, you can use *The Old Farmer's Almanac* to convert Boston data to your location.

Lab 17.2 Inverse Functions and Parametric Curves

Prerequisites:

Ch. 1	A Brief Tour of *Joy*
Sec. 3.5	How to Graph 2D Parametrized Curves
Sec. 3.7.2	How to Modify a Graph

This lab shows how to use parametrized curves to graph a function $y = f(x)$ and its inverse $y = f^{-1}(x)$. Using this method, you can graph f^{-1} without finding its equation.

■ Before the Lab (Do by Hand)

The graph of $y = f(x)$ is a curve, and you can plot the curve if you know an equation for the function. But sometimes you can plot the curve even if you cannot solve for y in terms of x and even if the curve is not the graph of a function.

A curve is the path of a point whose x- and y-coordinates are functions of a quantity t, called a *parameter*. These functions define a *parametrization* of the curve. For example, $x = t^3 - 3t$, $y = t$ is a parametrization of the curve $x = y^3 - 3y$. It is often useful to think of t as representing a time variable and the curve as being traced out by a point (x, y) that moves as time passes. In this example, when $t = 1$ the point is at $(x, y) = (-2, 1)$. The curve is not the graph of a function $y = f(x)$ because two different points can have the same x-coordinate.

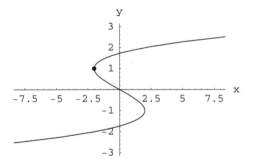

The point where $t = 1$ on the curve $x = t^3 - 3t$, $y = t$.

In the special case where the curve is the graph of a function $y = f(x)$, you can let $x = t$, $y = f(t)$ parametrize the curve. If f has an inverse function $y = f^{-1}(x)$, then their graphs are symmetric about the line $y = x$. This means that for each point $(t, f(t))$ on the graph of f, the mirror image point $(f(t), t)$ is on the graph of f^{-1}, i.e., $t = f^{-1}(f(t))$. So $x = f(t)$, $y = t$ is a parametrization of the graph of f^{-1}. If f does not have an inverse function, then reversing coordinates in this way produces a curve that is not a function.

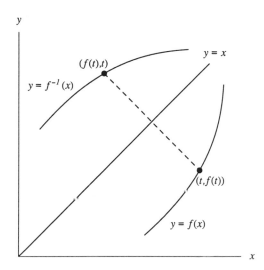

Symmetric points on the graphs of $y = f(x)$ and $y = f^{-1}(x)$.

1. What is a parametrization for the line $y = x$?

2. What is a parametrization for the curve $y = x + \sin x$?

3. Suppose f is any function with an inverse. If the graph of f crosses the line $y = x$, why must the graph of f^{-1} cross the line at the same point?

4. The function $f(x) = x + \sin x$ has an inverse, but you cannot find its equation because you cannot solve for x in terms of y. What is a parametrization for the graph of f^{-1}?

5. What are the values of x where the graphs of $f(x) = x + \sin x$ and its inverse cross the line $y = x$? Express your answer in terms of π.

■ In the Lab

1. Graph the parametrization that you found in Question 1 (Before the Lab) for $-8 \le t \le 8$. (**Graph ▷ Parametric Plot**)

2. Repeat for Questions 2 and 4 (Before the Lab).

3. a) Combine the three graphs (**Graph ▷ Combine Graphs**).

    b) Redraw the graph in a) with **Equal scales** and a **Grid**. (**Graph ▷ Modify Graph**)

    ϑ You need to choose **Equal scales** for the symmetry to appear when you combine the graphs.

    c) Verify that the graphs of f and f^{-1} appear to be symmetric about the line. If the symmetry doesn't appear, reexamine your parametrizations and make any necessary corrections.

4. Print your work.

■ After the Lab

1. a) In your graph for Question 3b) (In the Lab), estimate the x and y coordinates where the three curves cross.

    b) Compare your answer to a) with your predictions in Question 5 (Before the Lab).

2. Let $f(x) = x + \sin x$. Explain why the inverse function of f exists.

3. a) Let $f(x) = x + \sin x$. The graph of f^{-1} should appear to have a vertical tangent at alternate crossing points. What feature of the graph of f at those points causes this?

    b) Use a derivative to prove that the graph of f has the property in a).

Chapter 18
Limits and Continuity

Examples and Exercises

Prerequisites:

Ch. 1	A Brief Tour of *Joy*
Sec. 4.2,5	Tabulating expressions, factoring
Sec. 5.1	How to Solve Equations Algebraically

■ 18.1 Conjecturing Limits

18.1.1 Example: Limits via Graphs

When $x = 0$ the expression $\sin \frac{1}{x}$ is undefined. But does $\lim_{x \to 0} \sin \frac{1}{x}$ exist, and if so what is its value? You can visualize the behavior of this function with a graph.

- Graph $\sin \frac{1}{x}$ for $-5 \leq x \leq 5$.

Mathematica graphs this expression:

▷ Graph $\text{Sin}\left[\frac{1}{x}\right]$ on $-5 \leq x \leq 5$.

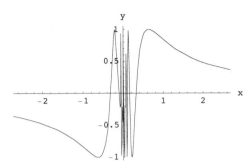

Out[2]= - Graphics -

It's hard to see what is happening near $x = 0$, so we'll zoom in.

- On the *Joy* palette, click **Last Dialog**.
- Change the interval to $-0.05 \le x \le 0.05$ and graph again.

The result is:

▷ Graph $\mathrm{Sin}\left[\dfrac{1}{x}\right]$ on $-0.05 \le x \le 0.05$.

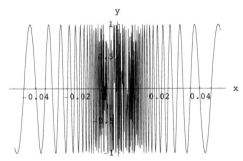

Out[3]= - Graphics -

It appears from the graphs that the function continues to oscillate between $y = \pm 1$ as x gets closer and closer to 0. The oscillations are so close together that *Mathematica* cannot show them all accurately. In fact, it can be proved that $\sin \dfrac{1}{x}$ does not approach any limit as $x \to 0$.

18.1.2 Example: Limits via Tables

In this example we will use numerical data to conjecture whether this limit exists, and if so with what value:

$$\lim_{x \to 2} \frac{x^5 + x^4 - 2x^3 - 2x^2 - 8x - 8}{x - 2} \ .$$

Substituting $x = 2$ would give $\frac{0}{0}$, which is indeterminate, so you cannot find the limit by substitution. First, we'll express this as a function in *Mathematica* so we can refer to it more easily.

- Create the function $f(x) = \frac{x^5 + x^4 - 2x^3 - 2x^2 - 8x - 8}{x-2}$. (**Create ▷ Function**)

Mathematica creates the function:

▷ Define f[x] to be $\dfrac{x^5 + x^4 - 2x^3 - 2x^2 - 8x - 8}{x - 2}$

Next, we'll graph the function.

- Graph $f(x)$ for $-5 \le x \le 5$.

The graph is:

▷ Graph f[x] on $-5 \le x \le 5$.

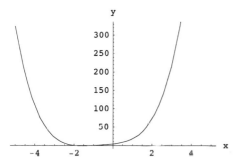

Out[3]= - Graphics -

The graph suggests that near $x = 2$ the function f is increasing and that $\lim_{x \to 2} f(x)$ exists and is roughly equal to 75. *We'll assume that f is increasing in an interval around $x = 2$.*

⚠ Although the graph doesn't show it, $f(2)$ is undefined. *Mathematica* graphs curves by plotting a certain number of points and connecting them, and didn't include $x = 2$ as one of these. The output looks like the graph of a continuous function, but the actual graph has a hole where $x = 2$ and f is not continuous there.

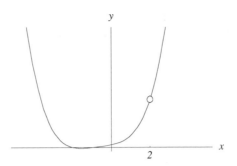

We can get a better idea of how $f(x)$ behaves near $x = 2$ by making a table of values.

- From the *Joy* menus, choose **Create ▷ List**.
- If necessary, set the popup menu to **x Equally Spaced**.
- Fill in **x, f[x]**.
- Fill in $1.5 \le x \le 2$ with increments of 0.05.
- Make sure **Display in decimal form** and **Display as a table** are checked and click **OK**.

Mathematica creates a table of values and warns that $f(2)$ is indeterminate.

▷ Create the list with values {x, f[x]} for 1.5` ≤ x ≤ 2 in steps of size 0.05.

    Power::infy : Infinite expression $\dfrac{1}{0.}$ encountered.

    ∞::indet : Indeterminate expression 0. ComplexInfinity encountered.

Out[4]//TableForm=

1.5	37.1875
1.55	39.8536
1.6	42.6816
1.65	45.6784
1.7	48.8511
1.75	52.207
1.8	55.7536
1.85	59.4984
1.9	63.4491
1.95	67.6136
2.	Indeterminate

The table shows that for the selected values of $x < 2$, the values of $f(x)$ increase and are less than 75. Here's a way to home in on the limit, taking values on both sides of $x = 2$.

- On the *Joy* palette, click **Last Dialog** to bring back the **List** dialog.
- Create a new table, using the interval $1.95 \le x \le 2.05$ and increments of 0.01.

The resulting table shows values of $f(x)$ as x approaches 2 from either side (we omit *Mathematica*'s warning message):

▷ Create the list with values {x, f[x]} for 1.95` ≤ x ≤ 2.04999999999999982`
  in steps of size 0.01.

Out[5]//TableForm=

1.95	67.6136
1.96	68.4729
1.97	69.3411
1.98	70.2183
1.99	71.1046
2.	Indeterminate
2.01	72.9046
2.02	73.8185
2.03	74.7417
2,04	75.6743
2.05	76.6164

Since f is increasing near $x = 2$, the limit is between $y = 71.1046$ and $y = 72.9046$. You can continue this way to achieve any desired precision. For example, the following table shows the limit is between 71.91 and 72.09.

▷ Create the list with values {x, f[x]} for
   1.9950000000000001` ≤ x ≤ 2.00499999999999989` in steps of size 0.001 .

Out[6]//TableForm=

1.995	71.5511
1.996	71.6407
1.997	71.7304
1.998	71.8202
1.999	71.91
2.	Indeterminate
2.001	72.09
2.002	72.1802
2.003	72.2704
2.004	72.3607

The tables show the limit is near 72. Example 18.1.3 shows how to use algebra to establish that the limit is *exactly* 72.

18.1.3 Example: Limits via Algebra

Example 18.1.2 above suggests that

$$\lim_{x \to 2} \frac{x^5 + x^4 - 2x^3 - 2x^2 - 8x - 8}{x - 2} = 72 .$$

Here's how to show this is true.

- If you didn't define this function in Example 18.1.2, do it now:

$$f(x) = \frac{x^5 + x^4 - 2x^3 - 2x^2 - 8x - 8}{x - 2}$$

- From the *Joy* menus, choose **Algebra ▷ Simplify**.
- Fill in `f[x]`, choose **Factor**, and click **OK**.

The result is:

▷ Factoring f[x] gives

   Out[7]=   (1 + x) (2 + x) (2 + x²)

Thus as $x \to 2$, we see $f(x) \to 3 \cdot 4 \cdot 6 = 72$.

This factorization shows we can rewrite the function as

$$f(x) = \begin{cases} (1+x)(2+x)(2+x^2), & x \neq 2 \\ \text{undefined}, & x = 2 \end{cases}.$$

Because $f(2)$ is undefined, f isn't continuous at $x = 2$ even though $\lim_{x \to 2} f(x)$ exists.

Exercises

1. a) Graph this function on $-5 \le x \le 5$. Should the graph have a hole where $x = 4$? Should it have a hole where $x = 3$? Why?

    $$f(x) = \frac{x^3 - 4x^2 + 5x - 20}{x - 4}.$$

    b) Zoom in until you can estimate the value of $\lim_{x \to 4} f(x)$.

2. a) Use algebra to find

    $$\lim_{x \to 4} \frac{x^3 - 4x^2 + 5x - 20}{x - 4}.$$

    b) Is f continuous at $x = 4$? Why?

3. a) Use tables to conjecture whether $\lim_{x \to 0} \sin \frac{1}{x}$ exists.

    b) Is $\sin \frac{1}{x}$ continuous at $x = 0$? Why?

4. Use a graph to conjecture the value of $\lim_{x \to 0} \frac{\sin x}{x}$. Should the graph have a hole where $x = 0$? Why?

5. a) Graph on $-10 \le x \le 10$:

    $$f(x) = \frac{x^2 - 5x + 6}{x^2 - 4x - 5}.$$

b) *Mathematica* tries to connect points on different parts of a graph and sometimes does this inaccurately. What lines appear on the graph of $f(x)$ but do not belong there? How could you have predicted these lines from the equation?

c) Use tables to conjecture whether $\lim_{x \to -1} f(x)$ exists and if so, what its value is.

d) Repeat c) for $\lim_{x \to 2} f(x)$.

■ 18.2 The Definition of Limit

18.2.1 Example: Epsilons and Deltas Graphically and Numerically

The definition of $\lim_{x \to a} f(x) = L$ is: no matter what number $\epsilon > 0$ you choose, you can always find a number $\delta > 0$ such that

whenever $0 < |x - a| < \delta$ then $|f(x) - L| < \epsilon$.

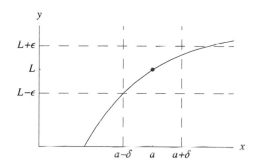

This means that given any pair of horizontal lines $y = L \pm \epsilon$ you can always find a pair of vertical lines $x = a \pm \delta$ with the following property:

whenever $x \neq a$ and (x, y) is on the graph and between the vertical lines, then it is also between the horizontal lines.

For a given ϵ, if one value for δ satisfies the definition then so does any smaller value. If there are horizontal lines $y = L \pm \epsilon$ for which no choice of vertical lines $x = a \pm \delta$ has this property, then L is not the limit of the function. Here $f(x)$ may or may not be defined for $x = a$, and if it is defined then $f(a)$ may or may not equal L.

In Example 18.1.2 we let

$$f(x) = \frac{x^5 + x^4 - 2x^3 - 2x^2 - 8x - 8}{x - 2}$$

and showed algebraically in Example 18.1.3 that $\lim_{x \to 2} f(x) = 72$. We assumed that f is an increasing function near $x = 2$, based on its graph. The table in Example 18.1.2

```
1.995      71.5511
1.996      71.6407
1.997      71.7304
1.998      71.8202
1.999      71.91
2.         Indeterminate
2.001      72.09
2.002      72.1802
2.003      72.2704
2.004      72.3607
```

thus shows that when $x \neq 2$ and $1.999 < x < 2.001$, we have $71.91 < f(x) < 72.09$. So if we take $\epsilon = 0.1$ then we can choose $\delta = 0.001$ (or any smaller number):

whenever $0 < |x - 2| < 0.001$ then $|f(x) - 72| < 0.1$.

Graphing $y = f(x)$ together with the horizontal lines $y = 72 \pm 0.1$ illustrates the choice of $\delta = 0.001$:

- If you didn't define this function in Example 18.1.2, do it now.

$$f(x) = \frac{x^5 + x^4 - 2x^3 - 2x^2 - 8x - 8}{x - 2}$$

- Graph $f(x)$ and the constant functions 71.9, 72.1 simultaneously, for $1.999 \leq x \leq 2.001$.

▷ Graph

```
f[x]       Black      Solid
71.9       Red        . . . . .
72.1       Blue       _ _ _ _
```

on 1.999 ≤ x ≤ 2.001.

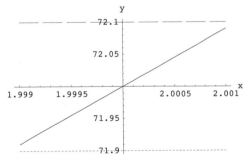

Out[8]= - Graphics -

The graph of $y = f(x)$ is nearly a straight line because we have zoomed down onto a very narrow x-interval. The horizontal lines $y = L \pm \epsilon = 72 \pm 0.1$ are at the top and bottom of the picture. The vertical lines at $x = a \pm \delta = 2 \pm 0.001$ aren't drawn but their x-coordinates are marked near the left and right edges. The graph shows that when $x \neq 2$ and (x, y) is on the graph and between the vertical lines, then it is also between the horizontal lines. That is, for the given $\epsilon = 0.1$ the choice of $\delta = 0.001$ satisfies the requirements. The graph suggests a slightly larger δ will also satisfy them.

By reducing the size of ϵ you can bring $f(x)$ closer to the limit $L = 72$, but you'll need to bring x closer to 2.

18.2.2 Example: Epsilons and Deltas Algebraically

In Example 18.2.1, we took $\epsilon = 0.1$ and used a table of values to find a number $\delta > 0$ such that whenever $0 < |x - 2| < \delta$ then $|f(x) - 72| < \epsilon$. Here we'll see how to find δ algebraically.

First, we'll find where the graph of $y = f(x)$ crosses the horizontal lines $y = L \pm \epsilon = 72 \pm 0.1$.

- If you didn't define this function in Example 18.1.2, do it now.

$$f(x) = \frac{x^5 + x^4 - 2\,x^3 - 2\,x^2 - 8\,x - 8}{x - 2}$$

- Solve the equation $f(x) = 72.1$ for x. (**Algebra** ▷ **Solve**)
- Repeat for $f(x) = 71.9$.

The solutions are:

▷ Solve the equation f[x] = 72.1 for x algebraically.

Out[9]= {{x → -3.7008}, {x → -0.650153   2.96191 I}, {x → -0.650153 + 2.96191 I},
         {x → 2.00111}}

▷ Solve the equation f[x] = 71.9 for x algebraically.

Out[10]= {{x → -3.69886}, {x → -0.650013 - 2.95991 I}, {x → -0.650013 + 2.95991 I},
         {x → 1.99889}}

We see that the real solutions closest to $x = 2$ are $x = 1.99889$ and $x = 2.00111$. To within the six significant digits that *Mathematica* reports, these are each 0.00111 units from $x = 2$. Since f is increasing in this interval, we can say that whenever $0 < |x - 2| < 0.00111$ then $|f(x) - 72| < 0.1$. So for $\epsilon = 0.1$ we can take $\delta = 0.00111$. This value is the *largest* value we can choose for δ. (It is slightly larger than the δ we found in Example 18.2.1.)

> If the two solutions were not equidistant from $x = a$, we would let δ be the distance from $x = a$ to the *nearer* of the two solutions.

- Graph $f(x)$ and the constant functions $y = 71.9$, $y = 72.1$ simultaneously, for $1.99889 \leq x \leq 2.00111$.

The result is:

▷ Graph

f[x]	Black	Solid
71.9	Red	.....
72.1	Blue	_ _ _ _

on 1.99889 ≤ x ≤ 2.00111.

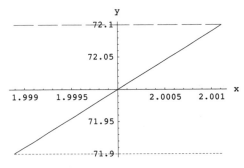

Out[11]= - Graphics -

The output shows the points where the graph of $y = f(x)$ crosses the horizontal lines $y = 72 \pm 0.1$. Since f is increasing, all points on the graph between these intersection points lie between the horizontal lines.

18.2.3 Example: Limits at Infinity

We say $\lim_{x \to \infty} f(x) = L$ if no matter what number $\epsilon > 0$ you choose, you can always find a number $N > 0$ such that whenever $x > N$, then $|f(x) - L| < \epsilon$.

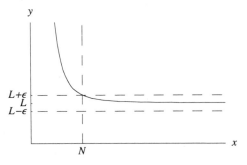

This means that given any pair of horizontal lines $y = L \pm \epsilon$ you can always find a vertical line $x = N$ such that whenever (x, y) is on the graph of f and to the right of the vertical line, it is also between the horizontal lines.

For example, consider

$$\lim_{x\to\infty} 1 + \frac{x^2 + 5x}{x^3 - x}.$$

- Create the function $f(x) = 1 + \frac{x^2 + 5x}{x^3 - x}$. (**Create ▷ Function**)

▷ Define f[x] to be $1 + \dfrac{x^2 + 5x}{x^3 - x}$

- Graph $f(x)$ for $0 \le x \le 25$.

The result is:

▷ Graph f[x] on $0 \le x \le 25$.

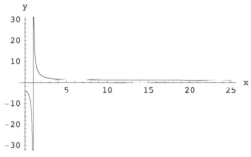

Out[3]= - Graphics -

The graph has a vertical asymptote at $x = 1$ and for $x > 1$ the function appears to be decreasing. We can change the graphing interval to focus on larger values of x.

- Change the x-interval to $5 \le x \le 100$.

The result is:

▷ Graph f[x] on $5 \le x \le 100$.

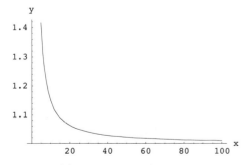

Out[4]= - Graphics -

Here, *Mathematica* places the horizontal axis at $y = 1$. The graph suggests that $\lim_{x \to \infty} f(x) = 1$ but we can't be sure. In verifying this using the definition of limit, we will assume that for $x > 1$, the function f is decreasing and $f(x) > 1$.

We can show by hand that in fact $\lim_{x \to \infty} f(x) = 1$. As $x \to \infty$,

$$1 + \frac{x^2 + 5x}{x^3 - x} = 1 + \frac{\frac{1}{x} + \frac{5}{x^2}}{1 - \frac{1}{x^2}} \to 1 + \frac{0 + 0}{1 + 0} = 1.$$

Here's how to find the value for N for a given value of ϵ as required by the definition of limit. For example, take $\epsilon = 0.02$. The graph suggests that $1 < f(60) < 1.02$. We can check this by evaluating $f(60)$:

In[5]:= **f[60]**

Out[5]= $\dfrac{3664}{3599}$

Use the **Numeric** button on the *Joy* palette to express this in decimal form:

In[6]:= $\mathbf{N}\left[\dfrac{3664}{3599}\right]$

Out[6]= **1.01806**

Since f is decreasing and $f(x) > 1$, we can make $f(x)$ come within 0.02 of $L = 1$ by taking $x > 60$. So for $\epsilon = 0.02$ we can let $N = 60$.

The smallest possible N we can use is a solution to the equation $f(x) = 1.02$, and we can use *Joy* to solve for it:

- Solve $f(x) = 1.02$ for x. (**Algebra** ▷ **Solve**)

The result is:

▷ Solve the equation f[x] = 1.02 for x algebraically.

Out[7]= $\{\{x \to -4.5973\}, \{x \to 54.5973\}\}$

The relevant solution here is the positive one, so the smallest N we can take is $N = 54.5973$.

If we let ϵ be arbitrary, we can find N similarly:

▷ Solve the equation f[x] = 1 + ε for x algebraically.

Out[8]= $\left\{\left\{x \to \dfrac{1 - \sqrt{1 + 20\,\epsilon + 4\,\epsilon^2}}{2\,\epsilon}\right\}, \left\{x \to \dfrac{1 + \sqrt{1 + 20\,\epsilon + 4\,\epsilon^2}}{2\,\epsilon}\right\}\right\}$

The relevant solution is the second one, which is greater than 1. That is, we can let $N = \frac{1+\sqrt{1+20\,\epsilon+4\,\epsilon^2}}{2\,\epsilon}$ or any larger number. Assuming f is decreasing for $x > 1$, this proves $\lim_{x\to\infty} f(x) = 1$.

> ☿ You can enter ϵ using the **Greek** button on the *Joy* palette. When you are done, close the palette by clicking in the close box in its upper-left corner (Macintosh) or on the **X** in its upper-right (Windows).

Exercises

1. a) Use a graph to estimate how close (δ) a number x must be to 0 for $\frac{\sin x}{x}$ to come within $\epsilon = 0.01$ of its limit as $x \to 0$.

    b) Is $f(x) = \frac{\sin x}{x}$ continuous at $x = 0$? Explain.

2. Use a table to estimate: How close (δ) must x be to 4 for you to believe that
    $$\frac{x^3 - 4\,x^2 + 5\,x - 20}{x - 4}$$
    is within $\epsilon = 0.001$ of its limit L as $x \to 4$?

3. Let
    $$f(x) = \frac{x^3 - 4\,x^2 + 5\,x - 20}{x - 4}$$
    and let $L = \lim_{x\to 4} f(x)$. Use **Algebra ▷ Solve** as in Example 18.2.2 to find the largest value of δ for which $0 < |x - 4| < \delta$ implies that $|f(x) - L| < 0.001$. What must you assume about f for this to be valid?

4. In the United States, the percentage of income that you must pay in taxes depends on your income level and family status. For example, in 1998 single taxpayers whose taxable income was between \$25,350 and \$61,400 paid \$3802.50 + 28% of the amount over \$25,350.

    The 28% rate is less for lower income levels and more for higher ones. Taxable income is the amount of your total income minus certain adjustments and deductions.

    a) Let x be the amount of taxable income and $f(x)$ be the corresponding amount of income tax. Write an equation for $f(x)$ when $25350 \le x \le 61400$, using the 1998 formula.

    b) Suppose your taxable income is \$30,000. How much tax do you have to pay?

    c) Graph the income tax function. Based on the graph (zoom in if necessary), what do you think is the value of $L = \lim_{x\to 30000} f(x)$? Do you think f is continuous when $x = 30000$? Why?

    d) Use a table of values to estimate how close (δ) an income level x must be to \$30,000 for the tax $f(x)$ to be within \$1 of the limit L?

5. a) Graph $\frac{\sin x}{|x|}$ for $-10 \le x \le 10$. (Use the **Absolute value** button on the *Joy* palette to enter $|x|$.)

    b) Based on the graph, do you think $\lim_{x \to 0} \frac{\sin x}{|x|}$ exists? Why?

    c) Let $L = 1$. Find a number $\epsilon > 0$ for which there is no number $\delta > 0$ that satisfies the definition of $\lim_{x \to 0} \frac{\sin x}{|x|} = L$.

    d) Repeat for $L = 0$.

6. a) Graph on $-10 \le x \le 10$:

    $$f(x) = \frac{x^2 - 5x + 6}{x^2 - 4x - 5} \ .$$

    b) Let $L = 0$. Find a number $\epsilon > 0$ for which there is no number $\delta > 0$ that satisfies the definition of $\lim_{x \to 5} f(x) = L$.

    c) Repeat for $L = 2$.

    d) Explain why there is no number L such that $\lim_{x \to 5} f(x) = L$.

7. a) Graph $1 - 10\,e^{-x}$ for $0 \le x \le 15$. What do you think is the value of $L = \lim_{x \to \infty} 1 - 10\,e^{-x}$?

    b) Suppose $\epsilon = 0.01$. Graph $1 - 10\,e^{-x}$ together with the horizontal lines $y = L - \epsilon$ and $y = L + \epsilon$, for $0 \le x \le 15$.

    c) From your graph, find a value of N such that whenever $x > N$ we have $|(1 - 10\,e^{-x}) - L| < \epsilon$.

    d) Use **Algebra ▷ Solve** as in Example 18.2.3 to solve for the smallest N in c). Repeat for an arbitrary value of ϵ.

8. a) Graph $\frac{\sin x}{x^3}$ for $-10 \le x \le 10$. Based on the graph, do you think $\lim_{x \to 0} \frac{\sin x}{x^3}$ exists (i.e., equals some number)? Evaluate the limit by hand to confirm this.

    b) The graph in a) doesn't show the influence of the factor $\sin x$ in the numerator. Experiment with different intervals $a \le x \le b$ with $a, b > 0$ until the graph shows the effect of $\sin x$.

    c) Based on the graph, what do you think is the value of $L = \lim_{x \to \infty} \frac{\sin x}{x^3}$? Find a number N such that whenever $x > N$ we have $|\frac{\sin x}{x^3} - L| < 0.001$. (You cannot use **Algebra ▷ Solve** to find the smallest N that satisfies this inequality, but you can use the properties of $\sin x$ to find some value of N that satisfies it.)

Lab 18.1 Visualizing the Definition of Limit

Prerequisites:
Ch. 1	A Brief Tour of *Joy*
Sec.18.2	The Definition of Limit

This lab helps you visualize the definition of limit illustrated in Example 18.2.1: $\lim_{x \to a} f(x) = L$ means that no matter what number $\epsilon > 0$ you choose, you can always find a number $\delta > 0$ such that

$$\text{whenever } 0 < |x - a| < \delta \text{ then } |f(x) - L| < \epsilon. \tag{*}$$

■ Before the Lab (Do by Hand)

1. If you haven't read Examples 18.2.1 and 18.2.2, read them now.

2. a) Draw a picture like the one in Example 18.2.1 for $f(x) = 3x + 5$, where $a = 1$ and $\epsilon = 0.15$. What should L be?

    b) Find the points in a) where the graph of $f(x) = 3x + 5$ intersects the horizontal lines.

    c) Are the intersection points the same distance from $x = 1$, measured horizontally?

    d) Determine the largest possible δ that satisfies the condition (*) when $\epsilon = 0.15$.

3. Repeat Question 2 without specifying a numerical value for ϵ. That is, find a relation between the permissible values of δ and an arbitrary given value of ϵ.

4. The graph of $f(x) = 10 - 4x^2$ is a parabola:

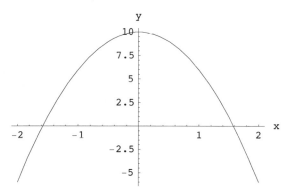

    a) Draw a picture like the one in Example 18.2.1 for this parabola, where $a = 1$ and $\epsilon = 0.1$. What should L be?

b) Find the points in the first quadrant where the parabola intersects the horizontal lines.

c) Are the intersection points the same distance from $x = 1$ (measured horizontally)?

d) Determine the largest possible δ that satisfies the condition (*) when $\epsilon = 0.1$.

5. Repeat Question 4b) without specifying a numerical value for ϵ. That is, find a relation between the permissible values of δ and an arbitrary given value of ϵ.

■ In the Lab

1. Create the function $f(x) = 2x^3 - 12x^2 + 22x - 10$. (**Create ▷ Function**)

2. Graph $f(x)$ for $-5 \le x \le 5$. What do you think will be the value of $L = \lim_{x \to 1} f(x)$?

3. a) In your *Mathematica* notebook, type $\epsilon = 0.1$ and press $\boxed{\text{SHIFT}} - \boxed{\text{RET}}$. (You can type ϵ using the **Greek** button on the *Joy* palette.)

   b) Type L = and the value you decided for $\lim_{x \to 1} f(x)$. Press $\boxed{\text{SHIFT}} - \boxed{\text{RET}}$.

4. Graph $f(x)$, $L + \epsilon$, and $L - \epsilon$ simultaneously, making the x-interval just wide enough to include the points where the curve crosses the two horizontal lines. Estimate the x-coordinates of these points to three decimal places.

5. Estimate to three decimal places the largest value of δ that satisfies condition (*) for the given value of ϵ.

6. a) Solve for the x-coordinate of the point where the curve $y = f(x)$ meets the line $y = L + \epsilon$. (**Algebra ▷ Solve**)

   b) Repeat for $y = L - \epsilon$.

   c) Use a) and b) to calculate the value of δ that you estimated in Question 5.

7. Print your work.

■ After the Lab

1. In Questions 2 and 4 (Before the Lab) and Question 5 (In the Lab), you needed to decide whether to calculate the largest δ by using an interval to the left of $x = 1$ or to the right. How did you decide?

2. a) Explain how the way the steepness of the curve changes indicates whether the largest δ is found to the left of $x = 1$ or to the right.

   b) For $f(x) = 10 - 4x^2$ from Question 4 (Before the Lab), express your answer to a) using the values of $f'(1)$ and $f''(1)$.

   c) Repeat with $f(x) = 2x^3 - 12x^2 + 22x - 10$ from Question 5 (In the Lab).

3. a) In your graph from Question 2 (In the Lab), suppose $L = 2.5$ and $\epsilon = 1$. Illustrate with a sketch that you can find a $\delta > 0$ that satisfies condition (*). Why doesn't this mean that $\lim_{x \to 1} f(x) = 2.5$?

    b) With $L = 2.5$, find a number $\epsilon > 0$ for which there is no $\delta > 0$ that satisfies condition (*). What does this tell you about $\lim_{x \to 1} f(x)$?

Lab 18.2 Numerical Evidence and Limits

Prerequisites:
  Ch. 1   A Brief Tour of *Joy*
  Sec. 4.2   How to Tabulate an Expression

This lab points out some of the pitfalls in using evidence from numerical experiments to decide the value of a limit. It is based on "Introduction to Limits, or Why Can't We Just Trust the Table?" by Allen J. Schwenk, *The College Mathematics Journal* (28) 1997, 51.

■ Before the Lab (Do by Hand)

In this experiment, you're going to investigate

$$\lim_{x \to 0} \sin \frac{\pi}{x}$$

by substituting various values of x into $\sin \frac{\pi}{x}$. These questions ask you to do this by hand for some simple values of x.

1. What is the value of $\sin \frac{\pi}{x}$ when $x = 0.1, 0.01, 0.001, 0.0001, \dots$?

2. Repeat for $x = 0.2, 0.02, 0.002, 0.0002, \dots$.

3. Repeat for

   $x = 0.4, 0.04, 0.004, 0.0004, \dots;$

   $x = 0.5, 0.05, 0.005, 0.0005, \dots;$

   $x = 0.8, 0.08, 0.008, 0.0008, \dots.$

4. Based on your experiments so far, what do you think the value of $\lim_{x \to 0} \sin \frac{\pi}{x}$ will be and why?

■ In the Lab

1. Follow these steps to make a table of values for $\sin \frac{\pi}{x}$ with $x = 0.3, 0.03, 0.003, \dots, 3 \times 10^{-10}$.

   - From the *Joy* menus, choose **Create ▷ List**.
   - Fill in x, $\texttt{Sin}\left[\frac{\pi}{x}\right]$.
   - From the popup menu, choose **x=f(n)**.
   - Fill in $\texttt{x = 3} \_10^{-n}$ and $1 \le n \le 10$.

The symbol ⌴ denotes a blank space to indicate multiplication. Don't forget the minus sign in $-n$.

- Choose **Display in Decimal Form** and **Display as a table**.
- Click **OK**.

2. Repeat for $x = 6 \times 10^{-n}$, $x = 7 \times 10^{-n}$, and $x = 9 \times 10^{-n}$.

3. Graph $\sin \frac{\pi}{x}$ for $0 \le x \le 1$. Repeat for $0 \le x \le 0.1$.

4. Print your work.

■ After the Lab

1. Based on your numerical data, what can you now say about $\lim_{x \to 0} \sin \frac{\pi}{x}$?

2. Use your graphs to illustrate the values taken by $\sin \frac{\pi}{x}$ for the different values of x that you substituted (use different colors or markings for different categories of x-values).

3. Without a calculator, derive the values of $\sin \frac{\pi}{x}$ that you obtained before the lab.

4. Without a calculator, derive the values you found in the lab when $x = 3 \times 10^{-n}$ and 6×10^{-n}. You may leave them in terms of square roots.

5. Without a calculator, explain why the values of $\sin \frac{\pi}{x}$ are all the same when $x = 9 \times 10^{-n}$, even if you don't know what that value is.

6. a) On your graphs, find x-values x_1, x_2, x_3, ... that approach 0 and satisfy
$\sin\left(\frac{\pi}{x_1}\right) = \sin\left(\frac{\pi}{x_2}\right) = \sin\left(\frac{\pi}{x_3}\right) = \ldots = 1$.

   b) Repeat for $\sin\left(\frac{\pi}{x_1}\right) = \sin\left(\frac{\pi}{x_2}\right) = \sin\left(\frac{\pi}{x_3}\right) = \ldots = -1$.

   c) Repeat for $\sin\left(\frac{\pi}{x_1}\right) = \sin\left(\frac{\pi}{x_2}\right) = \sin\left(\frac{\pi}{x_3}\right) = \ldots = 0.5$.

   d) For what y-values besides 1, -1, 0.5 can you find such x-values x_1, x_2, x_3, ...?

7. a) Find a number $\epsilon > 0$ for which there is no number $\delta > 0$ such that
   whenever $0 < |x - 0| < \delta$ then $|\sin \frac{\pi}{x} - 0| < \epsilon$.

   b) What does a) say about $\lim_{x \to 0} \sin \frac{\pi}{x}$?

Chapter 19
Derivatives in One Variable

Examples and Exercises

Prerequisites:	
Ch. 1	A Brief Tour of *Joy*
Sec. 5.1	How to Solve Equations Algebraically
Secs. 7.1–4	Differentiating functions and graphing derivatives

■ 19.1 Derivatives without Equations

19.1.1 Example: Predicting the Graph of a Derivative

This example shows how you can predict the graphs of the derivatives $y = f'(x)$ and $y = f''(x)$ from the graph of $y = f(x)$. We can make these predictions without having equations for $f'(x)$ and $f''(x)$. We'll illustrate this with the function

$$f(x) = \frac{x^2}{1 + x^2}.$$

- Create the function $f(x) = \frac{x^2}{1+x^2}$. (**Create ▷ Function**)

Mathematica creates the function:

▷ Define f[x] to be $\frac{x^2}{1 + x^2}$

- Graph $f(x)$ for $-5 \le x \le 5$.

The result is:

▷ Graph f[x] on -5 ≤ x ≤ 5.

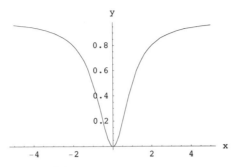

Out[3]= - Graphics -

We see that $f(x)$ decreases for $x < 0$, so $f'(x) < 0$ for those values. It increases for $x > 0$, so $f'(x) > 0$ there. The tangent is horizontal at the origin, so $f'(0) = 0$.

Also, $f(x)$ changes at a *decreasing* rate (the slope decreases) until about $x = -0.6$, and then changes at an *increasing* rate (the slope increases) until $x = 0$. So $f'(x)$ decreases to a minimum at $x < -0.6$ (*approx*.) and then increases until $x = 0$.

For $x > 0$, the pattern reverses. We see $f(x)$ increases at an increasing rate (the slope is positive and increases) until $x \approx 0.6$ and then increases at a decreasing rate (the slope is positive and decreases). So $f'(x)$ increases to a maximum at $x \approx 0.6$ and decreases afterward.

As $x \to \pm\infty$, the graph of f "flattens out" toward a horizontal asymptote at $y = 1$ ($\frac{x^2}{1+x^2} \to 1$), so we expect $f'(x) \to 0$.

So the graph of f' should have the general shape:

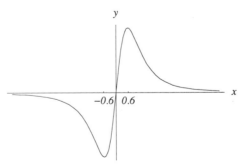

Now we'll check this:

- Graph $f'(x)$ for $-5 \le x \le 5$.

⚠ If you want to graph the derivative without having an equation for it, you must graph it as $f'(x)$. *Mathematica* will not graph it in the form $\frac{d}{dx}\left(\frac{x^2}{1+x^2}\right)$. This is why we created the function $f(x)$ at the start.

The graph is:

▷ Graph f'[x] on -5 ≤ x ≤ 5.

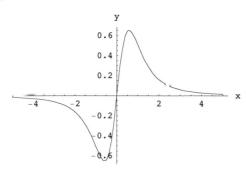

Out[4]= - Graphics -

We can also predict the graph of f''. The graph of f is concave down until $x \approx -0.6$, then concave up until $x \approx 0.6$, and concave down thereafter. So $f''(x) > 0$ for $-0.6 < x < 0.6$ (*approx.*). At $x \approx \pm 0.6$, $f''(x) = 0$. For all remaining x, we have $f''(x) < 0$.

Since f'' measures the rate of change of f', we can predict the graph of f'' by using the graph of f', just as we predicted the graph of f' from that of f. This procedure suggests that $f''(x)$ decreases until $x \approx -1$, increases up to $x \approx 0$, decreases until $x \approx 1$, and then increases. As $x \to \pm \infty$, we should have $f''(x) \to 0$. We'll show below that $x = 0, \pm 1$ are the exact points where these changes occur.

Thus the graph of f'' should have the form:

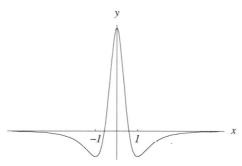

Finally, we check this with *Joy*:

- Graph $f''(x)$ for $-5 \le x \le 5$.

The result is:

▷ Graph (f′)′[x] on -5 ≤ x ≤ 5.

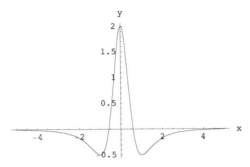

Out[5]= - Graphics -

The graph of f has inflection points where its concavity changes, which is also where $f'(x)$ reaches a maximum or minimum and $f''(x)$ changes sign. We estimated these from the graph at $x \approx \pm 0.6$. We can find them exactly by solving for the roots of $f''(x)$, as follows.

- Solve $f''(x) = 0$ for x. (**Algebra** ▷ **Solve**)

The solution is:

▷ Solve the equation f″[x] = 0 for x algebraically.

Out[6]= $\left\{\left\{x \to -\dfrac{1}{\sqrt{3}}\right\}, \left\{x \to \dfrac{1}{\sqrt{3}}\right\}\right\}$

Use the **Numeric** button on the *Joy* palette to express this in numeric form:

In[7]:= $\mathbb{N}\left[\left\{\left\{x \to -\dfrac{1}{\sqrt{3}}\right\}, \left\{x \to \dfrac{1}{\sqrt{3}}\right\}\right\}\right]$

Out[7]= $\{\{x \to -0.57735\}, \{x \to 0.57735\}\}$

We estimated the minimum of $f''(x)$ as $x \approx \pm 1$, and the maximum as $x \approx 0$, using the graph of f'. We can show these are the exact values by solving for the roots of $f'''(x)$, as follows.

- Solve $f'''(x) = 0$ for x.

The result is:

▷ Solve the equation f‴[x] = 0 for x algebraically.

Out[8]= $\{\{x \to -1\}, \{x \to 0\}, \{x \to 1\}\}$

Exercises

In these exercises use the methods in Example 19.1.1 to answer the questions and to check your answers without finding an equation for the derivative.

1. a) Create the function $f(x) = \sin^3 x$ and graph it for $0 \le x \le 2\pi$. (Enter as `Sin[x]`^{3}.)

    b) Where does the graph suggest that $f'(x)$ will be zero in this interval? positive? negative? Express your answers in terms of π. Confirm by graphing $f'(x)$ for $0 \le x \le 2\pi$.

2. Create and graph $f(x) = \sin^3 x$ for $0 \le x \le 2\pi$. (Enter as `Sin[x]`^{3}.) Estimate the x-coordinates where $f''(x)$ will be positive, negative, and zero. Check your answer by graphing $f''(x)$

3. a) Create and graph $f(x) = \ln(\ln x)$ for $1 \le x \le 20$. (Enter as `Log[Log[x]]`.) Predict the graph of the derivative. Then check your answer by graphing $f'(x)$.

    b) Predict the graph of the second derivative. Then check your answer by graphing $f''(x)$.

4. a) Create and graph $f(x) = e^{x+\sin x}$ for $-3 < x \le 3$. Where does the graph suggest that $f(x)$ will be increasing most rapidly in this interval? Check your answer by graphing $f''(x)$.

    b) Use *Joy* to solve for the value of x in $[-3, 3]$ where $f(x)$ is increasing most rapidly. (**Algebra ▷ Solve**)

5. a) Create and graph $f(x) = x^x$ for $0 \le x \le 2$. The graph of f has a vertical tangent at $x = 0$. Does $f'(0)$ exist?

    b) According to the graph, does $f'(x)$ have a minimum value for $0 \le x \le 2$? Check your answer by graphing $f'(x)$.

■ 19.2 Estimating Derivatives Numerically

Additional Prerequisite:
        Sec. 4.2   How to Tabulate an Expression

19.2.1 Example: Average and Instantaneous Rates of Change

The derivative of a function $y = f(x)$ is given by the *instantaneous* rate of change

$$f'(x) = \lim_{h \to 0} \frac{f(x+h) - f(x)}{h} .$$

So you can estimate a derivative by taking the *average* rate of change for a small value of h:

$$f'(x) \approx \frac{f(x+h) - f(x)}{h} \text{ when } h \text{ is small}.$$

For example, when $f(x) = x^x$ we can estimate the derivative at $x = 2$ by

$$f'(2) \approx \frac{f(2.001) - f(2)}{0.001}.$$

Here's how to do it:

- Create the function $f(x) = x^x$. (**Create ▷ Function**)

Mathematica creates the function:

▷ Define f[x] to be x^x

- If necessary, click in your *Mathematica* notebook below the last cells so that a horizontal line appears.
- Type $\frac{f[2.001]-f[2]}{0.001}$ and press ⌜SHIFT⌝-⌜RET⌝.

The result is:

In[3]:= $\dfrac{\mathbf{f[2.001] - f[2]}}{\mathbf{0.001}}$

Out[3]= 6.77933

So $f'(2) \approx 6.77933$. Here's how to get a better estimate by taking smaller and smaller values of h.

- From the *Joy* menus, choose **Create ▷ List**.
- Fill in h, $\frac{f[2+h]-f[2]}{h}$.
- From the popup menu, choose **x=f(n)**.
- Fill in h = 10^{-n} and $0 \le n \le 10$.

  This chooses the values $h = 1, 0.1, 0.01, \ldots, 10^{-10}$.

- Make sure **Display in decimal form** and **Display as a table** are checked.
- Click **OK**.

Mathematica produces the list:

▷ Create the list with values $\left\{ h, \dfrac{f[2+h] - f[2]}{h} \right\}$ for h = 10^{-n} with $0 \le n \le 10$.

Out[4]//TableForm=

1.	23.
0.1	7.49638
0.01	6.8404
0.001	6.77933
0.0001	6.77326
0.00001	6.77266
$1. \times 10^{-6}$	6.7726
$1. \times 10^{-7}$	6.77259
$1. \times 10^{-8}$	6.77259
$1. \times 10^{-9}$	6.77259
$1. \times 10^{-10}$	6.77259

This suggests that $f'(2) = 6.77259$ to six significant digits.

Here's how to find an equation for the derivative using *Mathematica* (or you can find it by hand using logarithmic differentiation).

- If necessary, click in your *Mathematica* notebook below the last output so that a horizontal line appears.
- Type f'[x].
- Press SHIFT – RET .

In[5]:= f'[x]

Out[5]= $x^x + x^x \, \text{Log}[x]$

So $f'(2) = 4 + 4 \ln 2$. You can let *Mathematica* express this in numerical form:

- Click the **Numeric** button on the *Joy* palette.
- Type f'[2] and press SHIFT – RET .

In[6]:= N[f'[2]]

Out[6]= 6.77259

Another way. You can also apply the **Derivative** button on the *Joy* palette to x^x:

In[7]:= $\dfrac{d}{dx}(x^x)$

Out[7]= $x^x + x^x \, \text{Log}[x]$

Exercises

In Exercises 1–4:

   a) Estimate $f'(x)$ by calculating

$$\frac{f(x + h) - f(x)}{h}$$

for the given values of x and h. Use the methods of the example in this section, Example 19.2.1.

b) If you know how to differentiate f, find $f'(x)$ for the given value of x by hand. If you don't know the derivative, let *Mathematica* find $f'(x)$ as in the example.

1. $f(x) = \cos x, x = 0, h = 0.000001$.

2. $f(x) = e^x, x = 1, h = 0.00001$.

3. $f(x) = \cos x, x = \frac{\pi}{4}, h = 10^{-n}, n = 0, 1, 2, \ldots, 10$.

4. $f(x) = \ln x, x = 1, h = 10^{-n}, n = 0, 1, 2, \ldots, 10$.

5. a) Create the function $f(x) = \tan x$. Graph $f(x + 1) - f(x)$ and $\sec^2 x$ simultaneously for $-\pi/2 \le x \le \pi/2$. (Enter $\sec^2 x$ as `Sec[x]`^{2}. Even though $\sec x$ is undefined for $x = \pm\pi/2$, you can use this interval for graphing.)

    b) Repeat with

    $$\frac{f(x + 1) - f(x)}{0.5}.$$

    Repeat again, replacing the denominator by 0.3 and then by 0.1.

    c) Summarize what the graphs suggest about the derivative of $\tan x$.

6. This exercise uses animation, which is discussed in Section 3.9.

    a) Create the function $f(x) = \sin x$. (**Create ▷ Function**)

    b) Animate the graphs of

    $$\frac{f(x + 2^{-k}) - f(x)}{2^{-k}}$$

    with x between $\pm 2\pi$. Let $-1 \le k \le 8$ and take the common plot frame to be $-2\pi \le x \le 2\pi$, $-1 \le y \le 1$. (**Graph ▷ Animation**)

    c) What curve do the graphs seem to be approaching as $h = 2^{-k}$ gets closer to 0? What does this suggest about the derivative of $\sin x$?

■ 19.3 Differentiation Practice

> Additional Prerequisite:
>   Sec. 7.5,6   Differentiating expressions

Exercises

In these exercises:

    a) Find the derivative by hand

    b) Check your answer using the **Derivative** button on the *Joy* palette. If *Mathematica* gives the answer in a different form, make sure it is equivalent to yours.

1. $y = (x^3 - x + 1)(2x^5 - x^2)$; find $\frac{dy}{dx}$.

2. $u = \frac{v^4 + v}{3v - 2}$; find $\frac{du}{dv}$ when $v = 0$.

3. $p = \sqrt{r^2 - 2r + 1}$; find $\frac{dp}{dr}$.

4. $y = \cos(x^3 - 3x)$; find $\frac{dy}{dx}$.

5. $p = \ln(q^2 + \sin q)$; find $\frac{dp}{dq}$ when $q = \pi$.

6.   a) $s = e^{t^2}$; find $\frac{ds}{dt}$.

     b) $s - (e^t)^2$; find $\frac{ds}{dt}$.

7. $y = \sin^2(x^2 + 2^x)$; find $\frac{dy}{dx}$ when $x = 3$.

■ 19.4 Derivatives, Tangents, and Graphs

Exercises

1.   By hand, find the equation of the line that is tangent to the curve $y = x^4 - 3x^3 + 2x^2 - x + 1$ at $x = 2$. Check your answer by graphing the curve and the tangent line simultaneously.

2.   a) Graph $y = 2x^5 - x^4 - 4x^3 - 2x$ for $-5 \le x \le 5$.

     b) Find the critical points and decide which are maxima and which are minima. (Differentiate by hand and use **Algebra ▷ Solve** to find where the derivative is zero.)

     c) Find the x-coordinates of the inflection points on the graph.

3. a) Find the first derivative of $y = x^4 - 6x^2 - 8x + 5$ by hand and then factor it using *Joy*. (**Algebra** ▷ **Simplify**)

    b) Predict the x-coordinates of the relative (local) maxima and minima of the function in a). Then graph the curve and check your answer. Explain any differences.

4. a) Find the second derivative of $y = x^5 + 5x^4 - 40x^2 + 3x + 10$ by hand and then factor it using *Joy*. (**Algebra** ▷ **Simplify**)

    b) Predict the number and location of the inflection points for the graph of the function in a). Then graph the curve and check your answer. Explain any differences.

■ 19.5 Implicit Differentiation

Additional Prerequisites:
Sec. 3.2	How to Graph Implicit Functions
Sec. 7.10	How to Differentiate an Implicit Function

19.5.1 Example: Finding the Derivative of an Implicit Function

When y is not given explicitly as a function of x, we can find the derivative $\frac{dy}{dx}$ by implicit differentiation. Here's how to visualize an implicit function and differentiate it using *Joy*, with the example $x^2 y^3 + y^4 = 1$.

- Graph $x^2 y^3 + y^4 = 1$ for $-2 \le x \le 2$. Choose the method of **solving for one variable**. (**Graph** ▷ **Implicit Function**)

Mathematica displays the graph, which is not the graph of a function. We see that the equation implicitly defines y as a function of x in two ways.

▷ Graph of $x^2 y^3 + y^4 = 1$ for $-2 \le x \le 2$.

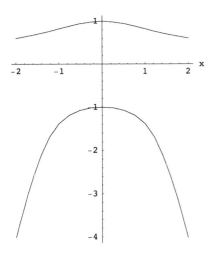

Out[2]= - Graphics -

When we find the derivative $\frac{dy}{dx}$, we are differentiating both functions at the same time. You can differentiate by hand, or use *Joy* as follows:

- If necessary, click in your *Mathematica* notebook below the last output so that a horizontal line appears.
- Type $x^2 \ y[x]^3 + y[x]^4$.
- Click the **Derivative** button on the *Joy* palette.
- Type x for the independent variable.
- Press SHIFT – RET .

The result is:

In[3]:= $\dfrac{d}{d \ x} \ (x^2 \ y[x]^3 + y[x]^4)$

Out[3]= $2 \ x \ y[x]^3 + 3 \ x^2 \ y[x]^2 \ y'[x] + 4 \ y[x]^3 \ y'[x]$

Since the right side of the equation $x^2 y^3 + y^4 = 1$ is constant, this derivative is 0. So we can set the result equal to 0 and solve for $y'(x)$:

$$\frac{dy}{dx} = y'(x) = \frac{-2 \, x \, y^3}{3 \, x^2 \, y^2 + 4 \, y^3} = \frac{-2 \, x \, y}{3 \, x^2 + 4 \, y}.$$

This is the slope of the tangent to the upper part of the graph when $y > 0$ and to the lower part when $y < 0$.

Exercises

In Exercises 1–3 use the method of the example in this section, Example 19.5.1.

a) By hand, find the derivative implicitly.

b) Check your answer using the **Derivative** button on the *Joy* palette and solving the resulting equation, as in the example. If *Mathematica* gives the answer in a different form, make sure it is equivalent to yours.

1. $x^2 y + x + x y^2 + y = 5$; find $\frac{dy}{dx}$.

2. $u e^v + v e^u = uv$; find $\frac{du}{dv}$ when $v = 0$.

3. $x \sin(xy) = 1$; find $\frac{dy}{dx}$.

4. a) Graph the curve $x^2 y^3 - y^4 = 1$ for $-3 \le x \le 3$ and estimate the slope of the tangent at the two points where $y = 2$. At these points, $x = \pm \sqrt{17/8} \approx 1.5$. (You may want to include a grid, as shown in Section 3.7.2, to make this easier.)

   b) Differentiate by hand to find the slopes and check them against your estimates.

5. a) Graph the curve $x^3 - y^4 + 8 y = 1$ for x between ± 3.

   b) By hand, find the y-coordinates of the points on the curve where $x = 1$. Then find the equation of the tangent line to the curve at each point.

   c) Graph the two tangent lines simultaneously. Then combine your two graphs and check that the lines appear tangent to the curve. (**Graph ▷ Combine Graphs**)

6. a) By hand, find any points on the curve $x^3 - y^4 + 8 y = 1$ where $\frac{dy}{dx}$ doesn't exist.

   b) Graph the curve for x between ± 3. Explain how you can tell from the graph where $\frac{dy}{dx}$ doesn't exist.

■ 19.6 Using Derivatives

Exercises

1. (**Minimization**) In this exercise, you can find how close a given curve comes to a specified point.

   a) Create $f(x) = x^2 - 3 x + 4$.

   b) Let $g(x)$ be the *square* of the distance between the origin and a point $(x, f(x))$ on the graph of f. Find an equation for $g(x)$.

It's easier to work with the square of the distance rather than the distance itself because this avoids square roots. The distance and its square are minimized by the same value of x.

c) Find $g'(x)$ by hand. Use *Joy* to show that g has only one real critical point, which will give the nearest point to the origin. (**Algebra ▷ Solve**)

d) Set a equal to the critical point you found by following these steps:

- If necessary, click below the last output so that a horizontal line appears.
- Type a = and fill in the value of the real critical point from c). You can copy and paste in this value, or type it.
- Press SHIFT – RET.

e) What slopes do $f'(a)$ and $\frac{f(a)}{a}$ represent?

f) Multiply $f'(a)$ and $\frac{f(a)}{a}$ together in your *Mathematica* notebook by typing the product and pressing SHIFT – RET. Then simplify the last output. (**Algebra ▷ Simplify**) What does the result prove?

2. (**Related rates**) A farmer wants to raise a bale of hay from the ground up to a hayloft in her barn. She attaches a rope from the hay up to a pulley at the loft and back down to a tractor. She then drives away from the barn at constant velocity and the hay is pulled up to the loft. The goal of this exercise is to describe the motion of the hay.

a) Make a sketch showing the hay partway up to the loft. Label the distance from the tractor to the barn $x = x(t)$ and the height of the hay $y = y(t)$.

b) Suppose the rope is 50 feet long, the pulley is 25 feet from the ground, and the tractor moves at 10 feet per second. Write an equation relating x and y. Where is the tractor when the bale reaches the top and when does that happen?

c) Use b) to find $\frac{dy}{dt}$ in terms of x, by hand. Graph the result for $0 \le x \le 50$. (The vertical axis will automatically be labeled y, but indicates $\frac{dy}{dt}$.) Does the hay rise at constant velocity? According to the graph, where is the tractor when the hay is moving fastest? Where is the hay at that time?

d) What information does the concavity of the graph give about the motion of the hay?

3. (**Mean Value Theorem**) The Leaning Tower of Pisa is about 185 feet high. If you drop a ball from the top, its height after t seconds is $h = -16 t^2 + 185$ feet.

a) (By hand) How long does the ball take to reach the ground? What is the average velocity of the ball during its flight?

b) Graph h from $t = 0$ to the time you found in a). How can you picture the average velocity on the graph?

c) Estimate the moment during the flight when the ball's instantaneous velocity equals the average velocity in b).

d) Solve for the time in c) by hand and check it against your estimate.

4. (**Mean Value Theorem**) a) Create the functions $f(x)$ and $g(x)$, where $f(x) = 2x + x \sin x$ and $g(x)$ is the height of the chord of the graph of f between $x = 0$ and $x = 2\pi$.

b) Graph $f(x)$, $g(x)$, and $f(x) - g(x)$ simultaneously, for $0 \le x \le 2\pi$.

c) Estimate the x-coordinates of the two points on the graph of f where the tangent is parallel to the chord. What seems to be the relationship between these points and the graph of $f - g$?

d) Choose one of the points in c) and solve for its x-coordinate. (Choose **Algebra** ▷ **Solve** and fill in an equation that determines this point. Select **Solve numerically**, using the estimate you made in c)).

> *Mathematica* cannot solve the equation exactly. **Solve numerically** gives a very good approximation, starting from an initial estimate.

e) Find the equation of the tangent at the point in d), and graph $f(x)$, $g(x)$, and the tangent for $0 \le x \le 2\pi$. Check that the tangent appears parallel to the chord.

f) Repeat d) and e) for the other point in c).

Lab 19.1 Local Linearity and Derivatives

Prerequisites:
     Ch. 1   A Brief Tour of *Joy*

What does a differentiable function look like? In this lab, you can explore the connection between the shape of a function's graph and the rate at which the function changes.

■ Before the Lab (Do by Hand)

1. a) Graph $\sin x$ by hand between $x = \pm 2\pi$, making the graph about 4 inches wide by 2 inches high. Plot the maxima, minima, and x-intercepts, and at least one point between each of these. (Use radians on your calculator.)

    b) Graph $\sin x$ for x between $\pm\pi$, adjusting the scales on the axes so that the curve fills the same size rectangle. Repeat for x between $\pm\pi/2$ and then $\pm\pi/4$. What shape does the curve appear to take as you do this?

2. From the last graph in Question 1, estimate the slope of the line that is tangent to the curve $y = \sin x$ at the origin. Also, estimate the equation of the tangent line.

3. Repeat Question 1 for $y = |x|$ between $x = \pm 1$ and then over progressively smaller intervals.

4. You cannot repeat Question 2 for $y = |x|$, because the graph does not have a tangent at the origin. Instead, estimate the slopes and equations of the two lines that are tangent at the origin from the left and from the right.

■ In the Lab

1. a) For x between $\pm 2\pi$, graph $\sin x$ simultaneously with the tangent whose equation you estimated in Question 2 (Before the Lab). Use the **Automatic** setting for the y-interval.

    b) Repeat a) over progressively smaller intervals centered at $x = 0$ until the shape of the graph appears not to change.

2. a) Graph $\sin x$ for $\frac{\pi}{2} - 1 \le x \le \frac{\pi}{2} + 1$. Use the popup menu to specify $-1 \le y \le 1$ as the y-interval.

    b) Repeat a) over progressively smaller x-intervals centered at $x = \frac{\pi}{2}$ with $-1 \le y \le 1$, until the shape of the graph appears not to change.

    If you don't specify the y-interval, then *Mathematica* will choose y-intervals that mask this behavior.

3. a) Graph $|\sin x|$ between $x = \pm 2\pi$, using the **Absolute value** button on the *Joy* palette to enter the function. (You can also type **Abs[Sin[x]]**). Use the popup menu to reset the y-interval to **Automatic**.

    b) Graph $|\sin x|$ over progressively smaller intervals centered at $x = 0$ until the shape of the graph appears not to change.

4. Repeat Question 3b) using intervals centered at $x = \pi$, beginning with $\pi - 1 \le x \le \pi + 1$.

5. Print your work.

■ After the Lab

If zooming in on the graph of $y = f(x)$ around some point $x = a$ makes it indistinguishable from a nonvertical line, we say that f is *locally linear* at a. The line is tangent to the graph when $x = a$, and its slope is the derivative $f'(a)$. The function f is differentiable at $x = a$ if there is a nonvertical line tangent to the graph at that point.

Answer the following questions based on your experiments before and in the lab.

1. a) Does $f(x) = \sin x$ appear to be locally linear at $x = 0$? If so, what is $f'(0)$? Are there separate slopes from the left and right?

    b) Repeat for $x = \pi/2$.

2. Does $f(x) = |x|$ appear to be locally linear at $x = 0$? If so, what is $f'(0)$? Are there separate slopes from the left and right?

3. a) Does $f(x) = |\sin x|$ appear to be locally linear at $x = 0$? If so, what is $f'(0)$? Are there separate slopes from the left and right?

    b) Repeat for $x = \pi$.

    c) Do you think $f(x) = |\sin x|$ is locally linear at $x = \pi/2$? Why?

Lab 19.2 Optimization

Prerequisites:
    Ch. 1   A Brief Tour of *Joy*
    Sec. 6.3   How to Tabulate a Function

What's the best design for a cylindrical juice can? Is one shape more efficient than the others in its use of material or in the volume that it holds? This lab shows how you can use *Joy* to solve optimization problems, where you must choose the value of some variable that produces the best outcome.

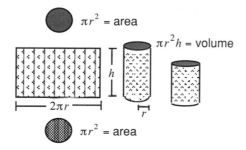

πr^2 = area

$\pi r^2 h$ = volume

h

$2\pi r$

r

πr^2 = area

■ Before the Lab (Do by Hand)

1. Suppose the radius of the can is r centimeters (cm) and the height is h cm. How much metal A does the can require?

2. A common size juice can holds 1 liter, or 1000 cubic centimeters (cc), of juice. If the radius of the can is 6 cm, how tall should the can be and how much metal will it require?

3. a) In general, let V be the volume of juice the can should hold. Express h in terms of r and V.

   b) Use your answer to a) to express A in terms of r and V.

   c) Check your answers to a) and b) with the numerical example in Question 2. Resolve any discrepancy before continuing.

■ In the Lab

1. First, you'll define the amount of metal as a function $A(r, V)$ of both the radius r and the volume V.

   • From the *Joy* menus, choose **Create ▷ Function**.
   • Fill in A[r,V] in the first input field.

- In the second input field, fill in your answer to Question 3b) (Before the Lab) for the amount of metal.

  Be sure to type a blank space between symbols where you need to indicate multiplication, e.g., $\pi\_\_r^2$.

- Click **OK**.

2. Next, you'll make a table showing how the amount of metal varies with the radius when the volume is 1000 cc.
   - Choose **Create ▷ List**.
   - Fill in `r,A[r,1000]`.
   - Make sure the popup menu reads **x Equally Spaced**, and fill in $1 \leq r \leq 15$ in increments of 1. Be sure to use the variable `r`.
   - Make sure **Display in decimal form** and **Display as a table** are checked.
   - Click **OK**.

3. Here's how to graph the amount of metal as a function of the radius when the volume is 1000 cc:

   - Choose **Graph ▷ Graph f(x)**.
   - Fill in `A[r,1000]` and $1 \leq r \leq 15$.

     The vertical axis is labeled *y* by default, but here stands for $A(r, 1000)$.

   - Click **OK**.

4. Here's how you can create graphs like these for different volumes, simultaneously:

   - Click the **Last Dialog** button on the *Joy* palette, to bring back the **Graph f(x)** dialog.
   - Fill in `A[r, 100]`, `A[r, 500]`, `A[r, 1000]`, `A[r, 1500]`, `A[r, 2000]`.
   - Make sure the interval is still $1 \leq r \leq 15$.
   - Click **OK**.

5. Print your work.

■ After the Lab

1. Why are very narrow and very wide 1000 cc cans not good choices from the point of view of the manufacturer?

2. a) Use your table in Question 2 (In the Lab) to estimate the radius of the 1000 cc can that uses the least amount of metal.

   b) Repeat a) using the graph in Question 3. If your answers to a) and b) aren't consistent, review your reasoning.

3. a) Let *A* be the expression you found for the amount of metal in Question 3b) (Before the Lab). Find $\frac{dA}{dr}$ by hand, assuming *V* is constant.

b) Find the value of r that minimizes A. Your answer should include V without specifying its numerical value.

c) In b), what is r when $V = 1000$? Compare this to your answer to Question 2 (After the Lab). Are your results consistent?

4. a) Estimate the minimum amount of metal for cans holding 500 cc and 1000 cc, using your graph in Question 4 (In the Lab). What are the units?

b) Let A be the expression you found for the amount of metal in Question 3b) (Before the Lab). When the radius minimizes A, how much metal do you need to use? Your answer should be a function of V.

c) Compare your answers to a) and b) and review your reasoning if they are not consistent.

d) Fill in the blanks:

The minimum amount of metal you need is _____ (*directly* or *inversely*) proportional to the _____ power of V.

If you double V, the minimum amount _____ (*increases* or *decreases*) by a factor of _____.

5. The ratio $\frac{h}{r}$ gives the proportions of the can. When the least amount of metal is being used, what is the value of this ratio?

■ For Further Exploration

A second approach is to design a can that holds as much liquid as possible, using a given amount of metal.

1. a) Create the function $V(r, A)$ that expresses the volume V in terms of the radius r and the amount A of metal.

b) For $1 \le r \le 15$, graph $V(r, 100)$, $V(r, 250)$, $V(r, 500)$, $V(r, 750)$, and $V(r, 1000)$ simultaneously. What do these curves represent and what do they show about the volume?

c) Does a wide can hold more than a narrow one that uses the same amount of metal?

2. a) Find $\frac{dV}{dr}$ by hand, treating A as a constant.

b) What radius maximizes the volume for an arbitrary amount A of metal? Compare your result with the graphs.

3. a) When the radius maximizes the volume, how many cc will the can hold? Your answer should include A as an unspecified constant.

b) Fill in the blanks:

The maximum volume of the can is _____ (*directly* or *inversely*) proportional to the _____ power of A.

If you double A, the maximum volume _____ (*increases* or *decreases*) by a factor of _____.

4. By hand, find $\frac{h}{r}$ when r maximizes the volume for a given value of A. Compare your answer to what you found in Question 5 (After the Lab).

Lab 19.3 Newton's Method for Estimating Roots

Prerequisites:
    Ch. 1   A Brief Tour of *Joy*
    Sec. 5.4   How to Visualize Newton's Method

This lab explains Newton's method for estimating roots of functions. In the lab, you can explore how the method works and see some of the phenomena that can occur. Section 5.4 gives a brief illustration of Newton's method.

Newton's method seeks a solution to an equation of the form $f(x) = 0$, i.e., a point where the graph of f crosses the x-axis. We begin with an initial estimate x_0 for such a root. The point x_1 where the tangent line at x_0 meets the x-axis is often a better estimate of the root than x_0.

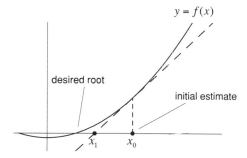

The equation of the tangent to the graph at x_0 is

$$y = f(x_0) + f'(x_0)(x - x_0).$$

The tangent crosses the x-axis at $(x_1, 0)$, so $0 = f(x_0) + f'(x_0)(x_1 - x_0)$, or

$$x_1 = x_0 - \frac{f(x_0)}{f'(x_0)}.$$

We move to $(x_1, f(x_1))$, the point on the graph where $x = x_1$, and repeat the procedure. In general, we calculate each point from the preceding one by

$$x_{n+1} = x_n - \frac{f(x_n)}{f'(x_n)}.$$

Under appropriate assumptions, it can be proved that if the initial estimate x_0 is sufficiently close to a root of f, then $x_0, x_1, x_2, \ldots$ will converge to that root.

■ Before the Lab (Do by Hand)

Here is a graph of $y = 2x^3 - 6x^2 + x + 1$.

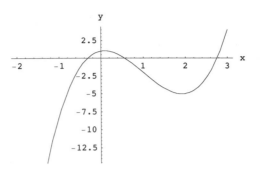

1. Let $x_0 = 0$. Calculate the Newton estimates x_1 and x_2.

2. On a copy of the graph, begin with $x_0 = 0$ and draw the first three tangent lines in Newton's method. Estimate the root to which you think the method will converge.

3. Repeat Question 2a) , starting with $x_0 = 1.5$.

4. What do you think will happen if you start at $x_0 = 2$? The page isn't large enough for you to draw the tangents.

5. What do you think will happen if you start at $x_0 = 1.8$?

■ In the Lab

We start with the function you studied before the lab.

1. Create the function $f(x) = 2x^3 - 6x^2 + x + 1$. (**Create ▷ Function**)

    This will make it easier to refer to this function several times during the lab.

2. a) From the *Joy* menus:

    • Choose **Calculus ▷ Newton's Method**.
    • Fill in the function **f[x]** and the initial estimate **x = 0**.
    • Check the option **Illustrate graphically**. Use the default setting 10 iterations.
    • Click **OK**.

    b) Compare the result with your answers to Questions 1 and 2 (Before the Lab) and review your reasoning if there are any inconsistencies.

3. a) Your output should predict the root $x = -0.320012$, to six significant digits, which is *Mathematica*'s default number of digits. Here's how you can check this result:

    • In your *Mathematica* notebook, click below the last output so that a horizontal line appears.
    • Type **f[-0.320012]**.
    • Press SHIFT – RET .

b) Since $x = -0.320012$ isn't exact, your result in a) won't be exactly 0. *Mathematica* calculated the root to many more significant digits than the six it displayed. Here's how you can get closer by using more of these digits.

- Drag across -0.320012 in the output for Question 2b) and copy it by pressing ⌘-C (Macintosh) or [CTRL]-C (Windows).
- Click below the last output so that a horizontal line appears.
- Type f[…], pasting the root between the brackets with ⌘-V (Macintosh) or [CTRL]-V (Windows).

   Mathematica automatically inserts all the digits it carries behind the scenes:

   f[-0.320011733446322343`]

   The mark ` may or may not be included, depending on the settings in your copy of *Mathematica*. It indicates the method *Mathematica* uses to determine how many digits are significant.

- Press [SHIFT] – [RET] .

   The result should be 0., which means that the first six significant digits are zero (not necessarily that the value is exactly 0). Entering f[Out[…]], using the output number for -0.320012, would give the same result. However, this approach lets you see the additional digits.

4. a) Click **Last Dialog** on the *Joy* palette and repeat Newton's Method with $x_0 = 1.5$.

   b) What root does the method predict? Check by evaluating $f(x)$ for this value of x. (It will not be exactly zero. Newton's method gives an excellent estimate but not the exact root.)

5. Repeat with $x_0 = 2$.

6. Repeat with $x_0 = 1.8$.

7. a) Create the function (**Create ▷ Function**)

$$f(x) = \frac{1 - 2x^2}{2(1 + x^4)} \quad .$$

   b) Graph $f(x)$ for $-5 \le x \le 5$.

   c) Apply **Newton's Method** to $f(x)$ starting with the initial point $x_0 = 2$. Do the estimates $x_0, x_1, x_2, \ldots$ appear to converge to a root?

   d) Evaluate $f(x)$ for the value x predicted by Newton's method.

   e) Repeat b) and c) using 20 iterations, starting with $x_0 = 2$.

8. a) Graph $x^3 - 5x$ for $-5 \le x \le 5$.

   b) Apply **Newton's Method** to $x^3 - 5x$ with the initial point $x_0 = 1$. Do the estimates $x_0, x_1, x_2, \ldots$ appear to converge to a root?

c) Is the value predicted by Newton's method a root of $x^3 - 5x$? Change the number of iterations to 15 and apply the method again. What happens?

d) Repeat c) with 20 iterations.

9. Print your work.

■ After the Lab

1. In your examples, when did Newton's method converge to a root that wasn't the nearest one to the starting point? What characteristics of the starting point might cause this?

2. According to the equation for Newton's method, what should happen if you start at a point where the function has a relative or local maximum or minimum? Illustrate this on the graph in Question 2 (Before the Lab).

3. What is the effect of taking more iterations in Question 7 (In the Lab)? Did Newton's method find a root?

4. Calculate the estimates x_1, x_2, x_3 for $f(x) = x^3 - 5x$, starting with $x_0 = 1$ and using the equation that expresses x_{n+1} in terms of x_n. Use this to explain the behavior you observed in Question 8 (In the Lab).

■ For Further Exploration

These questions all refer to the function $f(x) = x^3 - 5x$ from Question 8 (In the Lab).

1. (By hand) Using the graph of f, try to predict how the initial point x_0 determines whether Newton's method converges and, if so, to which root of f.

2. Using *Joy*, create the function $f(x) = x^3 - 5x$. Experiment with different initial points for Newton's method and check your predictions. By hand, sketch the graph of f, color coding different segments according to the root that initial points in those segments determine.

3. a) You can use the *Newton function*

$$\text{newton}(x) = x - \frac{f(x)}{f'(x)}$$

to describe the behavior of Newton's method. For example, $x_1 = \text{newton}(x_0)$, $x_2 = \text{newton}(x_1)$, …. Create the function newton(x) and calculate newton(newton(x)).

b) Solve newton(newton(x)) = x. You will find nine solutions, only five of which are real numbers. Identify the five real solutions with points on the graph of f, and explain why they satisfy newton(newton(x)) = x.

Lab 19.4 Iteration and Fixed Points

Prerequisites:

Ch. 1	A Brief Tour of *Joy*
Sec. 6.5	How to Iterate a Function
Sec. 5.4	How to Visualize Newton's Method (optional)

Section 6.5 shows how to use *Joy* to give a graphical illustration of iterating a function. In this lab, you can explore how changing the initial value in an iteration affects the process and how iterates can converge or diverge. You can also learn about the connection between iteration and Newton's method.

We can visualize the iteration process with a *cobweb graph*. Here are the first steps of a typical cobweb graph, showing how the iteration algorithm proceeds from the initial value x_0 to the first iterate $x_1 = f(x_0)$. It then applies the same procedure to x_1 to find $x_2 = f(x_1)$, etc.

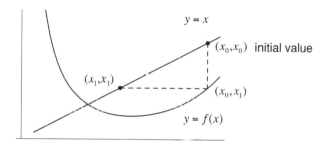

■ Before the Lab (Do by Hand)

1. Sketch the graphs of $y = \frac{x}{2} + 5$ and $y = x$ simultaneously. Starting with $x_0 = 0$, plot the points on $y = x$ corresponding to x_1, x_2, x_3. The iterates should appear to be converging to a limit. Find the coordinates of the limit.

2. Repeat for $x_0 = 15$.

3. Repeat Question 1 for $y = 2x - 3$, $y = x$, and $x_0 = 0$. Do the iterates approach a limit? What if $x_0 = 4$?

4. A value x^* for which $f(x^*) = x^*$ is called a *fixed point* of f. The fixed points correspond to where the graph of f meets the line $y = x$. What happens in an iteration if some value x_n is a fixed point?

■ In the Lab

1. From the *Joy* menus:

    - Choose **Calculus** ▷ **Iterate Function**.
    - Fill in the function $0.2\,x(1 - x)$ and the initial value $x_0 = 2$.
    - Fill in $n = 10$ iterates and select **Print all iterates**.
    - Check **Illustrate with cobweb graph** and click **OK**.

2. Repeat for several starting values: $x_0 = 5$, $x_0 = 6$, $x_0 = -2$, $x_0 = -5$.

3. Print your work.

■ After the Lab

1. a) In your cobweb graphs, to what values did the iterates of $0.2\,x(1 - x)$ appear to converge? Are these fixed points of the function?

    b) Are they the only fixed points of the function?

2. Imagine how the cobweb graph for $0.2\,x(1 - x)$ would be formed for all possible initial values x_0, not just the ones you used in the lab.

    a) To which values can the iterates of $0.2\,x(1 - x)$ converge, and to which initial values x_0 does each limit correspond?

    b) Which values of x_0 will produce iterates that don't converge to any limit?

3. Imagine how the cobweb graph for a line $y = m\,x + b$ would be formed for all possible values m, b, and x_0. Include positive, negative, and zero values.

    a) What are the fixed points of $m\,x + b$?

    b) For what kind of lines do you think the iterates converge? How does the initial value x_0 affect this?

    c) For what kind of lines do you think the iterates diverge? How does the initial value x_0 affect this?

4. Do the lines $y = \frac{x}{2} + 5$ and $y = 2\,x - 3$ that you worked with before the lab fit the pattern you found in Question 3 (After the Lab)?

5. What is the derivative of $0.2\,x(1 - x)$ at each of its fixed points? Does this correspond to the pattern you found for straight lines in Question 3 (After the Lab)?

■ For Further Exploration

Classifying Fixed Points

Based on the above examples, we say that a fixed point x^* of a function f is

an *attracting fixed point* if $|f'(x^*)| < 1$
a *repelling fixed point* if $|f'(x^*)| > 1$
a *superattracting fixed point* if $f'(x^*) = 0$ (because it attracts very fast)
a *neutral* fixed point if $|f'(x^*)| = 1$ (because this case is inconclusive).

1. Classify each fixed point of the functions $mx + b$ and $0.2\,x(1 - x)$ as attracting, repelling, superattracting, or neutral. Compare this to the behavior that you observed using *Joy* or by hand.

2. What are the fixed points of $c\,x(1 - x)$, where $c \neq 0$ is a constant? For which values of c is the nonzero fixed point attracting, repelling, superattracting, or neutral? Does the case $c = 0.2$ in the preceding question fit this pattern?

Iteration and Newton's Method

Newton's method for estimating the roots of a function f, i.e., a solution to $f(x) = 0$, is illustrated in Section 5.4. These questions help explain why Newton's method converges when it does. Newton's method starts with an initial value x_0 and computes the values

$$x_{n+1} = x_n - \frac{f(x_n)}{f'(x_n)},$$

which often converge to a root of f. Thus Newton's method amounts to iterating the *Newton function*

$$\text{newton}(x) = x - \frac{f(x)}{f'(x)}.$$

3. Show that the fixed points of the Newton function are the same as the roots of f.

4. Suppose x^* is a root of f. As a fixed point of the Newton function, is it attracting, repelling, superattracting, or neutral? Your answer is the reason for the efficiency of Newton's method in estimating solutions to equations.

Chapter 20

Integrals in One Variable

Examples and Exercises

Prerequisites:
      Ch. 1   A Brief Tour of *Joy*
      Ch. 8   Integrating Functions of One Variable

■ 20.1 Areas

These exercises ask for the *net area* (also called *signed area*) and the *total area* between a curve and the x-axis. The sketch below illustrates the difference between these concepts. Here A, B, C are the areas of three regions on either side of the x-axis.

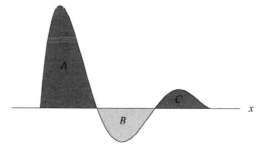

net area = $A - B + C$
total area = $A + B + C$.

Exercises

1. a) Graph $y = -4\,x^4 + 16\,x^3 - 16\,x^2 + 1$ for $0 < x \le 2.5$.

329

b) Use the graph to predict whether the integral $\int_0^{2.5} (-4 x^4 + 16 x^3 - 16 x^2 + 1) \, dx$ will be positive, negative, or zero. Calculate the integral to confirm your prediction.

2. a) Find the net area between the graph of $y = x^3 - 3 x + 2$ and the x-axis, for x between ± 3.

   b) Factor $x^3 - 3 x + 2$. (**Algebra** $\triangleright$ **Simplify**) Use the factors to find the total area between the graph and the x-axis, for x between ± 3.

3. a) Graph $y = \sin x \cos(2 x)$ for $0 \le x \le \pi$. In terms of π, where does the graph cross the x-axis?

   b) Find the net area between the graph and the x-axis.

   c) Find the total area between the graph and the x-axis.

4. a) Find the net area between the graph of $y = \sin(x^2)$ and the x-axis for $0 \le x \le 3$, *integrating numerically.*

   b) Express the points where the graph crosses the x-axis in terms of π. Now find the total area.

5. a) Plot $y = x^3 - x^2 + 1$ and $y = 3 x - 1$ simultaneously for $-5 \le x \le 5$. Find the points where the graphs intersect. (**Algebra** $\triangleright$ **Solve**)

   b) Find the total area between the two graphs.

■ 20.2 Accumulated Change and the Fundamental Theorem

Exercises

1. a) Graph $y = -4 x^4 + 16 x^3 - 16 x^2 + 1$ for $0 \le x \le 2.5$.

   b) The function $y = \int_0^x f(t) \, dt$ represents the net area accumulated under the graph of $f(t)$ between $t = 0$ and $t = x$. Use the graph in a) to predict approximately the values of x where the graph of $y = \int_0^x (-4 t^4 + 16 t^3 - 16 t^2 + 1) \, dt$ will lie above and below the x-axis as x ranges from 0 to 2.5. Explain your reasoning.

   c) Calculate $y = \int_0^x (-4 t^4 + 16 t^3 - 16 t^2 + 1) \, dt$ (by hand or with *Joy*) and graph the result for $0 \le x \le 2.5$. Use the graph to verify your prediction.

2. a) Graph $y = \cos(x^2)$ for $0 \le x \le \sqrt{2 \pi}$.

   b) Based on the graph in a), predict approximately where the graph of $y = \int_0^x \cos(t^2) \, dt$ will be increasing, decreasing, concave up, and concave down as x ranges from 0 to $\sqrt{2 \pi}$. Explain your reasoning.

c) Verify your predictions by graphing the *numerical integral* $\int_0^x \cos(t^2)\,dt$ for $0 \le x \le \sqrt{2\pi}$ and examining the result. (This function cannot be integrated exactly. Use the **Numeric Integral** button to enter the integral in the **Graph** ▷ **Graph f(x)** dialog.)

3. The number of hours of daylight each day varies according to the time of year and is different at different latitudes. At Boston, Massachusetts, a reasonable estimate is given by the function $f(t) = 12.2 + 3.1 \cos\left(\frac{2\pi}{365}(t - 172)\right)$. Here $f(t)$ is the number of hours of daylight per day at time t, where t is measured in days, $t = 0$ at 12:00 A.M., January 1st, and we ignore leap years.

    a) Explain why the area in a thin rectangle under the graph of $f(t)$

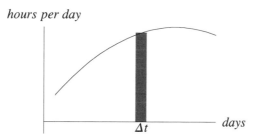

    is approximately the number of hours of daylight during the time interval of length Δt corresponding to the rectangle. Explain why the area under the curve in any interval $a \le t \le b$ is exactly the total amount of daylight during that interval.

    b) What is the total number of hours of daylight in Boston during January? What is the average number of hours of daylight per day during January, i.e., how long is the average January day?

    > If you integrate using *Mathematica* your answer may include a term $0.I$, which equals 0 and can be ignored.

    c) Graph $f(t)$ for $0 \le t \le 365$. Calculate the length of the average day for the months that appear to have the longest days. Which month has the longest days on average?

4. a) Evaluate $y = \int_a^x (t^2 - 1)\,dt$ by hand for $a = 0, 1, 2$.

    b) Graph the three results simultaneously for x between ± 3. For each graph, what is the value of y when $x = a$?

    c) What appears to be the effect of changing a on the graph of $y = \int_a^x (t^2 - 1)\,dt$? In particular, are the graphs related by shifting or scaling and how?

5. a) Evaluate $\int_a^x t \sin t\,dt$ for $a = 0, \pi/2, \pi$.

    b) Graph the three results simultaneously for x between $\pm 2\pi$. For each graph, what is the value of y when $x = a$?

c) Differentiate the three expressions in a) with respect to x. How does the value of a seem to affect $\frac{d}{dx}\left(\int_a^x t \sin t \; dt\right)$, and how can you predict this from the graphs in b)?

■ 20.3 Numerical Integration

20.3.1 Example: Custom Approximations for Integrals

When a function f has no elementary antiderivative or one cannot be found efficiently, we can estimate an integral $\int_a^b f(x)\, dx$ numerically. This example illustrates all the approximations in **Calculus ▷ Custom Integration** except for **Simpson's Rule**, which is described in Section 8.3.1 and in more detail in Lab 20.5.

If f is increasing on $[a, b]$, the sketch below shows we can use a **Left Endpoint Approximation** L to underestimate $\int_a^b f(x)\, dx$ and a **Right Endpoint Approximation** R to overestimate it. If f is decreasing, we can see the reverse is true.

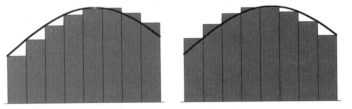

Left endpoint (left) and right endpoint (right) approximations.

So if f is *monotonic* on $[a, b]$, i.e., either increasing throughout $[a, b]$ or decreasing throughout, then $\int_a^b f(x)\, dx$ falls between L and R. If f is not monotonic, we can use a combination of left and right approximations to estimate the integral from below and above.

If the graph of f is concave down, the accompanying sketch shows we can use a **Trapezoidal Approximation** T to underestimate the integral. If the graph is concave up, T is an overestimate. It can be shown that the reverse is true for a **Midpoint Approximation** M: when the graph is concave down M is an overestimate, and when the graph is concave up M is an underestimate.

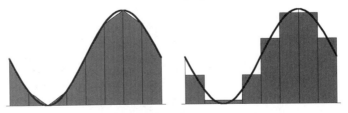

Trapezoidal (left) and midpoint (right) approximations.

So if the concavity of the graph doesn't change on $[a, b]$, then $\int_a^b f(x)\,dx$ falls between T and M. If the concavity changes, we can use a combination of trapezoidal and midpoint approximations to estimate the integral.

With these approximations, you can not only estimate the value of an integral whose exact value is unknown but also estimate the error in your approximation. For instance, *Mathematica* cannot find the value of $\int_1^2 (\frac{1}{10} + \ln(\sin x))\,dx$ in terms of elementary functions:

In[2]:= $\displaystyle \int_1^2 \left(\frac{1}{10} + \mathbf{Log[Sin[x]]} \right)\, d\!x$

Out[2]= $\frac{1}{10}$ ((1 + 15 I) + 10 Log[1 − E$^{2\,I}$] − 20 Log[1 − E$^{4\,I}$] − 10 Log[Sin[1]] +
  20 Log[Sin[2]] − 5 I PolyLog[2, E$^{2\,I}$] + 5 I PolyLog[2, E$^{4\,I}$])

You can graph the function to decide how to estimate the integral:

- Graph $y = \frac{1}{10} + \ln(\sin x)$ for $1 \le x \le 2$.

  Enter $\ln(\sin x)$ as **Log[Sin[x]]**, or use the **Functions** button on the *Joy* palette to enter it.

Mathematica displays the result:

▷ Graph $\frac{1}{10}$ + Log[Sin[x]] on 1 ≤ x ≤ 2.

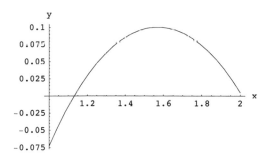

Out[3]= - Graphics -

Since the graph is concave down on $[1, 2]$, we will use the trapezoidal and midpoint approximations:

$$T \le \int_1^2 (\tfrac{1}{10} + \ln(\sin x))\,dx \le M .$$

If the concavity were to change but not the direction, i.e., if the function were either increasing throughout the interval or decreasing, we would use the left and right endpoint approximations instead.

The difference $M - T$ measures how precisely these approximations estimate the integral. It is called an *error bound*, since if we use either T or M to approximate the integral then $M - T$ is the largest error we could be making. The size of the error bound is determined by the number of subintervals in the approximations. You can find an error bound this way whenever you have an overestimate and an underestimate of the integral.

- Find the trapezoidal approximation to $\int_1^2 (\frac{1}{10} + \ln(\sin x))\, dx$ using 100 subintervals, assigning the name T to the result.

The output is:

▷ The trapezoidal approximation

for $\int_1^2 \left(\frac{1}{10} + \text{Log[Sin[x]]} \right) \text{d}x$ with 100 subintervals gives

T =

Out[4]= 0.0544887

- Repeat for the midpoint approximation, calling it M.

▷ The midpoint approximation

for $\int_1^2 \left(\frac{1}{10} + \text{Log[Sin[x]]} \right) \text{d}x$ with 100 subintervals gives

M =

Out[5]= 0.0545024

Next, find $M - T$:

- If necessary, click below the last output to make a horizontal line appear.
- Type M-T, and press ⌈SHIFT⌉ − ⌈RET⌉.

Mathematica calculates the error bound:

In[6]:= **M − T**

Out[6]= 0.0000137468

So the unknown value of the integral satisfies
$$T = 0.0544887 \le \int_1^2 (\tfrac{1}{10} + \ln(\sin x))\, dx \le 0.0545024 = M$$

and using either value to approximate it is accurate to within about $M - T = 1.37 \times 10^{-5}$. By using more subintervals, we can reduce the size of the error bound and thus get an approximation that is closer to the true value of the integral.

In general, the trapezoidal and midpoint methods are more efficient for estimating $\int_a^b f(x)\,dx$ than the left and right endpoint approximations. That is, they get closer to the value of the integral with the same number of subintervals. **Simpson's Rule** is the most efficient method in **Calculus ▷ Custom Integration**, but you generally can't predict whether it will overestimate or underestimate the integral. Lab 20.5 describes Simpson's Rule.

Exercises

1. a) Graph $y = \sin(x^2)$ for $0 \leq x \leq 1$. Based on the graph, will a left endpoint approximation L underestimate or overestimate $\int_0^1 \sin(x^2)\,dx$? Will a right endpoint approximation R be an underestimate or overestimate?

   b) Use $n = 100$ subintervals to find an overestimate, an underestimate, and an error bound for the integral.

2. a) Graph $y = 1/\sqrt{2 + x^3}$ for $-1 \leq x \leq 4$. Will the left and right approximations L and R overestimate or underestimate $\int_{-1}^4 1/\sqrt{2 + x^3}\,dx$?

   b) Using $n = 10, 20, 40, 80, 160, 320$ subintervals, find L for $\int_{-1}^4 1/\sqrt{2 + x^3}\,dx$. (The **Example button** in **Calculus ▷ Custom Integration** shows how to calculate these in one step.)

   c) Repeat b) for R. Then find $L - R$. In one step, this calculates the error bounds for each value of n.

   d) By about what factor does the error bound in a) seem to change when you double the number of subintervals? That is, fill in the blank to predict the general pattern:

   Error($2\,n$) ≈ ____ Error(n).

3. a) Graph $y = \sin(x^2)$ for $1 \leq x \leq \sqrt{\pi}$. Will a trapezoidal approximation T underestimate or overestimate

   $$\int_1^{\sqrt{\pi}} \sin(x^2)\,dx \ ?$$

   Will a midpoint approximation M underestimate or overestimate it?

   b) Use $n = 100$ subintervals to find an overestimate, an underestimate, and an error bound for the integral.

4. a) Graph $y = \ln(1 + \sin x)$ for $0 \leq x \leq \pi/2$. Based on the graph, will the trapezoidal approximation T and midpoint approximation M underestimate or overestimate

   $$\int_0^{\pi/2} \ln(1 + \sin x)\,dx ?$$

b) Use $n = 10, 20, 40, 80, 160, 320$ subintervals to calculate T. (The **Example** button in **Calculus** ▷ **Custom Integration** shows how to calculate these in one step.) Repeat for M.

c) Calculate $M - T$. In one step, this calculates the error bounds for each value of n.

d) By about what factor does the error bound seem to change when you double the number of subintervals? That is, fill in the blank to predict the general pattern:

$\text{Error}(2\,n) \approx \underline{\quad} \text{Error}(n)$.

5. The graph of $y = 1/\sqrt{1 + x^4}$ is both decreasing and concave up for $2 \le x \le 3$. Find under- and overestimates and the corresponding error bound for $\int_2^3 1/\sqrt{1 + x^4}\, dx$ using

   a) the left and right approximations with $n = 100$ subintervals;

   b) the midpoint and trapezoidal approximations with $n = 100$ subintervals.

   c) Which method gives a smaller error bound for the same number of subintervals?

6. a) Graph $y = \ln(2 + \sin(2\,x))$ for $0 \le x \le 2$. Where in this interval is the function increasing and where is it decreasing? Use exact values, e.g., express them in terms of π rather than numerically.

   b) This function is not monotonic on $[0, 2]$. What combination of left and right endpoint approximations will underestimate the integral $\int_0^2 \ln(2 + \sin(2\,x))\, dx$? What combination will overestimate the integral?

   c) Using $n = 100$ subintervals, find an underestimate for the integral. Also find an underestimate and an error bound for these estimates.

7. Repeat Exercise 6 for $\sin(x^2)$ and $1 \le x \le \pi/2$.

8. a) Graph $y = 1/\sqrt{1 + x^4}$ for $0 \le x \le 5$. Use the second derivative to decide where in this interval the graph is concave up and where it is concave down. (You can calculate the second derivative by hand or using the **Derivative** button, and factor it with **Algebra** ▷ **Simplify**.)

   b) This function changes concavity in $[0, 5]$. What combination of trapezoidal and midpoint approximations will underestimate $\int_0^5 1/\sqrt{1 + x^4}\, dx$? What combination will overestimate the integral?

   c) Using $n = 100$ subintervals, find an underestimate for the integral. Also find an overestimate and an error bound for these estimates.

9. Repeat Exercise 8 for $y = e^{-x^2}$ on $0 \le x \le 2$.

Lab 20.1 The Fundamental Theorem of Calculus and the Normal Curve

---

Prerequisites:
     Ch. 1   A Brief Tour of *Joy*
     Sec. 8.3   How to Integrate Numerically

In this lab, you can explore the relation between integration and differentiation in the context of the normal or Gaussian curve

$$g(x) = \frac{1}{\sqrt{2\pi}}\, e^{-x^2/2},$$

which plays an essential role in probability and statistics.

■ Before the Lab (Do by Hand)

It turns out that g has no antiderivative that can be expressed using standard functions. But this function is continuous, so it has an antiderivative. For example, one antiderivative is

$$G(x) = \int_0^x g(t)\, dt = \int_0^x \frac{1}{\sqrt{2\pi}}\, e^{-t^2/2}\, dt\,.$$

This integral can only be evaluated numerically. The Fundamental Theorem of Calculus implies that $G'(x) = g(x)$.

> Notice that $\int_0^x g(t)\, dt$ is a function of x, not of t. The variable t is a "dummy variable," i.e., you can replace it with another symbol without changing the value of the integral. When you evaluate the definite integral the result is expressed in terms of x, the upper limit of integration. This is similar to $\int_1^x t^2\, dt = \frac{x^3}{3} - \frac{1}{3}$, which is also a function of x, except that the integral of the Gaussian function cannot be evaluated by algebraic means.

Here is a sketch illustrating the function g and its antiderivative G. Notice the "bell" shape and the symmetry of the curve about the y-axis. For each $x \geq 0$, the value of $G(x)$ is the area accumulated under the graph of g from 0 to x.

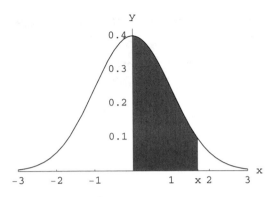

Normal curve $g(x) = \frac{1}{\sqrt{2\pi}} e^{-x^2/2}$ with shaded area equal to $G(x) = \int_0^x g(t)\, dt$.

1. Based on the graph of g, predict the following and explain your reasoning:

    a) Describe how each of the values $G(1)$, $G(0)$, $G(-1)$ represents ± an area for a curve over an interval, and specify the curve and interval.

    b) Estimate the values of $G(1)$, $G(0)$, $G(-1)$ by estimating areas from the graph.

    For what values of x will $G(x)$:

    c) be zero, positive, or negative?

    d) increase or decrease?

    e) increase or decrease fastest?

    f) have a local maximum or minimum?

    g) For what values of x will the graph of G be concave up or down? have an inflection point?

2. Combine your answers to Question 1 and make a rough sketch of the graph of G.

■ In the Lab

1. Create the function $g\,(x) = \frac{1}{\sqrt{2\pi}} e^{-x^2/2}$. Type **E** for the constant e (or use the e button on the *Joy* palette). (**Create ▷ Function**)

2. Graph $g(x)$ for $-3 \le x \le 3$. This is the Gaussian function, or normal curve.

3. Create the function $G(x)$ equal to the *numerical integral* $\int_0^x g(t)\, dt$.

    In the **Create ▷ Function** dialog, enter the integral using the **Numeric Integral** button on the *Joy* palette.

4. Graph $G(x)$ for $-3 \leq x \leq 3$.

5. Combine the graphs of g and G. (**Graph ▷ Combine Graphs**)

6. Print your work.

■ After the Lab

1. Describe the pattern you see in the graph of G:

   a) What range of values does $G(x)$ appear to take?

   b) For which x is $G(x)$ positive or negative, increasing or decreasing, concave up or down?

2. Does the rough sketch you made of G in Question 2 (Before the Lab) correspond to the graph you made with *Joy*? If not, review the reasoning you used to make the sketch and see where it is inconsistent with the numerical pattern you found.

3. What features of the graph of g determine whether:

   a) $G(x)$ is zero, positive, or negative?

   b) G is increasing or decreasing at x?

   c) the graph of G is concave up or down?

Lab 20.2 Animating the Fundamental Theorem of Calculus

Prerequisites:
Ch. 1   A Brief Tour of *Joy*
Sec. 3.9   How to Animate Graphs

This lab uses animation to illustrate the Fundamental Theorem of Calculus. The theorem says that for any continuous function g and constant a,

$$G(x) = \int_a^x g(t)\, dt$$

is an antiderivative of g.

■ Before the Lab (Do by Hand)

Let $g(x) = \frac{1}{\sqrt{2\pi}} e^{-x^2/2}$.

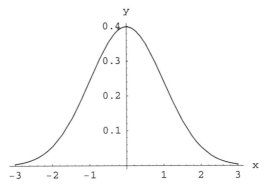

The normal curve $g(x) = \frac{1}{\sqrt{2\pi}} e^{-x^2/2}$.

The graph of g is the Gaussian or normal curve, which plays an essential role in probability and statistics. It turns out that g has no antiderivative that can be expressed in a finite number of terms using the standard functions. But the Fundamental Theorem says that

$$G(x) = \int_0^x g(t)\, dt = \int_0^x \frac{1}{\sqrt{2\pi}} e^{-t^2/2}\, dt$$

is an antiderivative of g, i.e., $G'(x) = g(x)$.

1.   On a copy of this graph, mark two points x and x + h, where h > 0.

2. a) Express $\Delta G = G(x + h) - G(x)$ as one integral.

    b) On the sketch, indicate the area that corresponds to the integral ΔG.

3. The ratio $\frac{\Delta G}{h}$ can be interpreted as the average value of some function over some interval. Which function and which interval?

4. In terms of the function $G(x)$, what is

    $\lim_{h \to 0} \frac{\Delta G}{h}$?

    What does the Fundamental Theorem of Calculus say about this limit and the function $g(x)$?

■ In the Lab

1. Create the function $g(x) = \frac{1}{\sqrt{2\pi}} e^{-x^2/2}$. (**Create ▷ Function**)

    Type E for the constant e (or use the e button on the *Joy* palette).

2. Graph $g(x)$ for $3 \le x \le 3$.

    This is the Gaussian function, or normal curve.

3. Now you'll define ΔG.

    - If necessary, click beneath the last output so that a horizontal line appears.
    - Type ΔG = and click the **Definite Integral** button on the *Joy* palette.

      ⚠ *Don't* type a space after Δ, since ΔG isn't a product. You can enter Δ using the **Greek** button on the *Joy* palette.

    - Fill in the integral $\int_{x}^{x+h} g[t] \, dt$.
    - Press ⎰SHIFT⎱ − ⎰RET⎱.

Mathematica expresses the integral in terms of Erf, a function whose values cannot in general be found exactly but can only be estimated by numerical techniques. The Erf function cannot be expressed in a finite number of terms using the standard functions.

The next step is to create an animation showing how the graph of $\frac{\Delta G}{h}$ behaves as h gets smaller and smaller. The animation will contain eight frames, corresponding to $h = \frac{1}{2^k}$, $k = -2, -1, 0, ..., 5$. Each frame will also show the graph of the normal curve g. Section 3.9 gives more detail about animation.

4. In your *Mathematica* notebook:

    - If necessary, click beneath the last output so that a horizontal line appears.
    - Type h $= \frac{1}{2^k}$.
    - Press ⎰SHIFT⎱ − ⎰RET⎱.

5. From the *Joy* menus:

    -   Choose **Graph ▷ Animation**.
    -   From the popup menu, choose **Graph f(x)** and click **OK**.
    -   In the **Graph f(x)** window, fill in $\frac{\Delta G}{h}$ and $-3 \le x \le 3$.

        Mathematica will substitute for *h* automatically, expressing it in terms of *k*.

    -   Click in the **Animation Info** window to make it active.
    -   Fill in $-2 \le k \le 5$, to specify the eight frames.
    -   Fill in the common plot frame, $-3 \le x \le 3$ and $0 \le y \le 0.4$.
    -   Check **Combine with** and fill in Out[...], using your output number for the graph of *g*.
    -   Click in the **Graph f(x)** window to make it active and click **OK**.

6. Open the animation frames by double-clicking on the cell bracket that contains them.

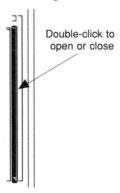

Double-click to
open or close

7. Print your work. Don't print all eight animation graphs, but cut out every other one and you will still see the pattern.

■ After the Lab

1. What are the largest and smallest values of *h* in the animation of $\frac{\Delta G}{h}$?

2. For the values of *h* shown in the animation, how is the graph of $\frac{\Delta G}{h}$ similar to or different from the normal curve? For example, does it have the characteristic bell shape as the normal curve, the same *y*-values, and the same symmetry property?

3. Describe the behavior of the animation in the language of limits and derivatives. How does the animation illustrate the Fundamental Theorem of Calculus?

Lab 20.3 Numerical Integration with Left and Right Endpoint Approximations

Prerequisites:
Ch. 1	A Brief Tour of *Joy*
Sec. 8.3	How to Integrate Numerically

In this lab you'll see how to use *Joy* to compute numerical approximations of integrals according to specific methods that often allow you to tell whether your estimate is too large or too small and how accurate it is. Unless you know the true value of the integral and therefore don't need to estimate it, you can't know the error in the estimate exactly. But in many cases, you can say how large the error can become in the worst case (an *error bound*).

This lab focuses on the left and right endpoint approximations, which are pictured here.

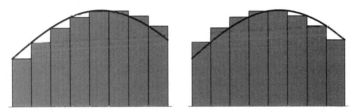

Left endpoint (left) and right endpoint (right) approximations.

Other approximating sums are more efficient than these, which means they converge to the true value of the integral faster as the number of subintervals increases, but in certain cases the left and right approximations are easier to apply. Lab 20.4 discusses the trapezoidal and midpoint methods and Lab 20.5 examines Simpson's Rule.

■ Before the Lab (Do by Hand)

You will be working with the integral of $\ln(1 + \sin x)$, a function for which no antiderivative is known that has only a finite number of terms and uses the standard functions in calculus. So its integral can only be approximated numerically. Here is a portion of its graph.

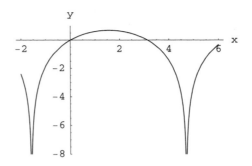

Graph of $y = \ln(1 + \sin x)$.

1. For the x-values shown in the graph, where does $\ln(1 + \sin x)$ appear to be increasing or decreasing? Then use derivatives to answer this question.

2. Indicate on the graph the area that corresponds to $\int_0^{\pi/2} \ln(1 + \sin x)\, dx$. Which endpoint approximation, left or right, do you predict will underestimate this integral? Which will overestimate it? Explain briefly.

3. By hand, find the left and right endpoint approximations to $\int_0^{\pi/2} \ln(1 + \sin x)\, dx$ with $n = 4$ subintervals. Be sure to use radians and the natural logarithm (ln) on your calculator. Keep at least four decimal places for all numbers, including the partition points.

4. If you use your left endpoint approximation to estimate the value of the integral, what is the largest error you could be making? This is called an *error bound* for the integral.

■ In the Lab

1. Find the left endpoint approximation to $\int_0^{\pi/2} \ln(1 + \sin x)\, dx$ with 10, 20, 40, 60, 80, 160, 320 subintervals, naming the output L. (**Calculus ▷ Custom Integration**)

    The default example shows how to enter several values of n at once.

2. Repeat for the right endpoint approximation, naming it R.

3. In your *Mathematica* notebook, click below the last output and type R-L. Press SHIFT – RET.

    This calculates the error bounds for all six values of n at once.

4. Repeat Questions 1–3 with 10, 100, 1000 subintervals.

5. Print your work.

■ After the Lab

1. What inequality gives under- and overestimates for $\int_0^{\pi/2} \ln(1 + \sin x)\, dx$ when the number of subintervals is $n = 10$? when $n = 320$? when $n = 1000$?

2. For $n = 10, 20, \ldots, 320$, by about what factor does the error bound change from each estimate to the next?

3. By about what factor does the error bound change when $n = 10, 100, 1000$?

4. The actual error in an approximation is unknown unless you know the true value of the integral. However, the error bound tells you the most the error can possibly be and is usually a conservative estimate of the error. Based on your results so far, predict the general pattern for the left and right endpoint approximations by filling in the blanks below:

    Error($2\,n$) $\approx$ ___ Error (n)   (express in terms of 2)

    Error($10\,n$) $\approx$ ___ Error (n)   (express in terms of 10)

    Error ($c\,n$) $\approx$ ___ Error (n)   (express in terms of c)

    With 10 times as many subintervals, you gain $\approx$ ___ decimal places of accuracy.

Lab 20.4 Numerical Integration with Trapezoidal and Midpoint Approximations

Prerequisites:
    Ch. 1   A Brief Tour of *Joy*
    Sec. 8.3   How to Integrate Numerically

In this lab you'll see how to use *Joy* to compute numerical approximations of integrals according to specific methods that often allow you to tell whether your estimate is too large or too small and how accurate it is. Unless you know the true value of the integral and therefore don't need to estimate it, you can't know the error in the estimate exactly. But in many cases, you can say how large the error can become in the worst case (an *error bound*).

This lab focuses on the midpoint and trapezoidal approximations, which are pictured here.

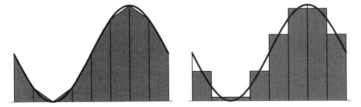

Trapezoidal (left) and midpoint (right) approximations.

Simpson's Rule is more efficient than these, which means it converges to the true value of the integral faster as the number of subintervals increases, but in certain cases the trapezoidal and midpoint approximations are easier to apply. Lab 20.3 discusses the left and right endpoint approximations and Lab 20.5 treats Simpson's Rule.

■ Before the Lab (Do by Hand)

You will be working with the integral of $\ln(1 + \sin x)$, a function for which no antiderivative is known that has only a finite number of terms and uses the standard functions. So its integral can only be approximated numerically. Here is a portion of its graph.

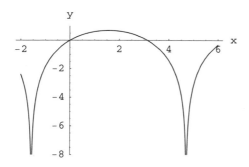

Graph of $y = \ln(1 + \sin x)$.

1. Where does the graph of $y = \ln(1 + \sin x)$ appear to be concave up or concave down? Use derivatives to answer this question.

2. Indicate on the graph the area that corresponds to $\int_0^{\pi/2} \ln(1 + \sin x)\, dx$. From the graph, do you predict that the trapezoidal approximation will underestimate or overestimate the integral? Explain briefly.

It can be shown that if the concavity of the graph doesn't change on some interval, then the integral of the function over that interval falls between the trapezoidal and midpoint approximations. So if one approximation is an underestimate, then the other is an overestimate and vice versa.

3. By hand, find the trapezoidal and midpoint approximations to $\int_0^{\pi/2} \ln(1 + \sin x)\, dx$ with $n = 4$ subintervals. Be sure to use radians and the natural logarithm (ln) on your calculator. Keep at least four decimal places for all numbers, including the partition points.

4. If you use your trapezoidal approximation to estimate the value of the integral, what is the largest error you could be making? (This is called an *error bound* for the integral.)

■ In the Lab

1. First find the trapezoidal approximation to $\int_0^{\pi/2} \ln(1 + \sin x)\, dx$ with 10, 20, 40, 60, 80, 160, 320 subintervals, naming the output T. (**Calculus ▷ Custom Integration**)

    The default example shows how to enter several values of n at once.

2. Repeat for the midpoint approximation, naming it M.

3. In your *Mathematica* notebook, click below the last output and type M–T. Press SHIFT − RET.

    This calculates the error bounds for all six values of n at once.

4. Repeat Questions 1–3 with 10, 100, 1000 subintervals.

5. Print your work.

■ After the Lab

1. What inequality gives under- and overestimates for $\int_0^{\pi/2} \ln(1 + \sin x)\, dx$ when the number of subintervals is $n = 10$? when $n = 320$? when $n = 1000$?

2. For $n = 10, 20, \ldots, 320$, by about what factor does the error bound change from each estimate to the next?

3. By about what factor does the error bound change when $n = 10, 100, 1000$?

4. The actual error in an approximation is unknown unless you know the true value of the integral. However, the error bound tells you the most the error can possibly be and is usually a conservative estimate of the error. Based on your results so far, predict the general pattern for the trapezoidal and midpoint approximations by filling in the blanks below:

   Error($2\,n$) $\approx$ ____ Error (n)   (express in terms of 2)

   Error($10\,n$) $\approx$ ____ Error (n)   (express in terms of 10)

   Error ($c\,n$) $\approx$ ____ Error (n)   (express in terms of c)

   With 10 times as many subintervals, you gain $\approx$ ____ decimal places of accuracy.

Lab 20.5 Numerical Integration with Simpson's Rule

Prerequisites:

Ch. 1	A Brief Tour of *Joy*
Sec. 8.3	How to Integrate Numerically

In this lab you'll see how to use *Joy* to compute numerical approximations of integrals using Simpson's Rule. This is a method that is more efficient than the other approximations in **Calculus ▷ Custom Integration**, which means it converges to the true value of the integral faster as the number of subintervals increases, but it is more complicated to work out by hand. Lab 20.3 discusses the left and right endpoint approximations and Lab 20.4 covers the trapezoidal and midpoint approximations.

■ Before the Lab (Do by Hand)

Simpson's Rule, which is pictured on page 350, estimates an integral by using areas under approximating parabolas rather than rectangles or trapezoids. It can also be thought of as a combination of the trapezoidal and midpoint approximations. Generally, you can't predict whether Simpson's Rule will underestimate or overestimate an integral and you can't know the precise error in the estimate. However, by comparing the Simpson estimates to an integral that you can find exactly, you can see how the errors change as you take more and more subintervals. In this lab, you'll work with $\int_0^4 x^4 \, dx$, whose exact value you can find easily.

1. Find the trapezoidal approximation T to $\int_0^4 x^4 \, dx$ with four subintervals.

2. Repeat for the midpoint approximation M, and then find the weighted average
   $\frac{2}{3} M + \frac{1}{3} T$.

Simpson's approximation to $\int_a^b f(x) \, dx$ is given by the expression

$$S = \frac{1}{3} \left(f(x_0) + 4 f(x_1) + 2 f(x_2) + 4 f(x_3) + \ldots + 2 f(x_{n-2}) + 4 f(x_{n-1}) + f(x_n) \right) \Delta x.$$

This formula gives the sum of the areas under parabolas that approximate the curve, where each parabola spans two adjacent subintervals as illustrated in the accompanying sketch. In this version of the formula n stands for the total number of subintervals, which must be even since the subintervals are paired, and there are $\frac{n}{2}$ parabolas.

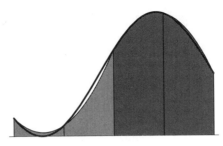

Simpson's Rule with $n = 4$ subintervals and $\frac{n}{2} = 2$ parabolas.

3. Apply this formula to find S for the integral $\int_0^4 x^4 \, dx$ with $n = 8$ subintervals.

It can be shown that the value of S for n subintervals, where n is even, always equals the value of $\frac{2}{3} M + \frac{1}{3} T$ for $\frac{n}{2}$ subintervals.

■ In the Lab

1. Find Simpson's approximation $\int_0^4 x^4 \, dx$ with 10, 20, 40, 80, 160, 320 subintervals, naming the result S. (**Calculus ▷ Custom Integration**)

2. On the *Joy* palette, click the **Numeric** button and fill in the command so that it reads N[S,15]. Press SHIFT – RET.

    This expresses the values given by Simpson's Rule to 15 significant digits.

3. Repeat Questions 1 and 2 with 10, 100, 1000 subintervals.

4. Print your work.

■ After the Lab

1. What is the exact value of $\int_0^4 x^4 \, dx$, expressed as a decimal?

2. In Question 3 (Before the Lab), you estimated $\int_0^4 x^4 \, dx$ by S with $n = 8$. What is the error in that approximation?

3. During the lab, you estimated $\int_0^4 x^4 \, dx$ with $n = 10, 20, 40, 80, 160, 320$ subintervals.

    a) Make a table showing the number of subintervals and the error for each estimate.

    b) By about what factor does the error change from each estimate to the next?

4. Repeat for $n = 10, 100, 1000$.

5. Based on your results so far, predict the general pattern for Simpson's Rule by filling in the blanks below:

   Error($2n$) $\approx$ ___ Error (n) (express in terms of 2)

   Error($10n$) $\approx$ ___ Error (n) (express in terms of 10)

   Error (cn) $\approx$ ___ Error (n) (express in terms of c)

   With 10 times as many subintervals, you gain $\approx$ ___ decimal places of accuracy.

■ For Further Exploration

Here's how you can derive the formula for Simpson's Rule. These steps will be described in detail below:

- Find the coefficients of the parabola $y = ax^2 + bx + c$ that fits through three points $(p, f(p))$, $(p + \Delta x, f(p + \Delta x))$, $(p + 2\Delta x, f(p + 2\Delta x))$ on the graph of $y = f(x)$.

  Here p represents any point in a partition of $[a, b]$ and Δx is the distance between any two consecutive partition points.

  This also shows that a parabola can be fitted to any three points with different x-coordinates.

- Find the area under the parabola for $p \leq x \leq p + 2\Delta x$.
- Add the areas for the $\frac{n}{2}$ parabolas approximating $y = f(x)$.

If you used the symbols f, p, a, b, c, or Δx earlier in this session, you will need to clear their values. Otherwise, skip to Question 2 below.

1. Choose **Create** ▷ **Clear**. Fill in `f,p,a,b,c,`Δ`x` and click **OK**.

2. From the *Joy* menus, choose **Algebra** ▷ **Solve**. Solve these equations algebraically for a, b, c, naming the result `coeffs`:

   $$f(p) \quad = \quad ap^2 + bp + c$$
   $$f(p + \Delta x) \quad = \quad a(p + \Delta x)^2 + b(p + \Delta x) + c$$
   $$f(p + 2\Delta x) \quad = \quad a(p + 2\Delta x)^2 + b(p + 2\Delta x) + c$$

   To enter Δx, use the **Greek** button on the *Joy* palette to enter Δ. Do *not* leave a space between Δ and **x**, since Δx does not represent a product. Close the **Greek** palette by clicking in the box in the upper left of the palette (Macintosh) or on the X in the upper right of the palette (Windows).

3. In your *Mathematica* notebook:

   - Click below the last output to make a horizontal line appear.
   - Type `area = `$\int_{p}^{p+2\Delta x}$`(a x`^{2}` + b x + c) d`x.
   - Press SHIFT − RET .

4. From the *Joy* menus:

    - Choose **Algebra ▷ Substitute**.
    - Fill in `area` in the first input field.
    - Click **make the substitutions** and fill in `coeffs` in the accompanying field.

    This substitutes the values you found for a, b, c into the expression for the area.

5. Choose **Algebra ▷ Simplify**. Choose **Last Output** from the popup menu and choose **Simplify**. Click **OK**.

The result should be

$$\left\{ \frac{1}{3}\, \Delta x \, (\mathtt{f[p]} + 4\, \mathtt{f[p+\Delta x]} + \mathtt{f[p+2\,\Delta x]}) \right\}$$

The first parabola, which extends from $a = x_0$ to $x_2 = x_0 + 2\,\Delta x$, thus contains the area

$$\frac{1}{3}\, (f(x_0) + 4\, f(x_1) + f(x_2))\, \Delta x \,.$$

The second parabola extends from $x_2 = x_0 + 2\,\Delta x$ to $x_4 = x_2 + 2\,\Delta x$ and contains the area

$$\frac{1}{3}\, (f(x_2) + 4\, f(x_3) + f(x_4))\, \Delta x \,,$$

and the areas under the remaining parabolas are found similarly.

6. By hand, add the areas under all the parabolas and show the sum has the value given in the section Before the Lab:

$$S = \frac{1}{3}\, (f(x_0) + 4\, f(x_1) + 2\, f(x_2) + 4\, f(x_3) + \ldots + 2\, f(x_{n-2}) + 4\, f(x_{n-1}) + f(x_n))\, \Delta x.$$

Lab 20.6 Improper Integrals and Numerical Integration

Prerequisites:	
Ch. 1	A Brief Tour of *Joy*
Sec. 8.3	How to Integrate Numerically

This lab shows how to estimate improper integrals such as

$$\int_{-\infty}^{\infty} \frac{1}{\sqrt{2\pi}} e^{-x^2/2} \, dx \, ,$$

the area under the entire normal or Gaussian curve. The function being integrated has no antiderivative that can be expressed in a finite number of terms using the standard functions. So we need to use numerical integration.

■ Before the Lab (Do by Hand)

Here is a graph of the normal curve $g(x) = \frac{1}{\sqrt{2\pi}} e^{-x^2/2}$. The curve never meets the *x*-axis, and the axis is an asymptote of the curve as $x \to \pm\infty$.

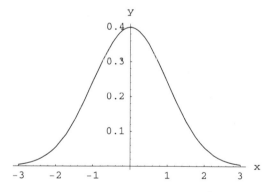

The normal curve $g(x) = \frac{1}{\sqrt{2\pi}} e^{-x^2/2}$.

Because $\int_{-\infty}^{\infty} g(x) \, dx = 2 \int_{0}^{\infty} g(x) \, dx$ by symmetry, we can concentrate on finding $\int_{0}^{\infty} g(x) \, dx$.

1. Using derivatives, show that the graph of *g* has inflection points at $x = \pm 1$.

Numerical estimates for $\int_{a}^{b} g(x) \, dx$, such as the trapezoidal and midpoint approximations, can only be carried out when *a* and *b* are finite. To estimate $\int_{-\infty}^{\infty} g(x) \, dx$, we will use the improper integral of a function *f* that *can* be integrated exactly.

2. Let $f(x) = \frac{1}{\sqrt{2\pi}} e^{-x}$.

   a) Explain why $g(2) = f(2)$ and why for $x > 2$, $g(x) < f(x)$.

   b) Explain why for any $c \geq 2$, we have $0 < \int_c^\infty g(x)\,dx \ < \int_c^\infty f(x)\,dx$.

   c) Verify that $\int_c^\infty f(x)\,dx = \frac{e^{-c}}{\sqrt{2\pi}}$.

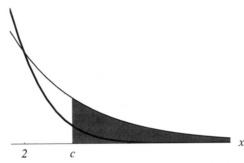

Graphs of $g(x) = \frac{1}{\sqrt{2\pi}} e^{-x^2/2}$ (**bold**) and $f(x) = \frac{1}{\sqrt{2\pi}} e^{-x}$.

Recall that on an interval where a curve is concave down, the trapezoidal approximation underestimates the integral and the midpoint approximation overestimates it. For example, between $x = 0$ and $x = 1$ the graph of g is concave down and we have

$$\int_0^1 g(x)\,dx \approx 0.341343 \ \text{(trapezoidal, } n = 100 \text{ subintervals)}$$

$$\int_0^1 g(x)\,dx \approx 0.341346 \ \text{(midpoint, } n = 100 \text{ subintervals)}$$

Where the curve is concave up, the reverse is true. For example, between $x = 1$ and $x = 2$ we have

$$\int_1^2 g(x)\,dx \approx 0.135906 \ \text{(trapezoidal, } n = 100 \text{ subintervals)}$$

$$\int_1^2 g(x)\,dx \approx 0.135905 \ \text{(midpoint, } n = 100 \text{ subintervals)}$$

Since for any c,

$$\int_0^\infty g(x)\,dx = \int_0^1 g(x)\,dx + \int_1^c g(x)\,dx + \int_c^\infty g(x)\,dx \ ,$$

we can estimate $\int_0^\infty g(x)\,dx$ by combining estimates for the three integrals on the right. The difference *overestimate − underestimate* for an integral is called an *error bound*. It measures how precisely these approximations estimate the integral.

3. a) Using $c = 2$, combine the above numerical estimates and your answers to Question 2b) and c) to find an overestimate of $\int_0^\infty g(x)\,dx$.

b) Repeat to find an underestimate of $\int_0^\infty g(x)\,dx$. Question 2b) gives an underestimate for $\int_c^\infty g(x)\,dx$.

c) Find the error bound for $\int_0^\infty g(x)\,dx$ corresponding to your underestimate and overestimate.

■ In the Lab

In these questions, you can improve the estimates of $\int_0^\infty g(x)\,dx$ that you found before the lab.

1   By hand or using *Joy* (**Algebra ▷ Solve**), find an *integer* c for which

$$0 < \int_c^\infty \frac{1}{\sqrt{2\pi}}\, e^{-x}\,dx \; < 0.00005.$$

To do this, use the expression $\int_c^\infty f(x)\,dx = \frac{e^{-c}}{\sqrt{2\pi}}$ that you established in Question 2 (Before the Lab).

2.   Create the function $g(x) = \frac{1}{\sqrt{2\pi}}\, e^{-x^2/2}$. (**Create ▷ Function**)

3.   Using 100 subintervals, find an underestimate and an overestimate for $\int_1^c g(x)\,dx$, where c is the integer in Question 1 (In the Lab). (**Calculus ▷ Custom Integration**)

4.   a) Combine your work before and during the lab to find an overestimate for $\int_0^\infty g(x)\,dx$, using your value for c.

    b) Repeat to find an underestimate for $\int_0^\infty g(x)\,dx$.

    c) Find the error bound for $\int_0^\infty g(x)\,dx$ given by these estimates.

5.   Print your work.

■ After the Lab

1.   What is the point of Question 1 (In the Lab)? That is, why do we want to choose a value of c that makes $\int_c^\infty \frac{1}{\sqrt{2\pi}}\, e^{-x}\,dx$ small?

2.   a) What will happen to the error bound for $\int_0^\infty g(x)\,dx$ if you increase the number of subintervals in your estimates for $\int_0^1 g(x)\,dx$ and $\int_1^c g(x)\,dx$?

    b) How small can you make the error bound without changing c?

3.   a) Describe a procedure for estimating $\int_0^\infty g(x)\,dx$ with an error bound of 10^{-6}. What should c be to accomplish this?

b) Why might you need to increase the number of subintervals for $\int_1^c g(x)\,dx$ in order to get good results?

4. Based on your work, what do you think is the area $\int_{-\infty}^{\infty} g(x)\,dx$ under the *entire* normal curve?

Chapter 21
Sequences and Series

Examples and Exercises

Prerequisites:

Ch. 1	A Brief Tour of *Joy*
Ch. 9	Working with Sequences and Series

■ 21.1 Conjecturing Limits of Sequences

21.1.1 Example: Limits Using Tables

Here's how to make a table of values for a sequence and conjecture the value of its limit. We will use the example

$$\lim_{k \to \infty} \frac{(-1)^k k}{k^3 - \ln k} \ .$$

- Create the sequence $a_k = \frac{(-1)^k k}{k^3 - \ln k}$. (**Create ▷ Sequence**)

Mathematica defines the sequence:

▷ Define a_k to be $\dfrac{(-1)^k k}{k^3 - \text{Log}[k]}$

- Choose **Create ▷ List** and fill in k, a_k.
- Make sure the popup menu is set to **x Equally Spaced**.
- Fill in the interval $1 \le k \le 100$, using increments of 1.
- Make sure that **Display in decimal form** and **Display as a table** are checked.

For such a long list you might want to *uncheck* **Display as a table**, but we include this here to make the pattern clearer.

- Click **OK**.

To save space, we show only part of *Mathematica*'s output. You can see that a_k alternates in sign and decreases slowly in absolute value.

▷ Create the list with values {k, a_k} for 1 ≤ k ≤ 100.

```
Out[3]//TableForm=
    1.        -1.
    2.        0.273716
    3.        -0.115824
    4.        0.0638838
    5.        -0.0405217
    6.        0.0280101
    7.        -0.0205246
    8.        0.0156887
    9.        -0.012383
   10.        0.0100231
   11.        -0.00827938


        . . .


   90.        0.000123458
   91.        -0.000120759
   92.        0.000118148
   93.        -0.000115621
   94.        0.000113174
   95.        -0.000110804
   96.        0.000108508
   97.        -0.000106282
   98.        0.000104124
   99.        -0.000102031
  100.        0.0001
```

Here's a way to take larger values of k without taking too much space. Since a_k has a factor of $(-1)^k$, we want k to have both even and odd values.

- Choose **Last Dialog** on the *Joy* palette to bring back the **Create ▷ List** dialog.
- Fill in 500 ≤ k ≤ 600 in increments of 25.
- *Uncheck* **Display as a table**.
- Click **OK**.

The result is:

▷ Create the list with values {k, a_k} for 500 ≤ k ≤ 600 in steps of size 25.

Out[4]= {{500., 4.×10⁻⁶}, {525., -3.62812×10⁻⁶}, {550., 3.30579×10⁻⁶},
{575., -3.02457×10⁻⁶}, {600., 2.77778×10⁻⁶}}

The numerical evidence now suggests that $\lim_{k\to\infty} a_k = 0$, but this is not a proof. To prove this you can use L'Hôpital's Rule to show $\lim_{k\to\infty} \frac{k}{k^3 - \ln k} = 0$, which implies that $\lim_{k\to\infty} a_k = 0$.

21.1.2 Example: Limits Using Graphs

You can also visualize $\lim_{k\to\infty} \frac{(-1)^k k}{k^3 - \ln k}$ by plotting the sequence.

- If you didn't create a_k in Example 21.1.1, do it now.
- Graph a_k for $1 \le k \le 100$. (**Graph ▷ Sequence**)

   If you like, you can check **Plot values on y-axis**. Section 9.4 illustrates its effect.

Mathematica graphs the sequence:

▷ Plot of the sequence a_k for $1 \le k \le 100$.

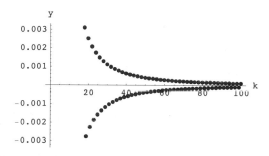

Out[5]= - Graphics -

- Repeat for $1 \le k \le 1000$.

Here is the result:

▷ Plot of the sequence a_k for $1 \le k \le 1000$.

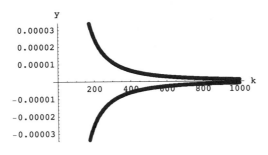

Out[6]=   - Graphics -

The *y*-values get closer to 0 as k grows, which suggests that $\lim_{k\to\infty} a_k = 0$. You can prove it using L'Hôpital's Rule.

Another way. If you created the list in Example 21.1.1, you can plot it directly using **Graph ▷ Points**. Just fill in the name or output number of the list, e.g., `Out[3]` for a plot with $1 \le k \le 100$.

Exercises

1. Graph the sequence $(1 + \frac{1}{k})^k$ for $1 \le k \le 500$. What familiar number appears to be the value of $\lim_{k\to\infty} (1 + \frac{1}{k})^k$?

2. Use a table to conjecture the value of $\lim_{k\to\infty} \frac{2k^3 + 3k^2 - k}{k^2 - 4}$. Find an algebraic explanation for why your conjecture is true.

3. Use a table to conjecture the value of $\lim_{k\to\infty} k \sin(\frac{2}{k})$. Use a familiar limit to explain why your conjecture is true.

4. Use a table to conjecture the value of $\lim_{n\to\infty} \sqrt[n]{n}$. Apply the logarithm function to both sides of $y = \sqrt[n]{n}$ and use L'Hôpital's Rule to prove your conjecture.

■ 21.2 The Definition of the Limit of a Sequence

21.2.1 Example: Epsilon-*N* Graphically, Numerically, and Algebraically

The definition of $\lim_{k\to\infty} a_k = L$, is: no matter what number $\epsilon > 0$ you choose, you can always find an integer $N > 0$ such that whenever $k > N$, then $|a_k - L| < \epsilon$.

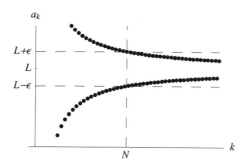

That is, given any pair of horizontal lines $y = L \pm \epsilon$ you can always find a vertical line $x = N$ such that whenever (k, a_k) is to the right of the vertical line, then it is also between the horizontal lines.

For example, consider $\lim_{k \to \infty} \left(\frac{4}{5}\right)^k$. First we'll create and graph the sequence.

- Create the sequence $a_k = \left(\frac{4}{5}\right)^k$. (**Create ▷ Sequence**)

The result is:

▷ Define a_k to be $\left(\dfrac{4}{5}\right)^k$

- Choose **Graph ▷ Sequence** and fill in a_k and $0 \le k \le 50$.
- Click **Assign name** and use the default name graph1.
- Click **OK**.

Mathematica plots the sequence:

▷ Plot of the sequence a_k for $0 < k < 50$.

graph1 =

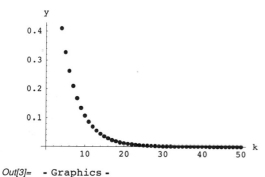

Out[3]=   - Graphics -

The sequence appears to converge to $L = 0$. We'll illustrate the definition with $\epsilon = 0.1$.

- Graph the lines $y = 0.1$ and $y = -0.1$ simultaneously for $0 \le k \le 50$, and name the result `graph2`. (**Graph ▷ Graph f(x)**)

The result is:

▷ Graph

-0.1	Black	Solid
0.1	Red	. . . . .

on 0 ≤ k ≤ 50.

`graph2 =`

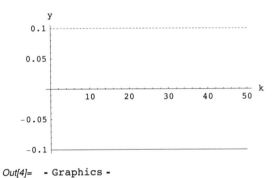

Out[4]=   - Graphics -

Now we can combine the graphs.

- Combine *graph1* and *graph2*. (**Graph ▷ Combine Graphs**)

We obtain:

▷ Combining the graphs {graph1, graph2} gives

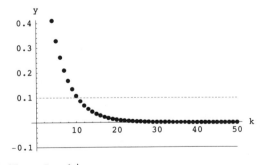

Out[5]=   - Graphics -

Since the sequence is decreasing, when k is an integer such that $k > 10$, then the points on the graph are all between the horizontal lines $y = \pm 0.1$. That is, $|a_k - 0| < 0.1$. So we can take $N = 10$ or any larger integer. By itself this isn't a proof, since we must show we can choose such an N no matter what $\epsilon > 0$ we take.

Similarly, we can illustrate the definition with a table of values.

- Make a table of values for k and a_k, for $0 \le k \le 50$ in steps of 1. (**Create ▷ List**)

To save space, we show only part of the table.

▷ Create the list with values {k, a_k} for 0 ≤ k ≤ 50.

Out[6]//TableForm=

0	1.
1.	0.8
2.	0.64
3.	0.512
4.	0.4096
5.	0.32768
6.	0.262144
7.	0.209715
8.	0.167772
9.	0.134218
10.	0.107374
11.	0.0858993

. . .

40.	0.000132923
41.	0.000106338
42.	0.0000850706
43.	0.0000680565
44.	0.0000544452
45.	0.0000435561
46.	0.0000348449
47.	0.0000278759
48.	0.0000223007
49.	0.0000178406
50.	0.0000142725

Since $\left(\frac{4}{5}\right)^k$ is decreasing, the table shows that for any integer $k > 10$ we have $|a_k - 0| < 0.1$. So, as with the graph, with $\epsilon = 0.1$ we can take $N = 10$. Similarly, with $\epsilon = 0.0001$ we can take $N = 41$ or any larger integer, since for any integer $k > 41$ we have $|a_k - 0| < 0.0001$.

To prove that $\lim_{k \to \infty} a_k = 0$, we must show we can find such a value of N for an arbitrary $\epsilon > 0$. To do this, we set

$$\left(\frac{4}{5}\right)^k = \epsilon$$

and solve for k by taking the logarithm of each side:

$$k \ln \frac{4}{5} = \ln \epsilon, \quad k = \frac{\ln \epsilon}{\ln 0.8} .$$

Now if we choose N to be any positive integer with $N \geq \frac{\ln \epsilon}{\ln 0.8}$, then for all integers k with $k > N$ we have $|a_k - 0| < \epsilon$. This proves $\lim_{k \to \infty} a_k = 0$.

For example, if $\epsilon = 0.0001$ then $\frac{\ln \epsilon}{\ln 0.8} \approx 41.3$ and we can take $N = 42$ or larger (as we saw in the table, we can also take $N = 41$).

21.2.2 Example: Replacing a Sequence with an Easier One

In this example, we consider

$$b_k = \frac{(-1)^k \, 8^k}{9^k + 10^k} .$$

- Graph the sequence $\frac{(-1)^k \, 8^k}{9^k + 10^k}$ for $0 \leq k \leq 50$, naming the result `graph3`. (**Graph ▷ Sequence, Assign name**)

    In this example there's no need to create the sequence first, but you can do that if you like. If you do, name it b_k.

Mathematica draws the graph:

▷ Plot of the sequence $\frac{(-1)^k \, 8^k}{9^k + 10^k}$ for $0 \leq k \leq 50$.

graph3 =

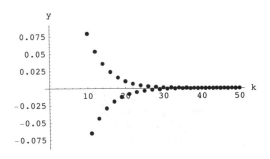

Out[7]=   - Graphics -

The graph suggests that $\lim_{k \to \infty} b_k = 0$. Here we'll show how to prove this using the definition.

Let $L = 0$. We must show that no matter what $\epsilon > 0$ we take, we can always find some $N > 0$ that satisfies the definition. In this example, we cannot solve for the integer N where the graph enters the region between the horizontal lines. (We cannot do this by hand, and *Mathematica* cannot do it using **Algebra ▷ Solve**.)

However, we can use an inequality to replace the sequence by one we can work with more easily:

$$|b_k| = \frac{8^k}{9^k + 10^k} < \frac{8^k}{10^k} = \left(\frac{4}{5}\right)^k = a_k,$$

i.e., $-a_k < b_k < a_k$.

Here a_k is the sequence of Example 21.2.1. Here is how to illustrate the relation between the two sequences in a graph. To make it easier to distinguish the sequences on screen and when printing, we will graph $\pm a_k$ in a contrasting color.

- Choose *Joy* ▷ **Graph Styles**.

Styles for multiple graphs:

	Color	LineStyle
Graph #1	Black ▾	Solid ▾
Graph #2	Red ▾	..... ▾
Graph #3	Blue ▾	___ ▾
Graph #4	Green ▾	___ ▾
Graph #5	Magenta ▾	_ . _ . ▾

Defaults Cancel OK

Graph Styles dialog (default).

- Click the **Color** popup menu for **Graph #1** and choose a different color. If you will be printing these graphs, **Gray** works well.

  This sets the colors and styles that *Joy* will use in all dialogs that create multiple graphs. This includes **Graph ▷ Sequence**, since you can plot sequences as functions of the parameter k and also along the y-axis.

- Choose **Graph ▷ Sequence**
- Graph a_k for $0 \le k \le 50$ and call it `graph1`.

  This replaces the original *graph1* with a new graph in the color you chose.

- Click **Last Dialog** and graph $-a_k$ for $0 \le k \le 50$, calling it `graph4`.
- Combine *graph3, graph4, graph1*. (**Combine Graphs**)

The result shows b_k in black and $\pm a_k$ in gray:

▷ Combining the graphs {graph3, graph4, graph1} gives

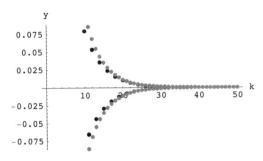

Out[10]=   - Graphics -

This plot illustrates the inequality $|b_k| < a_k$. It also shows that when the graph of a_k is between two horizontal lines of the form $y = L \pm \epsilon = \pm \epsilon$, then so is the graph of b_k. Example 21.2.1 showed that $\lim_{k\to\infty} a_k = 0$, so this proves $\lim_{k\to\infty} b_k = 0$.

For example, if $\epsilon = 0.1$, then we saw in Example 21.2.1 that we can take $N = 10$ or any larger integer for the sequence a_k. Here's how to add the lines $y = \pm 0.1$ to the graphs to illustrate that the same N satisfies the definition for b_k:

- Combine *graph3, graph4, graph1, graph2*. (**Combine Graphs**)

▷ Combining the graphs {graph3, graph4, graph1, graph2} gives

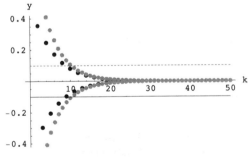

Out[10]=   - Graphics -

⚠ Depending on the order in which you list the graphs in the dialog, *Mathematica* may not display all of them. In this dialog, *graph2* (the horizontal lines) should be listed last.

When you're done, it's a good idea to restore *Joy*'s original colors and styles for graphing.

- Choose **Recent** ▷ **Graph Styles**, and then click **Default** and **OK**.

Exercises

In these exercises $k!$ stands for k *factorial*, the product

$$k != (k)\,(k-1)\,(k-2)\ldots(3)\,(2)\,(1)\,.$$

You can enter $k!$ by typing `k!`.

1. a) Graph the sequence $\frac{5^k}{k!}$ for $0 \le k \le 25$ and conjecture the value of $L = \lim_{k\to\infty} \frac{5^k}{k!}$.

    b) Let $\epsilon = 1$. Use the graph to find an integer $N > 0$ such that whenever k is an integer, $k > N$, we have $\left| \frac{5^k}{k!} - L \right| < \epsilon$.

    c) Redraw the graph over a suitable interval so you can find N for $\epsilon = 0.001$.

2. a) Graph the sequence $\frac{15^k}{k!}$ for the default interval $1 \le k \le 10$. Why can't you tell the value of $L = \lim_{k\to\infty} \frac{15^k}{k!}$ from the graph? For which value of k would you expect the pattern in the graph to change, and why?

    b) Graph $\frac{15^k}{k!}$ over an interval where you can conjecture the limit L. Use the graph to find an integer $N > 0$ such that whenever k is an integer, $k > N$, we have $\left| \frac{15^k}{k!} - L \right| < 0.01$.

3. a) Make a table to find how large n must be for $\frac{n^2-n}{2n^2+5}$ to come within 0.001 of its limit as $n \to \infty$.

    b) In a), you needed to assume that the sequence is increasing once n is sufficiently large. Differentiate $\frac{n^2-n}{2n^2+5}$ with respect to n to prove this is true (you may do this by hand or with *Joy*).

    c) Let $\epsilon > 0$ be arbitrary. Solve an equation to find an integer $N > 0$ that satisfies the definition of limit for this ϵ. (**Algebra ▷ Solve**)

    d) Substitute $\epsilon = 0.001$ to see if the value of N in c) is consistent with your answer to a). If it is not, review your reasoning.

4. In Example 21.2.2, suppose we had instead used the inequality

    $$\frac{8^k}{9^k + 10^k} < \frac{8^k}{9^k} = \left(\frac{8}{9}\right)^k$$

    to help prove $\lim_{k\to\infty} \frac{(-1)^k\,8^k}{9^k+10^k} = 0$.

    a) Create the graphs analogous to `Out[9]`, `Out[10]` that we would have obtained using this inequality.

b) Use the graphs of $\pm(\frac{8}{9})^k$ in a) to find an integer $N > 0$ that satisfies the definition of $\lim_{k \to \infty} \frac{(-1)^k 8^k}{9^k + 10^k} = 0$ when $\epsilon = 0.1$.

c) Let $\epsilon > 0$ be arbitrary. Find the value of N we would have obtained using this inequality instead, and check it against your answer to b).

d) Prove your value of N in c) satisfies the definition of $\lim_{k \to \infty} \frac{(-1)^k 8^k}{9^k + 10^k} = 0$ for arbitrary $\epsilon > 0$.

5. a) Graph the sequence $\frac{1}{k}$ for $1 \le k \le 500$.

    b) Let $L = 0$. For each of the following values of ϵ, use the graph to find an integer $N > 0$ such that whenever $k > N$ we have $|\frac{1}{k} - L| < \epsilon$:

    $$\epsilon = 0.02,\ 0.01,\ 0.005,\ 0.0025$$

    c) Prove that $\lim_{k \to \infty} \frac{1}{k} = 0$.

6. a) Graph the sequences $\frac{\cos k}{k}$, $\frac{1}{k}$, and $-\frac{1}{k}$ for $1 \le k \le 500$ and combine their graphs.

    b) What inequality does the combined graph represent? Why is the inequality true?

    c) Prove that $\lim_{k \to \infty} \frac{\cos k}{k} = 0$ by using the definition of limit and b). (Hint: use the result of Exercise 5.)

7. a) Use a graph to conjecture the value of $\lim_{n \to \infty} \frac{n^3 - n^2 + n}{5n^3 - 10}$. Find an algebraic explanation for why your conjecture is true.

    b) Use a derivative to show that the sequence $\frac{n^3 - n^2 + n}{5n^3 - 10}$ is increasing, at least after the first few terms.

    c) Solve an equation to decide how large n must be for $\frac{n^3 - n^2 + n}{5n^3 - 10}$ to come within 0.001 of its limit. (**Algebra ▷ Solve**)

    d) Why is b) needed to answer c) correctly?

■ 21.3 Infinite Series

21.3.1 Example: Terms vs Partial Sums of a Series

This example illustrates the difference between the individual terms of a series and its partial sums. It uses the geometric series

$$\sum_{k=0}^{\infty} \frac{1}{2^k}.$$

- From the *Joy* menus, choose **Create ▷ Sequence**.
- Fill in a_k and its value $\frac{1}{2^k}$.
- Check **Define a second sequence**.
- Fill in S_n and its value $\sum_{k=0}^{n} a_k$.

These are the *partial sums* of the series whose *terms* are a_k. Note that for this series, the terms start at $k = 0$ (the default example starts at $k = 1$).

- Click **OK**.

Mathematica creates the two sequences:

▷ Define a_k to be $\dfrac{1}{2^k}$ and define S_n to be $\displaystyle\sum_{k=0}^{n} a_k$

Now we'll make a table of values comparing the two sequences.

- Choose **Create ▷ List**.
- Fill in n, a_n, S_n.

In order to tabulate these sequences together, we must use the same index for each even though it's often clearer to use different indices k and n. The symbol we use for an index is a "dummy index," i.e., it can be replaced by any other symbol.

- Make sure the popup menu is set to **x Equally Spaced**.
- Fill in $0 < n \le 10$ in increments of 1.
- Keep the default settings **Display in decimal form** and **Display as a table**.
- Click **OK**.

Mathematica creates the table:

▷ Create the list with values $\{n, a_n, S_n\}$ for $0 \le n \le 10$.

```
Out[3]//TableForm=
  0       1.              1.
  1.      0.5             1.5
  2.      0.25            1.75
  3.      0.125           1.875
  4.      0.0625          1.9375
  5.      0.03125         1.96875
  6.      0.015625        1.98438
  7.      0.0078125       1.99219
  8.      0.00390625      1.99609
  9.      0.00195313      1.99805
 10.      0.000976563     1.99902
```

The partial sums (column 3) are the cumulative sums of the terms (column 2). For example, for $n = 3$ we have $a_3 = \frac{1}{2^3} = 0.125$ and $S_3 = \frac{1}{2^0} + \frac{1}{2^1} + \frac{1}{2^2} + \frac{1}{2^3} = 1.875$.

The table suggests that as we take more terms, the terms of the sequence approach 0 while the partial sums of the series approach 2. We'll graph the partial sums over a larger range of terms.

- Graph S_n for $0 \leq n \leq 50$. (**Graph ▷ Sequence**)

▷ Plot of the sequence S_n for $0 \leq n \leq 50$.

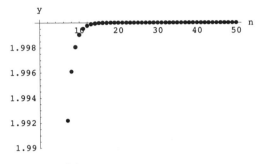

Out[4]= - Graphics -

The graph also suggests that the partial sums $S_n \rightarrow 2$ as $n \rightarrow \infty$, i.e., that the series converges to the sum $\sum_{k=0}^{\infty} \frac{1}{2^k} = 2$. This can be proved using the formula for the sum of a geometric progression (finite geometric series):

$$S_n = \frac{1 - (\frac{1}{2})^{n+1}}{1 - \frac{1}{2}} = 2 - \frac{1}{2^n},$$

so $\sum_{k=0}^{\infty} \frac{1}{2^k} = \lim_{n \rightarrow \infty} S_n = 2$. You can check that the values in the table on page 369 (Out[3]) satisfy $S_n = 2 - \frac{1}{2^n}$.

Another way. You can graph the partial sums of a series without first naming them as a sequence. For example, you can type $\sum_{k=0}^{n} \frac{1}{2^k}$ directly in the **Graph ▷ Sequence** dialog.

21.3.2 Example: Telescoping Series

The series

$$\sum_{k=3}^{\infty} \frac{1}{k\,(k-2)}$$

is called a *telescoping series* because its terms can be rewritten so that the partial sums telescope into a simpler expression. We can use this to find the sum of the series. To see that the sums telescope, we use the *partial fraction decomposition*

$$\frac{1}{k\,(k-2)} = \frac{1}{2}\left(\frac{1}{k-2} - \frac{1}{k}\right).$$

You can find the partial fractions either by hand or using *Joy*. Here's how to find them using *Joy*:

- From the *Joy* menus, choose **Algebra ▷ Simplify**.
- Fill in $\frac{1}{k\,(k-2)}$ and click **Find partial fractions**.
- Click **OK**.

Mathematica's result is:

▷ The partial fraction decomposition of $\frac{1}{k\ (k-2)}$ is

$$Out[2]= \quad \frac{1}{2\,(-2+k)} - \frac{1}{2\,k}$$

So the partial sums of the series are

$$\sum_{k=3}^{n} \frac{1}{k\,(k-2)} = \frac{1}{2}\left(\left(1-\frac{1}{3}\right)+\left(\frac{1}{2}-\frac{1}{4}\right)+\left(\frac{1}{3}-\frac{1}{5}\right)+\left(\frac{1}{4}-\frac{1}{6}\right)+ \dots +\right.$$
$$\left.\left(\frac{1}{n-3}-\frac{1}{n-1}\right)+\left(\frac{1}{n-2}-\frac{1}{n}\right)\right).$$

After cancelling, we obtain

$$\sum_{k=3}^{n} \frac{1}{k\,(k-2)} = \frac{1}{2}\left(1+\frac{1}{2}-\frac{1}{n-1}-\frac{1}{n}\right).$$

So the sum of the series is

$$\sum_{k=3}^{\infty} \frac{1}{k\,(k-2)} = \lim_{n\to\infty} \frac{1}{2}\left(1+\frac{1}{2}-\frac{1}{n-1}-\frac{1}{n}\right) = \frac{3}{4}.$$

Exercises

1. a) By hand, make a table for $1 \le n \le 5$ showing n, the nth term, and the nth partial sum of the series $\sum_{k=1}^{\infty} \frac{1}{k^2}$.

    b) Use the method of Example 21.3.1 to make the table in a) and check your work.

    c) Extend the table for $1 \le n \le 1001$ in increments of 100.

    d) Does c) prove that the series converges? What does it do? Explain.

2. a) Make a table showing n and the nth partial sum of $\sum_{k=1}^{\infty} \frac{1}{k^2}$, with $1 \le n \le 2001$ increasing in increments of 50. (**Create ▷ List, x Equally Spaced**) Using the **Numeric Sum** button to enter the partial sum will make this go faster.

    b) Repeat for the harmonic series $\sum_{k=1}^{\infty} \frac{1}{k}$. What does this suggest about the rates at which the partial sums of the two series increase?

3. a) Graph the partial sums $S_n = \sum_{k=1}^{n} \frac{1}{k^2}$ for $1 \le n \le 2001$ in increments of 50, using one of two methods. You can graph the table in Exercise 2 by entering its output number Out [...] in **Graph ▷ Points**. Or, you can use the **Numeric Sum** button to create the partial sum (**Create ▷ Sequence**) and then graph $S_{1+50\,m}$ for $0 \le m \le 40$ (**Graph ▷ Sequence**).

    b) Repeat for $S_n = \sum_{k=1}^{n} \frac{1}{k}$ and combine the graphs (**Graph ▷ Combine Graphs**). What does this suggest about the rates at which the partial sums of the two series increase?

4. a) Make a table of the partial sums S_n of $\sum_{k=0}^{\infty} e^{-k}$ from $n = 0$ to 20 and conjecture the sum of the series to five decimal places.

    b) This is a geometric series. By hand, find the exact value of the sum and express it in decimal form, rounded to five places.

5. a) Graph the partial sums of $\sum_{k=1}^{\infty} \frac{1}{(5\,k+1)\,(5\,k+6)}$ for $1 \le n \le 100$ and estimate the sum of the series.

    b) The series in a) is a telescoping series. Use the method of Example 21.3.2 to find its sum. Check that your answer is consistent with a), and review your reasoning if it is not.

6. Section 3.9 shows how to animate graphs. Here you'll animate the behavior of the partial sums of the geometric series $\sum_{k=0}^{\infty} x^k$. For this series, it can be shown that

    $$S_n = \sum_{k=0}^{n} x^k = \frac{1 - x^{n+1}}{1 - x} \quad (x \ne 1).$$

    a) Graph $\frac{1}{1-x}$ for $-1.5 \le x \le 1.5$ and assign the name *curve*. (**Graph ▷ Graph f(x)**)

    b) Animate

    $$\frac{1 - x^{5\,r+1}}{1 - x}$$

    for $-1.5 \le x \le 1.5$. In the **Animation Info** window, take $0 \le r \le 14$ to create 15 frames. For the common plot frame, take $-1.5 \le x \le 1.5$ and $-50 \le y \le 100$. Check **Combine with** to combine the animation with *curve*. (**Graph ▷ Animation, Graph f(x)**)

    c) Which partial sums S_n are included by animating $\frac{1-x^{5\,r+1}}{1-x}$ for $0 \le r \le 14$?

    d) Do the partial sums appear to converge for $-1 < x < 1$? If so, to what sum?

e) Repeat for $x > 1$ and $x < -1$. It is hard to tell from the animation what happens when $x = \pm 1$. For $x = \pm 1$, answer by finding the partial sums by hand from the definition $S_n = \sum_{k=0}^{n} x^k$.

■ 21.4 Convergence Tests

21.4.1 Example: The Integral Test and Remainder Estimates

Let $\sum_{k=1}^{\infty} a_k$ be a series where each term a_k is the value $f(k)$ of a function f that is continuous, positive, and decreasing for $1 \le x < \infty$. An example of such a series is $\sum_{k=1}^{\infty} \frac{1}{k^2}$, where $f(x) = \frac{1}{x^2}$.

The Integral Test says that the improper integral $\int_1^{\infty} f(x)\, dx$ converges if and only if the series $\sum_{k=1}^{\infty} a_k$ converges. That is, the integral and the series behave the same way.

These graphs give a method for using improper integrals to estimate the sum of the series when the series does converge. They suggest an underestimate and an overestimate for the "tail" $\sum_{k=n+1}^{\infty} a_k = a_{n+1} + a_{n+2} + \ldots$ of the series:

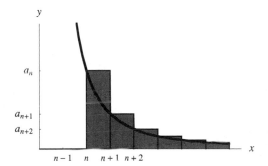 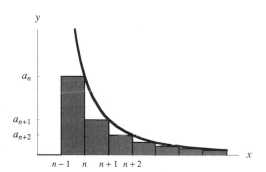

Underestimate (left): $\int_{n+1}^{\infty} f(x)\, dx \le \sum_{k=n+1}^{\infty} a_k$
Overestimate (right): $\sum_{k=n+1}^{\infty} a_k \le \int_n^{\infty} f(x)\, dx$.

We combine these inequalities, which are valid for every integer $n \ge 1$:

$$\int_{n+1}^{\infty} f(x)\, dx \le \sum_{k=n+1}^{\infty} a_k \le \int_n^{\infty} f(x)\, dx . \qquad (*)$$

Suppose the series converges to S, that is, $\sum_{k=1}^{\infty} a_k = S$. Then the sum in $(*)$ is the remainder or tail when we approximate S by its nth partial sum:

$$S = S_n + R_n \text{ , where } S_n = \sum_{k=1}^{n} a_k \text{ and } R_n = \sum_{k=n+1}^{\infty} a_k \text{ .}$$

The inequalities (*) tell us how closely S_n approximates S:

$$S_n + \int_{n+1}^{\infty} f(x)\, dx \le S \le S_n + \int_{n}^{\infty} f(x)\, dx.$$

The inequalities in (*) can also be used to prove that the statement of the Integral Test is true. Suppose $\int_{1}^{\infty} f(x)\, dx$ converges. Then $\int_{n}^{\infty} f(x)\, dx$ converges, and by (*) so does $\sum_{k=n+1}^{\infty} a_k$ and thus also $\sum_{k=1}^{\infty} a_k$.

Conversely, suppose $\sum_{k=1}^{\infty} a_k$ converges. Then $\sum_{k=n+1}^{\infty} a_k$ converges, and by (*) so does $\int_{n+1}^{\infty} f(x)\, dx$ and thus also $\int_{1}^{\infty} f(x)\, dx$.

For example, with $\sum_{k=1}^{\infty} \frac{1}{k^2}$ and $f(x) = \frac{1}{x^2}$, we have

$$\int_{n}^{\infty} f(x)\, dx = \int_{n}^{\infty} \frac{1}{x^2}\, dx = \lim_{a\to\infty} \left. -\frac{1}{x} \right]_{n}^{a} = \frac{1}{n} \text{ .}$$

So for any integer $n \ge 1$ we have

$$\frac{1}{n+1} \le R_n \le \frac{1}{n}$$

and

$$S_n + \frac{1}{n+1} \le S \le S_n + \frac{1}{n} \text{ .}$$

We'll use *Joy* to estimate S when $n = 100$. Since we'll be taking many terms and want to avoid fractions with large denominators, we will use numeric sums rather than exact sums.

- Choose **Create ▷ Sequence.**
- Fill in the sequence a_k given by $\frac{1}{k^2}$.
- If necessary, check **Define a second sequence** and fill in S_n.
- Type N to the left of $\sum_{k=1}^{n} a_k$, to create a numeric sum.
- Click **OK.**

  ♡ Creating S_n this way allows you to find the partial sum for any value of n, not just $n = 100$ as in this example.

Mathematica defines the numeric sum:

▷ Define a_k to be $\frac{1}{k^2}$ and define S_n to be NSum[a_k, {k, 1, n}]

⚠ *Joy* defines expressions of the form N $\sum_{k=1}^{n} a_k$ for *Mathematica*. If you use *Mathematica* without *Joy* and type N before a summation, the expression will be undefined and not produce the *Mathematica* function NSum shown in the paraphrase.

- If necessary, click below the cells defining the sum so that a horizontal line appears.
- Type S$_{100}$ and press ⌈SHIFT⌉ − ⌈RET⌉.

The result is:

In[3]:= **S$_{100}$**

Out[3]= 1.63498

Now we find estimates for the sum S.

- Type S$_{100}$ + $\frac{1}{101}$ and press ⌈SHIFT⌉ − ⌈RET⌉.

In[4]:= **S$_{100}$ + $\dfrac{1}{101}$**

Out[4]= 1.64488

Since $S_{100} + \frac{1}{100} = 1.64498$, our estimate for the sum is

$$1.64488 \leq S \leq 1.64498.$$

Another way to express this estimate is that $S = 1.64493 \pm 0.00005$. This is an excellent estimate, because it can be proved that the exact value of S is $\sum_{k=1}^{\infty} \frac{1}{k^2} = \frac{\pi^2}{6}$, which is approximately 1.644934.

Exercises

In Exercises 1–3:

a) By hand, use a comparison test to determine whether the series converges or diverges.

b) Using *Joy*, graph the partial sums of the series and of its comparison series, to illustrate the comparison in a). Use the graph to estimate the sum of the series if it converges or to illustrate that it diverges.

1. $\sum_{k=0}^{\infty} \frac{1}{5^k + 3}$

2. $\sum_{k=1}^{\infty} \frac{|\sin k|}{k^2}$ (Use the **Absolute Value** button to enter $|\sin k|$.)

3. $\sum_{k=1}^{\infty} \frac{\ln k}{k}$ (Enter $\ln k$ as Log[k] or use the **Functions** button on the *Joy* palette.)

4. a) Graph the sequence $a_k = \frac{1}{3^k}$ for $1 \leq k \leq 10$. Then graph $S_n = \sum_{k=1}^{n} a_k$ for $1 \leq n \leq 10$ and combine the two graphs.

b) Repeat for $a_k = \frac{1}{\sqrt{k}}$, $1 \le k \le 100$.

c) Conjecture from the graphs whether these series converge or diverge. Try to prove your answers.

5. a) In Example 21.4.1, how large must n be for the under- and overestimates of R_n in the inequalities (*) to be within 10^{-6} of each other, i.e., for $\frac{1}{n} - \frac{1}{n+1} < 10^{-6}$? (Solve by hand, putting the terms on the left over a common denominator, or use *Joy*.)

b) Using your answer to a), find an interval of length less than 10^{-6} that contains the sum S.

You will need to have *Mathematica* display more than six significant digits for the endpoints of the interval to be different from each other. For example, you can display 10 digits by using the command $N[\ldots,10]$ and filling in an expression or its name or output number. Section 1.3.1 illustrates this.

6. a) Use the Integral Test to prove that $\sum_{k=2}^{\infty} \frac{1}{k (\ln k)^2}$ converges. Note that the series begins with $k = 2$, since $\ln 1 = 0$.

b) Use the method of Example 21.4.1 to estimate the sum using the partial sum S_{100} (the sum from $k = 2$ to 100).

■ 21.5 Taylor Series

21.5.1 Example: Error Bounds for Taylor Series

It can be proved that the error in approximating $f(x)$ by its nth-degree Taylor polynomial $P_n(x)$ is related to the $(n+1)$st derivative $f^{(n+1)}$. This example illustrates that relationship.

We make suitable continuity and differentiability assumptions about the functions involved and take the Taylor polynomials centered at some value $x = a$:

$$P_n(x) = \sum_{k=0}^{n} \frac{f^{(k)}(a)}{k!} (x - a)^k$$

$$= f(a) + f'(a)(x - a) + \frac{f''(a)}{2!} (x - a)^2 + \ldots + \frac{f^{(n)}(a)}{n!} (x - a)^n.$$

To estimate the error $f(x) - P_n(x)$, we first let M be a bound on the values of the $(n+1)$st derivative of f between a and x:

$$|f^{(n+1)}(t)| \le M \text{ for all } t \text{ between } a \text{ and } x.$$

Then it can be proved that

$$\left| f(x) - P_n(x) \right| \leq \frac{M}{(n+1)!} \left| x - a \right|^{n+1}.$$ (*)

For example suppose we want to approximate $f(1.2)$, where

$$f(x) = \frac{\ln x}{x}.$$

There is no way to find the exact value of $f(1.2)$, since $\ln x$ can only be approximated for most values of x.

- Create the function $f(x) = \frac{\ln x}{x}$, entering $\ln x$ as Log[x]. (**Create ▷ Function**)

Mathematica gives.

▷ Define f[x] to be $\dfrac{\text{Log[x]}}{\text{x}}$

Since $\ln x$ is only defined for $x > 0$, we must choose some $a > 0$ as the center of the Taylor polynomials. We will choose $a = 1$.

- From the *Joy* menus, choose **Calculus ▷ Taylor Polynomials**.
- Fill in f[x] for the function and 1 for the point about which the polynomials are defined.
- Fill in degrees 1,2,3,4 and the interval $0 \leq x \leq 2$.
- Click **OK**.

Mathematica creates the polynomials:

▷ Define P_n[x] to be the Taylor polynomial for f[x] about the point 1.

For degrees n - {1, 2, 3, 4},
    graph the function f[x] and its Taylor polynomials on the interval $0 \leq x \leq 2$.

f[x]	Black	Solid
P_1[x]	Red	.....
P_2[x]	Blue	_ _ _ _
P_3[x]	Green	__ __
P_4[x]	Magenta	. _ . _ .

$-1 + x$

$-1 - \frac{3}{2} \, (-1 + x)^2 + x$

$-1 - \frac{3}{2} \, (-1 + x)^2 + \frac{11}{6} \, (-1 + x)^3 + x$

$-1 - \frac{3}{2} \, (-1 + x)^2 + \frac{11}{6} \, (-1 + x)^3 - \frac{25}{12} \, (-1 + x)^4 + x$

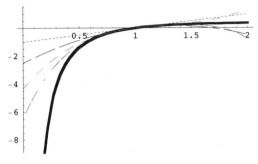

Out[3]=   - Graphics -

The polynomials appear to approximate $f(x)$ best when x is close to 1 and when the degree n is larger. So we'll approximate $f(1.2)$ by $P_4(1.2)$ and then estimate the error in the approximation.

- Click in your *Mathematica* notebook beneath the last output.
- Type P_4 [1.2] and press ⌈SHIFT⌉ − ⌈RET⌉.

In[4]:= P_4 [1.2]

Out[4]=   0.151333

To estimate the error in this approximation, we need a bound for the fifth derivative $f^{(5)}$ between 1 and 1.2.

- Graph $f^{(5)}(t)$ for $1 \le t \le 1.2$. (Enter $f^{(5)}(t)$ as f ' ' ' ' ' [t].)

▷ Graph $((((\texttt{f}')')')')'[\texttt{t}]$ on $1 \le \texttt{t} \le 1.2$.

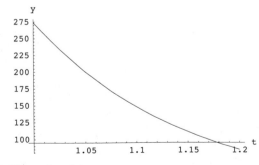

Out[5]=   - Graphics -

The graph suggests that $f^{(5)}$ is decreasing on the interval [1, 1.2], so we can take the bound M to be $f^{(5)}(1)$. (Alternatively, we can set M to be any larger number that is more convenient to find.)

- Click in your *Mathematica* notebook beneath the last output.
- Type **f'''''[1]** and press SHIFT − RET.

In[6]:= **f'''''[1]**

Out[6]= 274

> Rather than depend on the graph, you could calculate the sixth derivative $f^{(6)}$ to prove that $f^{(5)}$ is decreasing on [1, 1.2].

So we let $M = 274$ and calculate the error bound in inequality (*):

$$| f(1.2) - P_4(1.2) | \leq \tfrac{274}{5!} | 1.2 - 1 |^5 = \tfrac{274}{5!} (0.2)^5.$$

- Click in your *Mathematica* notebook beneath the last output.
- Type $\tfrac{274}{5!}$ **(0.2)**^{5} and press SHIFT − RET.

In[7]:= $\dfrac{274}{5!}$ **(0.2)**^{5}

Out[7]= 0.000730667

This tells us that our estimate $P_4(1.2) = 0.151333$ approximates $f(1.2) = \frac{\ln 1.2}{1.2}$ to within 0.000730667. In fact, if we use *Mathematica*'s value for the logarithm (also an approximation), we find

In[8]:= **f[1.2]**

Out[8]= 0.151935

The difference is

In[9]:= **f[1.2] - P$_4$[1.2]**

Out[9]= 0.000601297

This is within the error bound of 0.000730667 that we found.

Exercises

1. a) Define the Taylor polynomials for $\sin x$ about the point 0, and list and plot them for degrees 0, 2, 4, 6 with $-5 \leq x \leq 5$.

    b) The polynomials for degrees 1, 3, 5 are the same as for 2, 4, 6. Why?

    c) Repeat for degrees 8, 10, 12, 14.

d) Repeat for degrees 20, 22, 24, 26, letting $-10 \le x \le 10$. (To save space, *uncheck* **list the Taylor polynomials**.)

e) Write a paragraph summarizing how the behavior of the polynomials is related to the degree and to the graphing interval.

2. a) Define the Taylor polynomials of $\sqrt{x}$ about the point 1, and list and plot them for degrees 1, 2, 3, 4 with $0 \le x \le 3$.

   b) Repeat for degrees 8, 9, 10, 11.

   c) Repeat for degrees 20, 21, 22, 23, graphing for $1.9 \le x \le 2.1$. (To save space, *uncheck* **list the Taylor polynomials**.)

   d) Write a paragraph summarizing how the behavior of the polynomials is related to the degree and to the graphing interval. Also explain why we are not taking the Taylor polynomials about the point 0.

3. a) Find the 4th and 10th degree Taylor polynomials of $\sec x$ centered at 0.

   b) What are the numerical values of $\sec(0.5)$, $P_4(0.5)$, and $|\sec(0.5) - P_4(0.5)|$?

   c) Repeat b) for degree 10. By what factor does the error in the Taylor approximation of $\sec(0.5)$ decrease when n goes from 4 to 10?

4. a) Define the Taylor polynomials $P_n(x)$ of $\sqrt{x}$ about the point 1.

   b) Make a table showing n and $P_n(0.5)$, where $0 \le n \le 100$ in steps of 5. (**Create ▷ List, x Equally Spaced**) Does the Taylor series appear to converge to $\sqrt{0.5}$?

   c) Repeat for $P_n(1.5)$, $P_n(1.9)$, and $P_n(2.1)$. How are these cases different from each other?

5. a) Find the Taylor polynomials of degrees 0 through 6 about the point 0 for the function $5x^3 - 12x^2 - x + 4$.

   b) Repeat for $613x^5 - 27x^4 + 100x^3 - 15x^2 + 17x - 1$.

   c) Conjecture a theorem that describes the Taylor polynomials of a function that is itself a polynomial.

Exercises 6–9 refer to the Taylor error bound in inequality (*) of Example 21.5.1.

6. a) By hand, find the Taylor polynomial $P_7(x)$ of $\cos x$ about the point 0. Check your answer by using *Joy* to find $P_7(x)$.

   b) We know from trigonometry that $\cos\frac{\pi}{4} = \frac{1}{\sqrt{2}}$ exactly. Find the error in the approximation, $\frac{1}{\sqrt{2}} - P_7(\frac{\pi}{4})$, and express it in decimal form.

   c) Use inequality (*) to find an error bound for approximating $\cos\frac{\pi}{4}$ by $P_7(\frac{\pi}{4})$. Is the error bound consistent with the actual error in b)? If not, review your reasoning.

7. a) Define the Taylor polynomials $P_n(x)$ of $\sin x$ about the point 0.

   b) Make a table showing n and $P_n(\frac{\pi}{6})$, where $1 \le n \le 10$. To what value does $P_n(\frac{\pi}{6})$ appear to converge? **(Create ▷ List, x Equally Spaced, Display in decimal form)**

   c) We know from trigonometry that $\sin \frac{\pi}{6} = \frac{1}{2}$ exactly. Modify the table in b) to show n and the error in the approximation, $0.5 - P_n(\frac{\pi}{6})$.

   d) Use inequality (*) to find an error bound for approximating $\sin \frac{\pi}{6}$ by $P_n(\frac{\pi}{6})$. Make a table showing n and the error bound for $1 \le n \le 10$. Are the error bounds consistent with the actual errors in c)? If not, review your reasoning.

8. a) We can't find the exact value of $\tan 0.2$. Instead, find the fourth-degree polynomial $P_4(x)$ for $\tan x$ about the point 0, and then find $P_4(0.2)$.

   b) Plot the fifth derivative of $\tan x$ for $0 \le x \le 0.2$, as in Example 21.5.1. Use this to find an appropriate value of M in inequality (*).

   c) What is a bound for the error you would make in approximating $\tan 0.2$ by $P_4(0.2)$?

9. a) By hand, find the sixth degree Taylor polynomial $P_6(x)$ for $\sin x$ about the point 0. Check your answer by using *Joy* to find $P_6(x)$.

   b) Use inequality (*) to find an error bound for approximating $\sin x$ by $P_6(x)$. By hand, you can find a value of M that works for all x.

   c) Using your error bound in b), graph $\sin x$ together with $P_6(x) +$ the bound and $P_6(x) -$ the bound, for $-5 \le x \le 5$.

   d) The graph of $\sin x$ should lie between the other two curves. Why should it, and does it appear to do so in your output?

Lab 21.1 Alternating Series

Prerequisites:
    Ch. 1   A Brief Tour of *Joy*
    Secs. 9.5–8   Creating, tabulating, and graphing series
    Sec 3.9   How to Animate Graphs

This lab helps you visualize the behavior of the alternating harmonic series

$$\sum_{k=1}^{\infty} \frac{(-1)^{k+1}}{k} = 1 - \frac{1}{2} + \frac{1}{3} - \cdots$$

by calculating its partial sums and seeing how they change as the number of terms increases.

■ Before the Lab (Do by Hand)

1. Calculate the partial sums

$$S_n = \sum_{k=1}^{n} \frac{(-1)^{k+1}}{k}, \quad n = 1, 2, 3, 4, 5.$$

2. Do the partial sums increase, decrease, or neither? What if you consider the odd-numbered and even-numbered partial sums separately?

3. How do the values S_n and S_{n+1} compare in size?

■ In the Lab

1. Create the two sequences $a_k = \frac{(-1)^{k+1}}{k}$ and $S_n = \sum_{k=1}^{n} a_k$. (**Create ▷ Sequence**)

2. Create a table of values showing n and S_n for $1 \le n \le 10$. (**Create ▷ List, x Equally Spaced, Display in decimal form**)

3. Repeat for n ranging from 100 to 1000 in steps of 75.

4. Graph S_n for $1 \le n \le 10$. (**Graph ▷ Sequence**)

5. Here you'll animate the partial sums, plotting 15 frames in steps of 5.

   - Choose **Graph ▷ Animation**, choose **Sequence** from the popup menu, and click **OK**.
   - In the **Graph Sequence** window, fill in S_n and $1 \le n \le 5$ r.

- Check **Graph** and **Plot values on the y-axis**.
- In the **Animation Info** window, fill in $1 \le r \le 15$.

  Be sure to change the animation parameter to r.

- Fill in the common plot frame $0 \le n \le 75$, $0.5 \le y \le 1$.
- In the **Graph Sequence** window, click **OK**.

6. Double-click the bracket to open all the animation frames. Double-clicking opens and closes this group of cells.

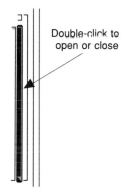

Double-click to
open or close

7. Print your work. Don't print all the animation graphs, but include at least the first four. You may cut out the rest before printing.

■ After the Lab

1. Estimate the sum S, to two decimal places, to which the series appears to converge. (It can be proved that the exact value is $S = \ln 2$.)

2. a) Using what you saw in the animation, describe the pattern that the partial sums take as they accumulate.

   b) Is this pattern consistent with what you observed in Questions 2 and 3 (Before the Lab)?

3. a) Choose a point (n, S_n) on your plot of the sequence. Draw a vertical line whose length is $|S - S_n|$, where $S = \lim_{n \to \infty} S_n$ is the sum of the series.

   b) Draw a vertical line whose length is $|S_{n+1} - S_n| = |a_{n+1}|$.

   c) Explain why these lines illustrate the inequality $|S - S_n| \le |a_{n+1}|$.

4. Use the inequality in Question 3c) to decide how many terms of the series are needed for the partial sum to be within 10^{-6} of S.

Lab 21.2 Definition of Convergence for a Series

Prerequisites:

Ch. 1	A Brief Tour of *Joy*
Sec 9.8	How to Plot the Partial Sums of a Series

In this lab you can illustrate the definition of convergence of a series. A series $\sum_{k=0}^{\infty} a_k$ converges to a sum S if its sequence of partial sums,

$$S_n = \sum_{k=0}^{n} a_k,$$

converges to S. That is, $\sum_{k=0}^{\infty} a_k = S$ means for every $\epsilon > 0$ there is an integer $N > 0$ such that whenever n is an integer and $n > N$, we have $|S - S_n| < \epsilon$.

■ Before the Lab (Do by Hand)

We'll use the example of the geometric series

$$\sum_{k=0}^{\infty} x^k .$$

The nth partial sum S_n is the sum of a finite geometric progression, and it can be proved that

$$S_n = \sum_{k=0}^{n} x^k = \frac{1 - x^{n+1}}{1 - x}, \quad \text{if } x \neq 1 . \tag{*}$$

Use equation (*) to answer the following questions.

1. a) Let $x = 0.5$. What is the value of $S = \lim_{n \to \infty} S_n$?

   b) How large must n be to guarantee that $|S - S_n| < 0.1$? Find the smallest such integer n and call it N.

   c) How large must n be to guarantee that $|S - S_n| < 0.001$? Find the smallest such integer n and call it N.

2. Repeat for $x = 0.99$.

■ In the Lab

1. a) From the *Joy* menus:

- Choose **Graph ▷ Sequence**.
- Fill in

$$\sum_{k=0}^{n} 0.5^k \quad \text{and} \quad 0 \le n \le 25.$$

- Click **OK**.

b) Graph $\sum_{k=0}^{n} 0.5^k$ again, letting n range between $N \pm 5$, where N is the value of n that you found in Question 1c) (Before the Lab). For example, if $N = 8$, then let $3 \le n \le 13$.

c) For $\sum_{k=0}^{\infty} 0.5^k$, graph (simultaneously) the horizontal lines $y = S + 0.001$ and $y = S - 0.001$, using the value of S you found in Question 1a) (Before the Lab). Again let n range between $N \pm 5$. (**Graph ▷ Graph f(x)**)

d) Combine the graphs in steps b) and c). (**Graph ▷ Combine Graphs**)

The result should show the partial sums crossing into the region between $y = S + 0.001$ and $y = S - 0.001$ at the midpoint of the graphing interval. If it doesn't do this, review your reasoning and try again.

2. Repeat for $\sum_{k=0}^{n} 0.99^k$, changing the value of N accordingly. In part a), use the interval $0 \le n \le 1500$.

3. Graph $\sum_{k=0}^{n} 1.01^k$ for $0 \le n \le 500$.

4. Print your work.

■ After the Lab

1. a) Let $0 < x < 1$ and $\epsilon > 0$ be arbitrary. Use equation (*) to decide how large n must be to guarantee that $|S - S_n| < \epsilon$. Let N be the smallest such integer; N will depend on x.

b) Is your answer valid for $x \ge 1$? Explain.

2. Check this value of N against the examples you did before and in the lab, with $x = 0.5$, 0.99 and $\epsilon = 0.1$, 0.001. Are your answers consistent with this value of N? If not, review your reasoning.

3. a) When $x = 0.9999$, what is the value of S?

b) How large would n have to be for S_n to come within $\epsilon = 0.1$ of S? Find the smallest such integer and call it N.

c) Repeat for $\epsilon = 0.001$.

4. For $x = 1$, the expression $S_n = \frac{1-x^{n+1}}{1-x}$ is indeterminate. Write out the terms of the series and find the value of S_n. Does the series converge?

5. a) For $x = 1.01$, what does your graph suggest about $\lim_{n \to \infty} S_n$?

    b) For any $x > 1$, what is $\lim_{n \to \infty} S_n$? Does the series converge or diverge?

6. Write a paragraph summarizing

    > for which values $x \geq 0$ the series converges and for which it diverges
    > for a given value x, $0 < x < 1$, how N appears to change as ϵ changes
    > for a given value of $\epsilon > 0$, how N appears to change as $x \to 1$ from the left

■ For Further Exploration

You can investigate the behavior of the series when $x < 0$ by answering these questions.

1. When $x = -0.5$, what is the value of S? Graph S_n, and experiment until you find a graphing interval for n that shows the pattern followed by S_n. How does S_n behave when n is odd? When n is even?

2. Repeat for $x = -0.9$, $x = -1$, and $x = -1.01$.

3. a) Let $\epsilon > 0$ and $-1 < x < 0$ be arbitrary. Suppose n is odd. Use equation (*) to decide how large must n be to guarantee that $|S - S_n| < \epsilon$. Let N be the smallest such integer.

    b) Repeat for n an even integer.

4. Write a paragraph summarizing

    > for which values $x < 0$ the series converges and for which it diverges
    > for a given value x, $-1 < x < 0$, how N appears to change as ϵ changes
    > for a given value of $\epsilon > 0$, how N appears to change as $x \to -1$ from the left

Lab 21.3 The Error in the Taylor Approximation

Prerequisites:

Ch. 1	A Brief Tour of *Joy*
Sec 9.10	How to Create a Taylor Series

In this lab, you can explore the convergence of the Taylor series for $\ln x$ and the error in the Taylor approximation.

■ Before the Lab (Do by Hand)

1. Sketch the graph of $\ln x$. Explain why you cannot take the Taylor series of $\ln x$ centered at the point 0.

2. Find the Taylor series for $\ln x$ about the point 1, up to the sixth-degree term.

3. Find a formula for $f^{(n+1)}(x)$, the $(n + 1)$st derivative of $f(x) = \ln x$.

■ In the Lab

1. a) Create the Taylor polynomials $P_n(x)$ for $\ln x$ about the point 1 and graph them for degrees 1, 2, 3, 4, 5, with $0 \le x \le 2$. (**Calculus ▷ Taylor Polynomials**)

    Enter $\ln x$ as **Log[x]** or use the **Functions** button on the *Joy* palette. Although $\ln 0$ is undefined, you can still use the interval $0 \le x \le 2$.

    b) Compare the polynomials with the ones you found before the lab and review your reasoning if there are any inconsistencies.

2. a) The difference $\ln x - P_n(x)$ is ± the vertical distance between the graphs of $\ln x$ and $P_n(x)$, and is the error in approximating $\ln(x)$ by its nth-degree Taylor polynomial. We'll call this Error(n, x) since it depends on both x and n.

    - From the *Joy* menus, choose **Create ▷ Function**.
    - Fill in **Error[n, x]** in the first input field and **Log[x]** − P$_n$**[x]** in the second.
    - Click **OK**.

    b) Graph Error(1, x), Error(2, x), Error(3, x), Error(4, x), Error(5, x) simultaneously, for $0 \le x \le 2$ and $-0.015 \le y \le 0.015$. (**Graph ▷ Graph f(x)**)

    c) Add a **Grid** to the graph in b). (**Graph ▷ Modify Graph**)

3. Print your work.

■ After the Lab

It can be proved that the Taylor series of $\ln x$ converges to $\ln x$ for $0 < x \le 2$. So the error $\ln x - P_n(x)$ is the sum of the terms in the tail of the series, beginning with power $n + 1$:

$$\ln x - P_n(x) = \frac{f^{(n+1)}(1)}{(n+1)!}(x-1)^{n+1} + \frac{f^{(n+2)}(1)}{(n+2)!}(x-1)^{n+2} + \dots .$$

For values of x near 1 we can approximate this sum by its first term,

$$\ln x - P_n(x) \approx \frac{f^{(n+1)}(1)}{(n+1)!}(x-1)^{n+1}.$$

1. a) Evaluate the right side of this estimate using the expression for $f^{(n+1)}(x)$ that you found before the lab.

    b) Use a) to explain why the graph of $\ln x - P_1(x)$ has the shape shown in Question 2 (In the Lab). For example, approximately what type of graph is it (line, parabola, cubic, etc.)?

    c) Repeat for $\ln x - P_2(x)$.

2. The graphs in Question 1 (In the Lab) show whether a Taylor approximation $P_n(x)$ is an underestimate or overestimate of $\ln x$, for a given x. How can you use the graphs in Question 2 (In the Lab) to show this?

3. We wish to approximate $f(1.5) = \ln 1.5$ by $P_n(1.5)$. According to the graphs in Question 2 (In the Lab), what is the approximate value of $|f(1.5) - P_n(1.5)|$ for degrees 3, 4, 5?

The error $f(x) - P_n(x)$ in a Taylor approximation always satisfies

$$\left| f(x) - P_n(x) \right| \le \frac{M}{(n+1)!} \left| x - a \right|^{n+1}.$$

The right-hand side is called an *error bound* for the approximation. Here the Taylor polynomials are centered at a and M is a bound on the values of the $(n + 1)$st derivative of f between a and x:

$$|f^{(n+1)}(t)| \le M \text{ for all } t \text{ between } a \text{ and } x.$$

4. We want to find the error bound given by the theory for the example of this lab.

    a) Use the expression for $f^{(n+1)}(x)$ that you found before the lab to find a value for M.

    b) Use M to find a bound for $|f(1.5) - P_n(1.5)|$.

5. a) What is the value of the error bound for degrees 3, 4, 5?

    b) Is your error bound consistent with the error you measured on the graph? That is, are the values of $|f(1.5) - P_n(1.5)|$ that you read off the graph at most equal to the error bounds in a)?

Chapter 22
Parameterized Curves

Examples and Exercises

Prerequisites:

Ch. 1	A Brief Tour of *Joy*
Sec. 3.5,8	Graphing 2D parameterized curves and polar coordinates
Sec. 10.6	How to Graph a Curve in Space

■ 22.1 Curves in the Plane

22.1.1 Example: How to Create a Function That Parameterizes a Curve

For example, here is how to create the parameterization $r(t) = (t \cos t, \sin t \cos t)$. For parameterized curves in space, use the same procedure with three coordinates.

- From the *Joy* menus, choose **Create ▷ Function**.
- Fill in `r[t]` in the first input field.
- Fill in the coordinates `{t␣Cos[t], Sin[t] Cos[t]}` in the second field.

    Here ␣ indicates that you must type a space between `t` and `Cos[t]` to indicate multiplication. *Mathematica* uses curly brackets `{ }` to denote lists. Vectors, such as these points in three-dimensional space, are written as lists in *Mathematica*.

    ☿ *Mathematica* recognizes that `Sin[t]` and `Cos[t]` are separate factors and so you don't need to type a space between them. However, extra spaces do no harm and if you're unsure how *Mathematica* will interpret an expression you may include a space.

- Click **OK**.

When you create a function, *Mathematica* does not display any output.

▷ Define `r[t]` to be `{t Cos[t], Sin[t] Cos[t]}`

You can use the $r(t)$ notation to evaluate this function at any t. For example:

- If necessary, click beneath the last cells so that a horizontal line appears.
- Type `r [π / 4]` and press SHIFT – RET.

The result is:

In[3]:= `r [π / 4]`

Out[3]= $\{\frac{\pi}{4\sqrt{2}}, \frac{1}{2}\}$

⚠ The graph of $r(t)$ is a curve in R^2. To draw the graph, choose **Graph ▷ Parametric Plot**. However, you will not be able to use the $r(t)$ notation in this dialog. Instead, fill in `x = t⌣Cos[t], y = Sin[t] Cos[t]`. Similarly, if $r(t)$ describes a curve in R^3, use **Graph 3D ▷ Space Curve** and fill in the x-, y-, z-coordinates.

Exercises

1. The graph of $x = \cos t$, $y = \sin t$ is a circle in R^2. Here you'll see the effect of multiplying x or y by a constant.

   a) Graph the circle $x = \cos t$, $y = \sin t$ for $0 \le t \le 2\pi$, choosing **Equal scales**. (**Graph ▷ Parametric Plot**)

   b) Predict the shape of the graph of $x = 2 \cos t$, $y = \sin t$. Check your prediction by graphing this curve and combining the two graphs. (**Graph ▷ Parametric Plot, Combine Graphs**)

   c) Repeat for $x = \cos t$, $y = 2 \sin t$.

2. a) Predict the shape of the graph of $x = 2 \cos t$, $y = 4 \sin t$. Check your prediction by graphing the curve for $0 \le t \le 2\pi$.

   b) Write an equation for this curve that doesn't contain a parameter, i.e., that contains only x and y.

3. Section 3.9 discusses animating graphs. Here you can use animation to tell the difference between the two parameterizations $x = \cos t$, $y = \sin t$ and $x = \cos 3t$, $y = \sin 3t$.

   a) Graph the two parameterizations for $0 \le t \le 2\pi$. How do the curves differ, if at all? (**Graph ▷ Parametric Plot**)

   b) Create $r(t) = (\cos t, \sin t)$. (**Create ▷ Function**)

   Then follow these steps to animate the motion of a point around the curve for $0 \le t \le 2\pi$. This will create 17 animation frames, one every $\frac{\pi}{8}$ radians.

   - Choose **Graph ▷ Animation** and then **Points**. Click **OK**.

- Fill in r[k__π / 8].
- In the **Animation Info** window, fill in 0 ≤ k ≤ 16, -1 ≤ x ≤ 1, -1 ≤ y ≤ 1.
- Check **Equal scales**.
- Check **Combine with** and fill in the name or output number of the graph of *r*(*t*) in a).
- Click **OK** in the **Plot Points** window.

c) Repeat for *p*(*t*) = (cos 3 *t*, sin 3 *t*) and describe the difference in the motion around the two curves.

■ 22.2 Curves in Space

22.2.1 Example: How to Differentiate along a Parameterized Curve

You can differentiate along a curve *r*(*t*) by entering the usual derivative notation *r'*(*t*) directly in your *Mathematica* notebook. We'll use the example $r(t) = (t^2 \cos t, \sin t \cos t, t^2 \sin t)$ in R^3. The same procedure applies to parametcrizations of curves in the plane.

- Choose **Create ▷ Function**.
- Fill in r[t].
- Fill in {t² Cos[t], Sin[t] Cos[t], t² Sin[t]}.

  Be sure to enclose the vector in curly brackets { }.

- Click **OK**.

Mathematica defincs the function:

▷ Define r[t] to be {t² Cos[t], Sin[t] Cos[t], t² Sin[t]}

- If necessary, click in your *Mathematica* notebook below the last cells so that a horizontal line appears.
- Type r'[t].
- Press [SHIFT] − [RET].

Mathematica differentiates the function one coordinate at a time:

In[3]:= **r'[t]**

Out[3]= {2 t Cos[t] - t² Sin[t], Cos[t]² - Sin[t]², t² Cos[t] + 2 t Sin[t]}

The derivative *r'*(*t*) is a vector tangent to the curve at the point where (*x*, *y*, *z*) = *r*(*t*). If *t* represents time, then we can interpret *r'*(*t*) as the velocity vector.

You can find *r"*(*t*) and higher-order derivatives similarly, for example:

- Click below the last output so that a horizontal line appears.
- Type r''[t] and press [SHIFT] − [RET].

⚠ Do not type double quotes " for the second derivative, but instead type two single quotes (apostrophes or primes) ' '.

The result is:

In[4]:= **r''[t]**

Out[4]= {2 Cos[t] − t² Cos[t] − 4 t Sin[t], −4 Cos[t] Sin[t],
4 t Cos[t] + 2 Sin[t] − t² Sin[t]}

Using the Derivative button. If you haven't created $r(t)$ as a function, here's how to use the **Derivative** button instead.

- Click in your *Mathematica* notebook beneath the last output, so that a horizontal line appears.
- Type {t² Cos[t], Sin[t] Cos[t], t² Sin[t]}

  Be sure to include the curly brackets { } that denote a vector.

- Drag across the expression you typed to select it.

$$\{t^2\,\mathrm{Cos[t]},\ \mathrm{Sin[t]\,Cos[t]},\ t^2\,\mathrm{Sin[t]}\}$$

- On the *Joy* palette, click the **Derivative** button.

Derivative button.

An expression for the derivative appears, with the vector function filled in and the input field for the independent variable selected.

$$\frac{d}{d\,\square}\left(\{t^2\,\mathrm{Cos[t]},\ \mathrm{Sin[t]\,Cos[t]},\ t^2\,\mathrm{Sin[t]}\}\right)$$

- Type t for the independent variable and press SHIFT − RET.

The result is the same as $r'(t)$ in Out[3] above:

In[5]:= $\dfrac{d}{d\,t}$ ({t² Cos[t], Sin[t] Cos[t], t² Sin[t]})

Out[5]= {2 t Cos[t] − t² Sin[t], Cos[t]² − Sin[t]², t² Cos[t] + 2 t Sin[t]}

Another way. If you click the **Derivative** button without typing and selecting the vector function first, the field for the independent variable will be selected.

$$\frac{d}{d\,\square}\,(\square)$$

In this case fill in the independent variable first, press TAB, and then fill in the expression to be differentiated.

Exercises

In these exercises, enter $\cos^2 t$ as `Cos[t]`^{2}.

1. a) Graph the curve $x = 1 + \cos t$, $y = \sin t$, $z = \cos^2 t$ for $0 \le t \le 2\pi$. On a printout of the graph, indicate the direction along the path that $r(t)$ follows as it moves for $0 \le t \le 2\pi$. (**Graph 3D ▷ Space Curve**)

    b) Find the velocity vector $r'(t)$ and draw it on the graph when $t = 0, \frac{\pi}{2}, \pi, \frac{3\pi}{2}, 2\pi$. Does $r'(t)$ appear to be tangent to the curve?

2. a) Section 10.8 discusses animating 3D graphs. Follow these steps to animate $x = 1 + \cos t$, $y = \sin t$, $z = \cos^2 t$ on $0 \le t \le 2\pi$. There will be 16 frames, at intervals of $\pi/8$ t-units.

    - Choose **Graph ▷ Animation** and then **Space Curve**. Click **OK**.
    - Let $x = 1 + \cos t$, $y = \sin t$, $z = \cos^2 t$.
    - Fill in $0 \le t \le \frac{\pi\_k}{8}$.

      The symbol _ means to leave a space to indicate multiplication.

    - Let $1 \le k \le 16$ and $0 \le x \le 2$, $-1 \le y \le 1$, $0 \le z \le 1$.

      This produces 16 frames. We choose the x-, y-, z-intervals to contain all the points on the curve.

    - In the **Space Curve** window, click **OK**.

    You can control the animation with the icons that appear at the bottom of the *Mathematica* window. Clicking anywhere in the notebook will stop the motion. If you stop the animation, you can restart it by double-clicking on the graphs.

    b) Based on the animation, approximately where along the path is the motion slowest and where is it fastest? You can see this best by slowing down the animation (click the control buttons or press a key from 1 (slowest) to 9 (fastest)).

    c) To verify your answer to b), calculate the speed, which is the length of the velocity vector $r'(t)$, and graph the speed for $0 \le t \le 2\pi$. Review your reasoning in b) and revise your answer if necessary. (**Graph ▷ Graph f(x)**)

Lab 22.1 Parametric Curves, Gravity, and Motion

Prerequisites:
Ch. 1	A Brief Tour of *Joy*
Secs. 3.5,9	Graphing parameterized curves and animation

The animation in this chapter is based on a famous physics experiment that has been implemented on a large scale by Robert Berg at Wellesley College. A tennis ball is launched from a small air cannon on the ground floor of the Science Center toward a falling target, e.g., a stuffed bunny that is dropped from a balcony. Can we aim the ball so that it hits the bunny?

■ Before the Lab (Do by Hand)

Let the balcony be h feet above the floor and the cannon be s feet away from the bunny's path. When we drop the stuffed bunny, it falls straight down.

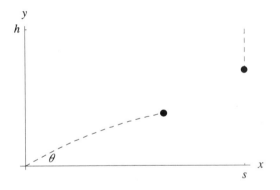

1. Find xbunny(t) and ybunny(t), the x- and y-coordinates of the bunny in terms of time t (assuming that air resistance is negligible).

We launch the ball at the exact moment when the bunny is dropped. Let θ be the angle at which we aim the cannon and v_0 ft/sec be the initial velocity of the ball, determined by how powerful the cannon is. The ball will travel along a parabolic arc given by the equations

$$\text{xball}(t) = (v_0 \cos \theta)\, t, \quad \text{yball}(t) = -16\, t^2 + (v_0 \sin \theta)\, t \,.$$

2. At what time t_s does the ball reach the vertical line along which the bunny falls? What are the y-coordinates of the ball and of the bunny for $t = t_s$?

3. Suppose the balcony is 30 feet high and the cannon can launch the ball with an initial velocity of 60 ft/sec. We place the cannon 50 feet away from the point where the ball will land.

   a) Find the functions xball(t), yball(t), xbunny(t), and ybunny(t) for these measurements. Some of these expressions will contain the angle θ, whose value is not specified.

b) Find the time t_s in terms of θ.

■ In the Lab

In the lab, you'll experiment with different values of the angle θ to see whether the ball reaches its target.

1. From the *Joy* menus:

    - Create the function `ybunny[t]`, using the value you found for it in Question 3a) (Before the Lab). (**Create ▷ Function**)
    - Repeat for the functions `xbunny[t]`, `xball[t]`, and `yball[t]`. Enter the symbol θ with the **Greek** button on the *Joy* palette.

      Mathematica may warn you of a possible spelling error because the names of the four functions are similar. You may ignore these messages if your spelling is correct.

      When you are done, close the Greek palette by clicking the close box in its upper-left corner (Macintosh) or the close button **X** in the upper-right corner (Windows).

2. In your *Mathematica* notebook:

    - If necessary, click below the last output so that a horizontal line appears.
    - Type t_s = and fill in the expression you found in Question 3b) (Before the Lab). You can enter t_s using the **Subscript** button on the Joy palette.
    - Press [SHIFT] − [RET] .

Try an Experimental Launch

Here's what happens if you launch the ball at an angle of $\theta = 0.75$ radians, which is about 43°.

3. In your *Mathematica* notebook:

    - If necessary, click below the last output so that a horizontal line appears.
    - Type θ = `0.75` and press [SHIFT] − [RET] .

4. Graph the parametric equations $x = \text{xball}(t)$, $y = \text{yball}(t)$ for $0 \le t \le 1.6$, assigning the name *ballpath* to the graph. (**Graph ▷ Parametric Plot**)

    This plots the path of the ball as it travels for 1.6 seconds.

    - Repeat for the path of the bunny, assigning the name *bunnypath*.
    - Combine the graphs *ballpath* and *bunnypath*. (**Graph ▷ Combine Graphs**)

Animating the Flight

Here's how to animate the experiment and show the bunny and ball in flight. There will be six frames, numbered $k = 0$ to $k = 5$, at $t = 0, 0.2\,t_s,\ 0.4\,t_s,\ \ldots,\ t_s$ seconds.

- From the *Joy* menus, choose **Graph ▷ Animation**.
- From the popup menu, choose **Points** and click **OK**.
- In the **Plot Points** dialog, fill in `{xbunny[0.2 ts ␣k], ybunny[0.2 ts ␣k]}, {xball[0.2 ts ␣k], yball[0.2 ts ␣k]}`.

  This plots the bunny and the ball at the same moment of time. The symbol ␣ means to leave a blank space to indicate multiplication.

- Click in the **Animate Info** dialog to make it active.
- Fill in `0 ≤ k ≤ 5`.
- Fill in the intervals for the common plot frame, `0 ≤ x ≤ 75, -15 ≤ y ≤ 30`.

  These intervals are large enough to show the flight paths of the ball and bunny.

- Click in the **Plot Points** dialog to make it active.
- Click **OK**.

Mathematica creates the six frames and collapses them, showing only the first, and sets the animation in motion.

You can control the animation with the icons that appear at the bottom of the *Mathematica* window. Clicking anywhere in the notebook will stop the motion. If you stop the animation, you can restart it by double-clicking on the graphs. You can change the speed with the icons or by pressing a number between 1 (slowest) and 9 (fastest).

5. What happened? Did the ball hit the bunny?

How Should You Aim the Cannon?

Depending on how you aim the cannon, the ball may get to $x = 50$ ft before or after the bunny passes the ball's flight path.

6. By hand and using the results you obtained before the lab, find the angle θ that makes the ball and the bunny meet each other on the vertical line.

7. In your *Mathematica* notebook:

   - If necessary, click below the last output so that a horizontal line appears.
   - Type `Θ =` and fill in the angle you found. Use the **Functions** button on the *Joy* palette to help enter the expression you found for the angle.
   - Press SHIFT – RET.

8. Repeat the animation by choosing **Recent** ▷ **Animation**. If you did not change the animation dialogs, the entries will be the same as before.

   Mathematica will automatically use the new value of θ and the resulting value of t_s.

   If the animation does not show that the ball meets the bunny, check your work and make any needed corrections.

9. Open the animation frames by double-clicking on the cell bracket that contains them. To save paper, you can cut out all but the last two frames before printing.

10. Print your work.

■ After the Lab

1. a) For $\theta = 0.75$, did the ball get to the bunny's path too soon or too late?

   b) Does this suggest that you should raise or lower the cannon? Why?

2. Is the angle you found in Question 6 (In the Lab) greater than or less than $\theta = 0.75$ radians?

3. What does the animation show that the parametric plot did not? Why can't the parametric plot show this, too?

■ For Further Exploration

Here are some other questions to consider. You can answer them by solving appropriate equations by hand, and then illustrate your answers by animating the flight paths of the ball and bunny.

1. Suppose that s is given but that you have a choice of cannons with different initial velocities. Show that if v_0 is large enough, you can aim the cannon (i.e., choose θ) so that the meeting height is arbitrarily close to h. That is, if you use a strong enough cannon and aim it correctly, then the ball will meet the bunny as close to the balcony as you want.

2. If the cannon is weak enough, can the bunny hit the floor before the ball gets to $x = s$ no matter what angle θ you use? How large must v_0 be to guarantee that the ball meets the bunny in time?

3. If the cannon is far enough away, can the ball hit the floor before it reaches the bunny's path? For a given velocity v_0, what is the maximum distance s at which you can place the cannon and still hit the bunny before it lands?

Lab 22.2 Two Parameterizations of an Ellipse

> Prerequisites:
>   Ch. 1   A Brief Tour of *Joy*
>   Sec. 3.3   How to Plot Points
>   Sec. 4.2   How to Tabulate an Expression

In this lab you can see how the parameterization of a curve describes motion along a path, by comparing two different parameterizations of the same ellipse.

■ Before the Lab (Do by Hand)

1. Show that the parametric equations $x = 2 \cos t$, $y = \sin t$ describe the ellipse $x^2 + 4y^2 = 4$.

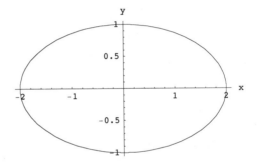

2. For which values of t will (x, y) be at the ends of the two axes of the ellipse?

The relations connecting polar coordinates (r, θ) and Cartesian coordinates (x, y) are $x = r \cos \theta$, $y = r \sin \theta$.

3. Show that the polar equation of the ellipse is

$$r = \frac{2}{\sqrt{1 + 3 \sin^2 \theta}}.$$

4. For which values of θ will (x, y) be at the ends of the two axes of the ellipse?

This sketch shows how the parameters t and θ determine (x, y) on the ellipse.

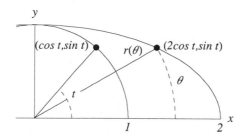

5. Do $t = \pi/4$ and $\theta = \pi/4$ produce the same point (x, y)?

■ In the Lab

1. Follow these steps to make a table of values for (x, y) as t moves from 0 to 2π in steps of $\pi/16$.

    *   From the *Joy* menus, choose **Create ▷ List**.
    *   Fill in 2 Cos[t], Sin[t].
    *   If necessary, set the popup menu to **x Equally Spaced**.
    *   Fill in $0 \leq t \leq 2\pi$ in increments of $\pi / 16$.

        Be sure to make the variable t, not x. Enter π by clicking the π button on the *Joy* palette.

    *   Check **Display in decimal form** and **Display as a table**.
    *   Check **Assign name** and fill in tlist.
    *   Click **OK**.

2. Create a polar function this way:

    *   Choose **Create ▷ Function**.
    *   Fill in r[θ] and $\dfrac{2}{\sqrt{1+3\,\text{Sin}[\theta]^2}}$. Click **OK**.

        Use the **Greek** button on the *Joy* palette to enter θ.

3. Make a table of values of $r(\theta)\cos\theta$, $r(\theta)\sin\theta$ for $0 \leq \theta \leq 2\pi$ in increments of $\pi/16$. Assign the name **polarlist**. (**Recent ▷ List**)

    Be sure to change the variable to θ. When you are done, close the Greek palette by clicking the close box in its upper-left corner (Macintosh) or the close button **X** in its upper-right corner (Windows).

4. Plot *tlist*. (**Graph ▷ Points**)

5. Plot *polarlist*.

6. Print your work. If you will be doing **For Further Exploration**, save your work so you can use it again.

■ After the Lab

1. Compare the coordinates in *tlist* and *polarlist*. Does $(r(\theta)\cos\theta,\ r(\theta)\sin\theta)$ seem to be moving ahead of $(2\cos t,\sin t)$ or behind it? As t and θ increase from 0 to 2π, which point appears to be in the lead?

2. According to the graph of *tlist*, does $(2\cos t,\sin t)$ appear to travel at constant speed? If not, where on the ellipse does it move more quickly and where more slowly?

3. Repeat for $(r(\theta)\cos\theta,\ r(\theta)\sin\theta)$ on the graph of *polarlist*.

■ For Further Exploration

1. By hand, find the velocity vector for the parameterization $(2\cos t,\sin t)$. Then find the length of this vector, which is the speed with which the point travels around the ellipse. Use *Joy* to graph the speed for $0 \le t \le 2\pi$. (**Graph ▷ Graph f(x)**)

2. Here's how to graph the speed with which $(r(\theta)\cos\theta,\ r(\theta)\sin\theta)$ moves around the ellipse.

   a) Create a function $s(\theta) = (r(\theta)\cos\theta,\ r(\theta)\sin\theta)$. (**Create ▷ Function**)

   Remember to use curly brackets { } since the value of $s(\theta)$ is a vector.

   b) In your *Mathematica* notebook, evaluate $s'(\theta)$. This is the velocity vector for the motion.

   c) In the notebook, type `speed` =, click the $\boxed{\sqrt{\blacksquare}}$ button on the *Joy* palette, and complete the equation. The right side should be the length of $s'(\theta)$. It's easier to copy and paste from the last output than to retype the coordinates of $s'(\theta)$. Be sure not to copy the curly brackets in the output. When you are done, press ⌗SHIFT⌗ − ⌗RET⌗.

   d) Graph *speed* for $0 \le \theta \le 2\pi$. Be sure to change the variable to θ. (**Graph ▷ Graph f(x)**)

3. Compare the graphs of the two speed functions with the plots of *tlist* and *polarlist* you made in the lab. Verify that the speeds behave in the way you predicted in the questions after the lab.

Lab 22.3 Numerical Integration and Halley's Comet

Prerequisites:

Ch. 1	A Brief Tour of *Joy*
Secs. 3.3,8	Plotting points, polar coordinates
Secs. 4.1,2	Simplifying and tabulating expressions
Secs. 5.1,2	Solving equations algebraically and numerically
Sec. 8.3	How to Integrate Numerically

What is the path of a comet? Does a comet travel with constant speed? In this lab, you can explore the answers to these questions. You'll see how to use *Joy* to plot the orbit of Halley's comet and solve an integral equation to determine where the comet is at different times.

■ Before the Lab (Do by Hand)

Kepler's First Law of Planetary Motion says that planets and those comets that return to the solar system after fixed periods of time travel in elliptical orbits with the sun at one focus. Halley's comet is perhaps the most famous returning comet.

The standard equation in Cartesian coordinates for an ellipse centered at the origin and with axes of length a and b is $\frac{x^2}{a^2} + \frac{y^2}{b^2} = 1$. We can think of the sun as being at the focus $(c, 0)$, where $c = \sqrt{a^2 - b^2}$. Let r and θ be the polar coordinates along the ellipse, with the origin moved to the sun (these are called *heliocentric* polar coordinates). The equations relating x, y to r, θ are

$$x = c + r\cos\theta, \quad y = r\sin\theta. \tag{*}$$

Here x, y are Cartesian coordinates relative to the center of the ellipse, but r, θ are polar coordinates relative to the sun.

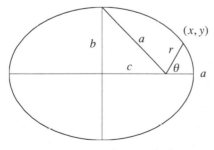

$$x = c + r\cos\theta, \, y = r\sin\theta$$

For Halley's comet, the dimensions of the orbit are $a = 2.686$, $b = 0.681$ billions of kilometers. The earth's orbit is much smaller, since our average distance from the sun is about 0.15 billion (150 million) kilometers.

1. The *eccentricity* of an ellipse is the number $e = c/a$, which is always between 0 and 1. Find e for the orbit of Halley's comet. Does the value of e imply that the orbit is almost a circle or very elongated?

To know how the comet travels around its orbit, we need Kepler's Second Law: as it moves around the sun, the comet sweeps out equal areas within the ellipse during equal periods of time.

The area of an ellipse is πab. Halley's comet travels once around its orbit approximately every $T = 76.1$ years. By Kepler's law, the area swept out during a given period of time is the same regardless of the location of the comet along the ellipse. This sketch illustrates the motion of the comet from its *perihelion* point P, where it is closest to the sun, to the point Q where it is one year later.

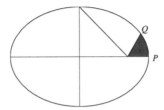

perihelion and one year later (not to scale).

Let area(θ) be the area swept out from P through an angle θ, and let dist(θ) be the distance traveled by the comet from P through that angle, i.e., the length of the arc.

2. Suppose the comet travels through an angle θ in t years. Explain why

$$\text{area}(\theta) = \pi ab \, \frac{t}{T} \, .$$

3. If necessary, review these concepts about area and arc length in a calculus text. Here $r = r(\theta)$ is the equation of the ellipse in heliocentric polar coordinates.

    a) The area from P (the perihelion point, on the x-axis) to the point with polar angle θ is

    $$\text{area}(\theta) = \int_0^\theta \frac{r^2(u)}{2} \, du \, .$$

    We use the dummy variable u to distinguish it from the particular value of θ that determines the area being calculated.

    b) The length of the arc from P to the point with polar angle θ is $\text{dist}(\theta) = \int_0^\theta \sqrt{x'(u)^2 + y'(u)^2} \, du$. Here $x = r(u) \cos u$, $y = r(u) \sin u$ are Cartesian coordinates relative to the sun, not to the center of the ellipse.

4. Show that

$$\text{dist}(\theta) = \int_0^\theta \sqrt{r(u)^2 + r'(u)^2} \; du.$$

■ In the Lab

1. Follow these steps to find the polar equation of an ellipse with the origin at one focus (the sun).

    *   Click in your *Mathematica* notebook beneath the last output, if any, to make a horizontal line appear.
    *   Type c = $\sqrt{a^2 - b^2}$ and press SHIFT – RET.
    *   From the *Joy* menus, choose **Algebra ▷ Solve**.
    *   Fill in

        $$\frac{(c + r\_Cos[\theta])^2}{a^2} + \frac{(r\_Sin[\theta])^2}{b^2} = 1$$

        The symbol __ means to leave a space to indicate multiplication. Use the **Greek** palette to enter θ, or use the keyboard shortcut ESC – q – ESC (don't type the dashes –). In this lab you'll need to type θ often, so the keyboard shortcut will be useful.

        If you use the Greek palette, close it by clicking the close box in the palette's upper-left corner (Macintosh) or the close button **X** in the upper-right (Windows).

    *   Fill in the variable r.
    *   Check **Solve algebraically** and click **OK**.

    It will help to simplify this result.

    *   Choose **Algebra ▷ Simplify**.
    *   Set the popup menu to **Last Output** and click the **Simplify** button.
    *   Click **OK**.

2. There are two solutions, but only one has $r > 0$. This is the solution we want:

    $$r(\theta) = \frac{2b^2 \left(a - \sqrt{a^2 - b^2} \cos\theta\right)}{a^2 + b^2 + (-a^2 + b^2)\cos(2\theta)}$$

    Create r as a function of θ. You can copy the above expression from the solution and paste it into the **Create ▷ Function** dialog.

3. In your *Mathematica* notebook, enter the values of a, b, T. That is:

    *   If necessary, click below the last output so that a horizontal line appears.
    *   Type a = 2.686 and press SHIFT – RET.
    *   Repeat for b = 0.681 and T = 76.1.

4. Graph $r = r(\theta)$ in polar coordinates, $0 \le \theta \le 2\pi$. This is the comet's orbit. (**Graph $\triangleright$ Polar Plot**)

To visualize the motion of the comet we need to express its position as a function of time, using Kepler's laws.

5. Here you'll create a function to represent the area swept out by the comet through an angle θ, beginning at perihelion. This is the function in Question 3a) (Before the Lab).

    *   Choose **Create $\triangleright$ Function**.
    *   In the first input field, fill in `area[`θ`]`.
    *   In the second field, fill in $\frac{r[u]^2}{2}$, select it, and click the **Numeric Integral** button on the *Joy* palette.

        Since the function is quite complicated, it's best to treat this as a numerical integral.

    *   Fill in the lower limit 0 and press $\boxed{\text{TAB}}$.
    *   Fill in the upper limit θ, $\boxed{\text{TAB}}$, and the dummy variable u.
    *   Click **OK**.

6. Similarly, create a function $\text{dist}(\theta) = \int_0^\theta \sqrt{r(u)^2 + r'(u)^2} \; du$, using a numeric integral. This function has no elementary antiderivative and must be integrated numerically.

7. By evaluating the functions area(θ) and dist(θ), find the area swept out by the comet and the distance it travels in one complete orbit.

    *   Click in your notebook below the last cells, so that a horizontal line appears.
    *   Type `area[...]`, filling in the appropriate value of θ, and press $\boxed{\text{SHIFT}}$ $-$ $\boxed{\text{RET}}$.
    *   Repeat for the distance.

8. Here's how to find the angle through which the comet travels during its first year after perihelion.

    *   Choose **Algebra $\triangleright$ Solve**.
    *   Fill in the equation `area[`θ`]` = ..., using the area that the comet must sweep out during the first year. You can find this area in Question 2 (Before the Lab).

        Remember to leave a space ⎵ to indicate multiplication.

    *   Fill in θ as the variable for which to solve.
    *   Click **Solve numerically**.
    *   For an initial estimate, fill in the angle through which the comet would travel if it passed through equal angles each year.
    *   Click **OK**.

        Mathematica will display a message about the limits of integration, because we defined a numerical integral using the symbol θ without giving it a value. You can safely ignore this message.

9. Use the angle you just found to calculate the distance the comet travels during the first year after perihelion.

The next few steps lead to plotting the position of Halley's comet at equally spaced *times*. This will enable you to picture how it moves along its orbit and how its speed behaves as it travels around the sun.

10. Here's how to find the angle of the comet at an arbitrary time in its orbit and then make a table of angles for various times.

- Choose **Recent ▷ Solve**.
- Fill in `area`$[\theta]$ = π `a` `b` $\frac{t}{T}$.

    This is the equation in Question 2 (Before the Lab).

- Change the initial condition to $\theta = \pi$.

    This estimate lets us solve for θ for values of *t* across the whole range of possible times.

- Check **Assign name** and fill in `angle`.
- Click **OK**.

    Mathematica will display an unevaluated command and various error messages because we defined a numerical integral using the symbol θ without giving it a value. You can safely ignore the messages.

- Choose **Create ▷ List**.
- Fill in `angle`.
- Make sure the popup menu is set to **x Equally Spaced**.
- Fill in $0 \le t \le T$ in increments of `T / 32`.

    This produces a table of angles, dividing the orbit into 32 equal time periods.

- Make sure **Display in decimal form** and **Display as a table** are checked.
- Check **Assign name** and fill in `angleData`.
- Click **OK**.

    Again, *Mathematica* will display messages that you can ignore.

11. Finally, here's how to plot the comet's travels.

- Choose **Algebra ▷ Substitute**.
- Fill in $\{$`r`$[\theta]$ `Cos`$[\theta]$`,` `r`$[\theta]$ `Sin`$[\theta]\}$.

    These are the *x*-, *y*-coordinates for the orbit with the sun at the origin.

- Click **make the substitutions** and fill in `angleData`.
- Check **Assign name** and fill in `xycoords`.
- Click **OK**.

- Choose **Graph ▷ Points**.
- Fill in `xycoords` and click **OK**.

  This shows the location of the comet at the 32 equally spaced times around its orbit, beginning at perihelion.

- Combine the plot of the entire orbit with the plot of the 32 points. (**Graph ▷ Combine Graphs**).

  If necessary, choose **Graph ▷ Modify Graph** and **Equal scales** so that the orbit has the same appearance as in Question 4 (In the Lab).

12. Print your work.

■ After the Lab

1. What are the units in your answers to Question 7 (In the Lab)? Is the area equal to πab?

2. Does the comet sweep out equal areas as it moves through equal *angles*? Explain your answer. (Hint: where is the sun in this coordinate system?)

3. Use your results in Questions 8 and 9 (In the Lab) to answer:

   a) Does the comet sweep out the same *angle* each year?

   b) Does the comet travel equal distances each year?

4. The table `angleData` in Question 10 shows how the angle θ changes with time.

   a) For which values of t does the table show the values of θ?

   b) What should θ be when $t = \frac{T}{2}$? Is that the value shown in the table? What might be a physical explanation for this?

5. According to the graph, where is the comet traveling quickly and where is it traveling slowly?

Chapter 23

Surfaces and Level Sets

Examples and Exercises

Prerequisites:
    Ch. 1   A Brief Tour of *Joy*
    Secs. 10.1–5   Graphing surfaces and level sets

■ 23.1 Surfaces

23.1.1 Example: Visualizing Vertical Cross-Sections

When you plot a surface such as

$$z = 10 - x^2 - \frac{y^2}{4},$$

the graph shows the cross-sections made by intersecting the surface with vertical planes where x or y is constant. (Example 23.2.1 shows how to visualize horizontal cross-sections.)

- Graph $z = 10 - x^2 - \frac{y^2}{4}$ over the intervals $-3.5 \leq x \leq 3.5$ and $-3.5 \leq y \leq 3.5$. **Assign the name** surface. (**Graph 3D ▷ Surface**)

Mathematica graphs the surface:

▷ Plot $10 - x^2 - \frac{y^2}{4}$ for $-3.5 \leq x \leq 3.5\char"0060$ and $-3.5 \leq y \leq 3.5\char"0060$

surface =

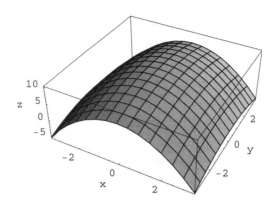

Out[2]= - SurfaceGraphics -

The cross-hatching on the surface indicates curves where x and y are constant. For example, one such curve is where $x = 2$ on the surface. We can visualize this curve by plotting $x = 2$ and combining the two graphs.

- If you are using *Mathematica* Version 4 with interactive 3D graphics, choose **Graph ▷ Modify Graph** and specify **Axis Labels "x","y","z"**.

    This displays scales along the coordinate directions and labels them. It also makes the graph noninteractive.

In R^3 the equation $x = 2$ does not have the form $z = g(x, y)$. Instead, it describes a plane that is a level surface for the function $g(x, y, z) = x$, i.e., a surface where this function is constant. We'll use that idea to graph the plane.

- From the *Joy* menus, choose **Graph 3D ▷ Level Surface**.
- Fill in **x** for the function and **2** for its constant value.
- Fill in the intervals $-3.5 \le x \le 3.5$, $-3.5 \le y \le 3.5$, and $-5 \le z \le 10$.

    We choose the intervals for x, y, z to match the values shown on the surface.

- Click **Assign name** and fill in `plane`. Click **OK**.

Mathematica graphs the plane:

```
▷ Graph of the level surface x = 2 for
  -3.5 ≤ x ≤ 3.5`,  -3.5 ≤ y ≤ 3.5`, -5 ≤ z ≤ 10.

plane -
```

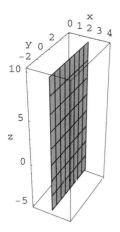

Out[4]=   - Graphics3D -

- Choose **Graph ▷ Combine Graphs**.
- Fill in `surface,plane`, or use output numbers if you didn't assign names.
- Click **OK**.

The output shows the plane $x = 2$ intersecting the surface.

▷ Combining the graphs {surface, plane} gives

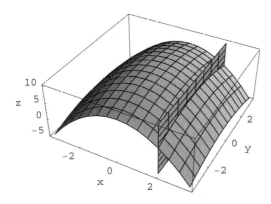

Out[5]=   - Graphics3D -

The plane $x = 2$ is parallel to the yz-plane. On the intersection of this plane with the surface $z = 10 - x^2 - \frac{y^2}{4}$, we have

$$z = 6 - \frac{y^2}{4}, \quad x = 2,$$

which describes a parabola in y and z facing downward. Similarly, all the vertical cross-sections

$$z = (10 - c^2) - \frac{y^2}{4}, \quad x = c$$

$$z = \left(10 - \frac{c^2}{4}\right) - x^2, \quad y = c$$

are downward-facing parabolas.

> ♡ *Mathematica*'s default setting divides the x- and y-intervals into 14 equal subintervals. Since x, y vary between ± 3.5, the cross-hatched curves are exactly 0.5 units apart and so $x = 2$ falls exactly on one of these curves.

Exercises

If you are using *Mathematica* Version 4 with interactive 3D graphics, you may want to choose **Graph ▷ Modify Graph** and specify **Axis Labels** in some of these exercises. That will help you estimate the coordinates of points on the surface.

1. a) Print a graph of $z = 10 - x^2 - 2y^2$ for x and y between ± 3.5. By hand, sketch in the curve where the surface intersects the plane $y = c$, for several values of c.

    b) Find the equation of the intersection of $z = 10 - x^2 - 2y^2$ and $y = c$, where c is an arbitrary constant. What kind of curve is the intersection, and how does it change as c changes?

2. a) Graph the surface $z = \sin x$ for x and y between $\pm \pi$. What is the effect on the graph of not having y in the equation?

    b) Describe the intersection of the surface with the plane $x = 3\pi/4$ and find its equation.

    c) Graph the plane $x = 3\pi/4$ and combine the two graphs. Does the intersection correspond to your description in b)? If not, check your reasoning. **(Graph 3D ▷ Level Surface)**

In Exercises 3 and 4:

    a) By hand, find the equation of the cross-section of the given surface with $x = c$. Then sketch the cross-section in three dimensions for enough values of c so that you see how it changes as c varies.

    b) Repeat for $y = c$.

    c) Use *Joy* to graph the surface and compare your sketches with the cross-sections shown in the graph. Check your reasoning if there are any inconsistencies.

3. $z = \sin x + \sin y$, with x and y between $\pm \pi$.

4. $z = y e^{-x}$ with x and y between ± 7. In c), *Mathematica* may cut off the surface where $|z|$ is very large. Use **Graph ▷ Modify Graph** to redraw the surface, choosing **Show All** in the **Range** popup menu.

In Exercises 5–7:

a) Predict what kind of symmetry the surface will have. Look for symmetry about the yz- and xz-planes, and about the plane $x = y$.

b) Graph the surface for x and y between ± 5 and test your prediction. You may wish to change your viewpoint of the surface to see the symmetry more clearly. In *Mathematica* Version 4 with interactive graphics, just drag the surface to change the viewpoint. In other cases, choose **Graph 3D ▷ 3D ViewPoint**.

5. $z = 10 x^2 - 4 y$.

6. $z = 10 x - 4 y^2$.

7. $z = 10 x^2 - 4 y^2$.

Exercises 8 and 9 are about the surface

$$z = \frac{x + y}{x^2 + y^2}.$$

8. a) Graph the surface for x and y between ± 1.

b) Change the viewpoint so you can look at the surface from underneath.

c) From the graphs, predict the limit of z as $(x, y) \to (0, 0)$ along the x- or y-axis from the left or right. Answer for all four possibilities.

d) Calculate the limits in c) from the equation of the surface.

9. a) Graph the surface for x and y between ± 2 and rotate it so that you can see more clearly where the function is undefined. *Mathematica* will display error messages saying that it cannot divide by 0.

b) *Mathematica* shows some points on the surface where $x^2 - y^2 = 0$, even though the function is undefined there. At which points where $x^2 - y^2 = 0$ can you define z so that the function is continuous there, and what should the value of z be? At which points can't you define z this way?

■ 23.2 Level Curves

23.2.1 Example: Visualizing Horizontal Cross-Sections

The horizontal cross-sections of a surface such as $z = 10 - x^2 - 2y^2$ are made by intersecting the surface with horizontal planes $z = c$, where c represents a constant. Example 23.1.1 showed how to graph vertical cross-sections.

For example, here is how to visualize the cross-section at $z = 1$.

- Graph $z = 10 - x^2 - 2y^2$ over the intervals $-3.5 \le x \le 3.5$ and $-3.5 \le y \le 3.5$, filling in the name `surface`. (**Graph 3D ▷ Surface, Assign name**)

Mathematica graphs the surface:

▷ Plot 10 - x² - 2 y² for -3.5 ≤ x ≤ 3.5` and -3.5 ≤ y ≤ 3.5`

surface =

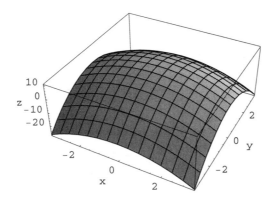

Out[2]= - SurfaceGraphics -

- If you are using *Mathematica* Version 4 with interactive 3D graphics, choose **Graph ▷ Modify Graph** and specify **Axis Labels "x", "y", "z"**.

  This displays scales along the coordinate directions and labels them. It also makes the graph noninteractive.

Now we graph the horizontal plane $z = 1$.

- Click **Last Dialog** to bring back the **Graph 3D ▷ Surface** dialog.
- Graph $z = 1$ over the same intervals, changing the name to `plane`.

Then we combine the two graphs.

- Combine *surface* and *plane*. (**Graph ▷ Combine Graphs**)

Mathematica combines the graphs:

▷ Combining the graphs {surface, plane} gives

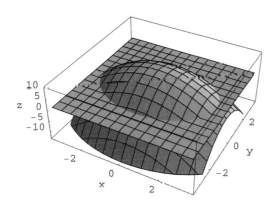

Out[4]= ▪ Graphics3D ▪

The equation of the cross-section at $z = 1$ is $1 = 10 - x^2 - 2y^2$, or $x^2 + 2y^2 = 9$, so the curve is an ellipse at height 1, centered at the origin with its major axis in the x-direction.

In general, the cross-section $z = c$ is an ellipse $x^2 + 2y^2 = 10 - c$, when $c < 10$. When $c = 10$ the cross-section is the point $(0, 0, 10)$ on the z-axis. When $c > 10$ the plane $z = c$ lies above the surface and the cross-section is the empty set.

The level curves, or contours, of the surface are the projections of the horizontal cross-sections onto the xy-plane. Here is what they look like for this surface.

- From the *Joy* menus, choose **Graph 3D ▷ Level Curves**.
- Fill in $10 - x^2 - 2y^2$ and the intervals $-3.5 \le x \le 3.5, -3.5 \le y \le 3.5$.
- Check **Contour Shading** and click **OK**.

The output shows a family of ellipses centered at the origin.

▷ Plot the level curves $10 - x^2 - 2y^2 = c$ for $-3.5 \le x \le 3.5$ ` and $-3.5 \le y \le 3.5$ `

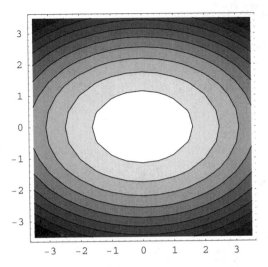

Out[5]= - ContourGraphics -

The shading is lighter where z is larger, so the ellipses closer to the origin correspond to cross-sections that are higher on the surface. They show curves $z = c$ for 10 (the default value) equally spaced values of c. The ellipses are closer together where the surface is steeper and farther apart where it is flatter, near the top of the surface.

For example, here's how to find the value of c corresponding to one of the ellipses by using *Mathematica* to read coordinates from the graph. Section 3.6 illustrates this in more detail.

- Click on the level curves to select the graph.
- Hold down the ⌘ (Macintosh) or CTRL (Windows) key and place the cursor on a point on some ellipse.
- Read the coordinates of the cursor in the lower-left corner of your notebook window.

For example, the next-to-smallest ellipse passes through (1.21, 1.55) (approximately). Now we'll substitute these coordinates into $z = 10 - x^2 - 2 y^2$.

- Click below the last output to make a horizontal line appear.
- Type $10 - 1.21^2 - 2 (1.55^2)$ and press SHIFT − RET.

In[6]:= $10 - 1.21^2 - 2 (1.55^2)$

Out[6]= 3.7309

So for this ellipse, $c \approx 3.7$.

Exercises

1. a) Graph the surface $z = x^2 - y^2$ for x and y between ±5.

b) Graph the plane $z = 5$ for x and y between ±5 and combine the two graphs. What are the equations and shape of the intersection curve?

c) Repeat for $z = -5$. You may need to change the viewpoint to see the intersection better, either by dragging (*Mathematica* Version 4, interactive mode) or by choosing **Graph 3D ▷ 3D ViewPoint**. How does the intersection differ from b)?

In Exercises 2–4:

a) Graph the level curves of the surface for x and y between ±2, with contour shading. What is the shape of the typical curve and what is its equation?

b) Approximately where are the highest and lowest points of the surface located, for x and y between ±2?

c) For which values of c are there no points on the level curve? For which values is there only one point?

2. $z = 4 - x^2 - y^2$.

3. $z = 4 - x^2 - 2y^2$.

4. $z = 4 - x^2 + 2y^2$.

5. a) Plot the level curves of
$$z = \frac{x + y}{x^2 + y^2}$$

for x and y between ±1. There will be some distortion near the origin.

b) Write the equation for the typical level curve $z = c$. Rewrite the equation in a way that lets you describe the shape of the curve (it will help to complete the square). Then compare your results with a) and review your reasoning if there are any inconsistencies.

c) Use the level curves to decide if this function has a limit as $(x, y) \to (0, 0)$.

■ 23.3 Level Surfaces

23.1 Example: Seeing Inside a Level Surface

Some level surfaces (contours) are easier to visualize if you "cut" through them by changing the domain over which you plot the surface. For example,

$$\frac{x^2}{4} + \frac{y^2}{2} + z^2 = w$$

is the equation of a function of x, y, z. We would like to see how the value of w affects the surface, e.g., in going from $w = 1$ to $w = 2$. It turns out that these are concentric ellipsoids, one inside the other. If you graph them simultaneously, you won't be able to see the inner ellipsoid. Here's how to see them both. We'll choose x-, y-, z-intervals that are large enough to show both surfaces.

- From the *Joy* menus, choose **Graph 3D ▷ Level Surface**.
- Fill in the expression $\frac{x^2}{4} + \frac{y^2}{2} + z^2$.
- Fill in the value 1.
- Fill in the intervals $-4 \le x \le 4$, $0 \le y \le 4$, $-4 \le z \le 4$.

Starting y at 0 instead of -4 lets you look inside the ellipsoids.

- Click **OK**.

Mathematica graphs the surface:

▷ Graph of the level surface $\frac{x^2}{4} + \frac{y^2}{2} + z^2 = 1$ for

$-4 \le x \le 4$, $0 \le y \le 4$, $-4 \le z \le 4$.

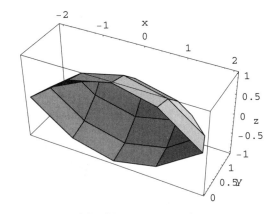

Out[2]= - Graphics3D -

Now we graph the function at level $w = 2$.

- Click **Last Dialog** on the *Joy* palette to reopen the **Level Surface** dialog.
- Change the value of the function to 2.
- Click **OK**.

Then we combine the graphs.

- Choose **Graph ▷ Combine Graphs**.
- Fill in Out[2],Out[3] or use your output names or numbers if they are different from these.
- Click **OK**.

Mathematica combines the two graphs (% is *Mathematica*'s abbreviation for `Out`):

▷ `Combining the graphs {%2, %3} gives`

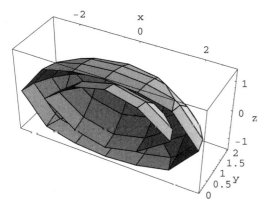

Out[4]= ‑ `Graphics3D` ‑

Exercises

1. a) Graph the level surfaces $x^2 + y^2 - z^2 = c$ for $c = -1$, 0, 1, with x, y, z between ±4. Don't combine the graphs.

    b) Find the equations of horizontal cross-sections of the three surfaces and use them to explain how the graphs differ.

2. a) Graph the level surface $x^2 + y^2 = 4$ with x, y, z between ±4.

    b) Why isn't the graph a circle?

3. a) By hand, predict the shape of the level surface $x - y^2 = 1$.

    b) Graph the surface for x, y, z between ±4 and check your answer to a). The graph will be smoother if you set the **Number of points in each direction** to 5.

    c) Repeat a) and b) for $x - z^2 = 1$.

Lab 23.1 Animating Cross-Sections and Level Curves

Prerequisites:
    Ch. 1   A Brief Tour of *Joy*
    Sec. 3.9   How to Animate Graphs
    Secs. 10.1,3 Graphing surfaces and level curves

This lab will help you visualize how the level curves (contours) of a function of two variables are related to the horizontal cross-sections of its graph.

■ Before the Lab (Do by Hand)

1. Sketch the level curve of $z = -2x^2 + y^2$ at level $z = 1$. (Hint: it's a hyperbola.)

2. Repeat for $z = -1$.

3. Predict the shape of the level curves of $z = -2x^2 + y^2$ for all contour levels.

■ In the Lab

1. Graph the surface $z = -2x^2 + y^2$ for x and y between ±2. (**Graph 3D ▷ Surface**)

2. Graph the level curves for this surface, for x and y between ±2. Include **Contour Shading**. (**Graph 3D ▷ Level Curves**)

Next, you'll animate horizontal planes $z = k$ intersecting the surface to produce cross-sections, with k changing from -5 to 5.

3. From the *Joy* menus:

- Choose **Graph ▷ Animation**.
- From the popup menu, choose **Surface** and click **OK**.
- In the **Graph Surface** dialog, fill in k.

    This is the horizontal plane $z = k$.

- Fill in the intervals $-2 \leq \mathbf{x} \leq 2$, $-2 \leq \mathbf{y} \leq 2$.
- In the **Animation Info** dialog, fill in $-5 \leq \mathbf{k} \leq 5$.

    This produces 11 animation frames, one for each integer k.

- For the common plot frame, fill in $-2 \leq \mathbf{x} \leq 2$, $-2 \leq \mathbf{y} \leq 2$, $-5 \leq \mathbf{z} \leq 5$.

    Since $z = k$, we know z will range between ±5.

- Click **Combine with** and fill in Out[2], or your output number for the surface if it is different from this.
- In the **Graph Surface** dialog, click **OK**.

You can control the animation with the icons that appear at the bottom of the *Mathematica* window. Clicking anywhere in the notebook will stop the motion and double-clicking on the graphs will restart it. You can change the speed with the icons or by pressing a number between 1 (slowest) and 9 (fastest).

4. Display all the animation frames by double-clicking on the cell bracket that holds them.

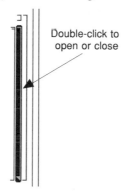
Double-click to
open or close

5. Before you finish this part of the lab, review the animation frames and make sure you can answer Question 1 (After the Lab).

6. Print your work, including the animation frames at least for $k = -1, 0, 1$. If you wish, you may cut out the other frames before printing

■ After the Lab

1. Write a paragraph describing the horizontal cross-sections in the animation and relating them to the plot of the level curves.

2. On your printout of the level curves, sketch the curves at levels $z = -1, 1$.

3. a) Find an equation for the level curve of $z = -2x^2 + y^2$ at level $z = 0$ and describe its shape.

   b) Add a sketch of this level curve to the printout.

Lab 23.2 Level Curves of Periodic Functions

Prerequisites:
        Ch. 1   A Brief Tour of *Joy*
        Secs. 10.1–3   Graphing surfaces and level curves

This lab will help you understand the behavior of the function $z = \sin x + \sin y$ using its graph, which is a surface in space, and its level curves.

■ Before the Lab (Do by Hand)

1. What are the maximum and minimum values of $z = \sin x + \sin y$ (you don't need to differentiate)?

2. At which points, expressed in terms of π, is $\sin x + \sin y$ a maximum? a minimum?

3. Find at least six points with $-\pi \leq x \leq \pi$ where $\sin x + \sin y = 0$.

■ In the Lab

1. Graph the surface $z = \sin x + \sin y$ for x and y between $\pm 3\pi$. To make it smoother, set the **Number of points in each direction** to 30. (**Graph 3D ▷ Surface**)

2. a) Graph the plane $z = 0$ for x and y between $\pm 3\pi$.

    b) Combine the graphs of the surface and plane. (**Graph ▷ Combine Graphs**)

    c) Rotate the combined graphs so you are looking down at them from above. (Drag the graph in *Mathematica* Version 4's interactive mode or choose **Graph 3D ▷ 3D ViewPoint**.)

    d) If you are using *Mathematica* Version 4 with interactive 3D graphics, choose **Graph ▷ Modify Graph** and specify **Axis Labels**. This displays scales along the coordinate directions and labels them.

3. Graph the level curves of $\sin x + \sin y$ for x and y between $\pm 3\pi$. Check **Contour Shading**. To make the plot smoother, set the **Number of points in each direction** to 60. (**Graph 3D ▷ Level Curves**)

4. Print your work.

■ After the Lab

1. a) On the graph of level curves, mark the points where $\sin x + \sin y$ has a maximum. Check the coordinates against the pattern you found for all maximum points in Question 2 (Before the Lab) and review your reasoning if there are inconsistencies.

    b) Repeat for the minimum points.

2. On the combined graph viewed from above, mark all the points where $\sin x + \sin y = 0$. Check that the points you found in Question 3 (Before the Lab) are among these and review your reasoning if they are not.

3. Find equations describing all the points where $\sin x + \sin y = 0$. You will need to consider two cases.

Lab 23.3 Continuity in Three Dimensions

Prerequisites:
      Ch. 1   A Brief Tour of *Joy*
      Secs. 10.1,4   Graphing surfaces, level surfaces

In this lab, you can explore the ideas of limit and continuity for functions of two variables.

■ Before the Lab (Do by Hand)

These questions concern the function

$$z = \frac{x^2 y}{x^4 + y^2},$$

which is undefined at the origin.

1. a) Find $\lim \frac{x^2 y}{x^4 + y^2}$ as $(x, y) \to (0, 0)$ along the x-axis.

    b) Repeat for the y-axis.

2. a) Repeat Question 1 as $(x, y) \to (0, 0)$ along the line $y = x$.

    b) Repeat for the line $y = -x$.

■ In the Lab

    ⚠ You will create several 3D graphics in this lab, which requires a lot of your computer's memory. Be sure to save your work as you go along. If you think you may run short of memory, see Section 2.5.5 for advice on how to manage memory.

1. Graph the surface $z = \frac{x^2 y}{x^4 + y^2}$ for x and y between ±1. (**Graph 3D ▷ Surface**)

    You can reduce the distortion near the origin by setting the **Number of points in each direction** to 30.

2. Graph the plane $y = x$ as a level surface $y - x = 0$ for x, y, z between ±1. (**Graph 3D ▷ Level Surface**)

3. a) Repeat for the plane $y = -x$, and then combine the graphs of the two planes and the original surface (**Graph ▷ Combine Graphs**)

    b) Use the combined graphs to illustrate the two limits you found in Question 2 (Before the Lab). You may need to rotate the graph in order to help do this. Review your reasoning if there are any inconsistencies. (Rotate the graph by dragging (*Mathematica* Version 4, interactive mode) or by choosing **Graph 3D ▷ 3D ViewPoint**.)

4. Graph the parabolic cylinder $y = x^2$ as a level surface, letting x, y, z vary between ± 1. (**Graph 3D ▷ Level Surface**)

5. Combine the graphs of $z = \frac{x^2 y}{x^4 + y^2}$ and $y = x^2$. (**Graph ▷ Combine Graphs**)

6. If you are using *Mathematica* Version 4 with interactive 3D graphics, choose **Graph ▷ Modify Graph** and specify **Axis Labels** on the graphs in Questions 3 and 5. This will help you read the coordinates on the graphs.

7. Print your work.

■ After the Lab

1. a) Calculate lim $\frac{x^2 y}{x^4 + y^2}$ as $(x, y) \to (0, 0)$ along the parabola $y = x^2$.

   b) Show how your answer can be seen in your graph for Question 5 (In the Lab).

2. a) Calculate lim $\frac{x^2 y}{x^4 + y^2}$ as $(x, y) \to (0, 0)$ along $y = mx$, where m is a given nonzero constant.

   b) Repeat for $y = mx^2$.

   c) Illustrate your answers to a) and b) on copies of your graph for Question 1 (In the Lab). In each case choose at least three values of m, including positive and negative values.

3. Does

$$\lim_{(x,y)\to(0,0)} \frac{x^2 y}{x^4 + y^2}$$

exist? Explain your answer.

4. Is it possible to choose a number a so that the function

$$f(x, y) = \begin{cases} \frac{x^2 y}{x^4 + y^2}, & (x, y) \neq (0, 0) \\ a, & (x, y) = (0, 0) \end{cases}$$

is continuous at the origin? Explain your answer.

Chapter 24
Derivatives in Several Variables

Examples and Exercises

Prerequisites:	
Ch. 1	A Brief Tour of *Joy*
Secs. 10.1–5	Graphing surfaces and level sets
Ch. 12	Differentiating Functions of Several Variables

■ 24.1 Visualizing Partial Derivatives

24.1.1 Example: Partial Derivatives as Slopes

Here is how to visualize partial derivatives as slopes. This example illustrates $\frac{\partial z}{\partial x}$ along the surface $z = \sin(x^2 + y)$ at the points where $y = -1$. We have

$$\frac{\partial z}{\partial x} = 2x\cos(x^2 + y) = 2x\cos(x^2 - 1).$$

This derivative measures the rate of change of z in the x-direction at each point along the curve where the surface $z = \sin(x^2 + y)$ meets the plane $y = -1$. Here is what the curve of intersection looks like.

- Graph $\sin(x^2 + y)$ for x and y between ± 2. (**Graph 3D ▷ Surface**)

Mathematica graphs the surface:

```
▷ Plot Sin[x² + y]  for  -2 < x ≤ 2   and  -2 ≤ y ≤ 2
```

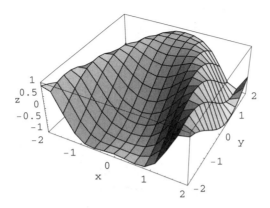

Out[2]= - SurfaceGraphics -

Now we graph the plane $y = -1$ and combine the two graphs.

- Graph $y = -1$ for x, y, z between ±2. **(Graph 3D ▷ Level Surface)**
- Combine `Out[2]`,`Out[3]`, using your output names or numbers if they are different. **(Graph ▷ Combine Graphs)**

The paraphrase shows the abbreviation %2 for `Out[2]`.

▷ Combining the graphs {%2, %3} gives

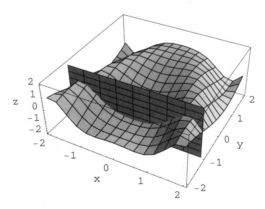

Out[4]= - Graphics3D -

The partial derivative $\frac{\partial z}{\partial x}$ can be seen as the slope of the tangent along this curve. The tangent is horizontal where $\frac{\partial z}{\partial x} = 2x\cos(x^2 - 1) = 0$. For the points shown in the graph, this happens at $x = 0$ and where $x^2 - 1 = \pi/2$, i.e., $x = \pm\sqrt{1 + \frac{\pi}{2}} \approx \pm 1.6$.

- If you are using *Mathematica* Version 4 with interactive 3D graphics, display the coordinates by choosing **Graph** ▷ **Modify Graph** and specifying **Axis Labels**.

The second partial $\frac{\partial^2 z}{\partial x^2}$ is reflected in the concavity of this curve. For example, this derivative should be positive at $x = 0$ because the curve is concave up there. You can check that $\frac{\partial^2 z}{\partial x^2} = 2 \cos(x^2 + y) - 4 x^2 \sin(x^2 + y)$ and that $\frac{\partial^2 z}{\partial x^2} > 0$ at $(x, y) = (0, -1)$.

Similarly, $\frac{\partial z}{\partial y}$ is the slope along the intersection of the surface with a plane perpendicular to this one, where x is constant and y varies. Also, the sign of $\frac{\partial^2 z}{\partial y^2}$ corresponds to the concavity of this intersection curve.

Another way. You can also visualize the partial derivatives without actually graphing the vertical planes. The cross-hatched curves that *Mathematica* draws on the surface are the intersections of the surface with vertical planes where x or y is constant. The slopes along these curves are the partial derivatives of z.

24.1.2 Example: Visualizing Partial Derivatives with Level Curves

We can also visualize the partial derivatives of a function by using its level curves to detect whether the function is increasing or decreasing in the x- or y-direction. Here's how to do that with $z = x^3 y$.

- Graph the level curves of $z = x^3 y$ for x and y between ±1, showing **Contour Shading**. (**Graph 3D** ▷ **Level Curves**)

The result is:

▷ Plot the level curves $x^3 y = c$ for $-1 \le x \le 1$ and $-1 \le y \le 1$

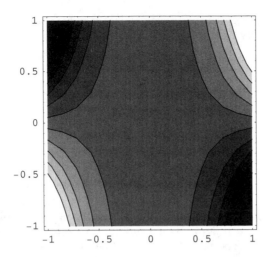

Out[2]= - ContourGraphics -

Choose a point (x, y) with $y > 0$ and increase x while keeping y constant. The shading appears to get lighter, implying that z increases. So the graph suggests that $\frac{\partial z}{\partial x} > 0$, which we can confirm from $\frac{\partial z}{\partial x} = 3 x^2 y$.

Begin where $x > 0$, $y > 0$ and increase only x. The contour lines appear to get closer together, implying that z increases at a faster rate. This suggests $\frac{\partial^2 z}{\partial x^2} > 0$, which we confirm from $\frac{\partial^2 z}{\partial x^2} = 6 xy$. You can imagine hiking along the surface. Moving in the x-direction you would be going uphill, with the hill getting steadily steeper.

However, if we start with $x > 0$, $y < 0$ and increase only x, then we see a different pattern. The shading appears to darken, suggesting that z decreases and $\frac{\partial z}{\partial x} < 0$. The contour lines get closer together, which implies that the negative numbers $\frac{\partial z}{\partial x}$ are themselves decreasing. This suggests that $\frac{\partial^2 z}{\partial x^2} < 0$. Now you are going downhill and the hill is getting steeper. We can confirm both conclusions using the above equations for the first and second partials.

You can graph the surface and check these results on the graph. Then examine the partials with respect to y similarly, and also see what happens with other starting points.

- Graph $z = x^3 y$ for x, y between ± 1. (**Graph 3D ▷ Surface**)

▷ Plot x³ y for -1 ≤ x ≤ 1   and -1 ≤ y ≤ 1

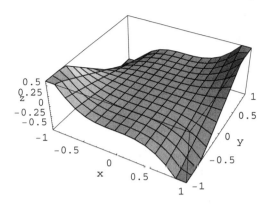

Out[3]=   - SurfaceGraphics -

Exercises

1. a) Combine the graphs of the surface $z = \sin(x^2 + y)$ and the plane $x = 1$. Print the result and mark the point on the intersection where it appears that $\frac{\partial z}{\partial y} = 0$.

   b) Find $\frac{\partial z}{\partial y}$ and then find the x-, y-, z-coordinates of this point.

2. a) Graph the level curves of $z = x^3 - 2y^2 + x - y$ for x and y between ± 2. From the graph, do you predict that $\frac{\partial z}{\partial x}$ is greater at $(-1, 0)$ or at $(1.5, 1.5)$?

   b) Find $\frac{\partial z}{\partial x}$ at the points in a) to confirm your prediction.

In Exercises 3–7:

   a) Graph the surface or level curves as indicated, for x and y between ± 2. (**Graph 3D ▷ Surface, Level Curves**)

   If you are using *Mathematica* Version 4 with interactive 3D graphics, you can display the coordinates on the surfaces by choosing **Graph ▷ Modify Graph** and specifying **Axis Labels**.

   With level curves, it will help to use **Contour Shading**.

   b) Print the graph and mark an example of a point, if any, where you think the graph shows that

   (i) $\frac{\partial z}{\partial x} > 0$ and $\frac{\partial z}{\partial y} > 0$; (ii) $\frac{\partial z}{\partial x} > 0$ and $\frac{\partial z}{\partial y} < 0$;

   (iii) $\frac{\partial z}{\partial x} < 0$ and $\frac{\partial z}{\partial y} > 0$; (iv) $\frac{\partial z}{\partial x} < 0$ and $\frac{\partial z}{\partial y} < 0$.

   c) Find the partial derivatives at the values (x, y) you chose and verify the signs you predicted. Review your reasoning if there are any inconsistencies.

3. $z = x^2 - y^2$ (surface).

4. $z = 10 x^3 + 25 x y^2 + 5 x + y$ (surface).

5. $z = \sin(x y + x)$ (surface).

6. $z = 2 x^2 - 2 y^2 - x + \frac{1}{8}$ (level curves).

7. $z = x^3 - y^3$ (level curves).

8. a) Graph the level curves of $z = x^3 - x^2 y + 4 y^2$ for x and y between ± 2. From the graph, what do the signs of $\frac{\partial^2 z}{\partial x^2}$ and $\frac{\partial^2 z}{\partial y^2}$ appear to be at $(-1, 1)$? at $(1, -1)$?

    b) Confirm your answers by calculating these derivatives.

■ 24.2 Estimating Partial Derivatives Numerically

Additional Prerequisite:
> Sec. 4.2   How to Tabulate an Expression

24.2.1 Example: Average and Instantaneous Rates of Change

The partial derivative of a function of two or more variables is a limit, where only one variable changes and the others are held constant. For example, if $f(x, y)$ is a function of two variables then $\frac{\partial f}{\partial x}$ is the *instantaneous* rate of change in the x-direction:

$$\frac{\partial f}{\partial x} = \lim_{h \to 0} \frac{f(x + h, y) - f(x, y)}{h} .$$

Another notation for $\frac{\partial f}{\partial x}$ is f_x. To emphasize that this derivative is a function of x, y, we may write $\frac{\partial f}{\partial x}\big|_{(x,y)}$ or $f_x(x, y)$.

We can estimate the value of $\frac{\partial f}{\partial x}$ by taking the *average* rate of change for a small value of h:

$$\frac{\partial f}{\partial x} \approx \frac{f(x + h, y) - f(x, y)}{h} \quad \text{when } h \text{ is small} .$$

For example, when $f(x, y) = (x + y)^x$ we can estimate $\frac{\partial f}{\partial x}$ at $(x, y) = (3, 1)$ by

$$\frac{\partial f}{\partial x}\bigg|_{(3,1)} = f_x(3, 1) \approx \frac{f(3.001, 1) - f(3, 1)}{0.001} .$$

Here's how to find this estimate using *Joy*:

*   Create the function $f(x, y) = (x + y)^x$. (**Create ▷ Function**)

Enter $f(x, y)$ as $\mathtt{f[x,y]}$.

Mathematica creates the function:

▷ Define $\mathtt{f[x, y]}$ to be $\mathtt{(x + y)}^{\mathtt{x}}$

- If necessary, click in your *Mathematica* notebook beneath the last cells, so that a horizontal line appears.
- Type $\frac{\mathtt{f[3.001,1]-f[3,1]}}{\mathtt{0.001}}$ and press $\boxed{\mathtt{SHIFT}}$ — $\boxed{\mathtt{RET}}$.

The result is.

$$In[3]:= \quad \frac{\mathtt{f[3.001, 1] - f[3, 1]}}{\mathtt{0.001}}$$

$$Out[3]= \quad 136.879$$

Here's how to get a better estimate by taking smaller and smaller values of h.

- From the *Joy* menus, choose **Create ▷ List**.
- Fill in $\mathtt{h}$, $\frac{\mathtt{f[3+h,1]-f[3,1]}}{\mathtt{h}}$.
- From the popup menu, choose **x=f(n)**.
- Fill in $\mathtt{h = 10^{-n}}$ and $0 \le \mathtt{n} \le 8$.

   This chooses the values $h = 1, 0.1, 0.01, \ldots, 10^{-8}$.

- Make sure **Display in decimal form** and **Display as a table** are checked.
- Click **OK**.

Mathematica produces the list:

▷ Create the list with values $\left\{ \mathtt{h}, \dfrac{\mathtt{f[3 + h, 1] - f[3, 1]}}{\mathtt{h}} \right\}$ for $\mathtt{h = 10^{-n}}$ with $0 \le \mathtt{n} \le 8$.

Out[4]//TableForm=

1.	561.
0.1	153.652
0.01	138.296
0.001	136.879
0.0001	136.738
0.00001	136.724
$1. \times 10^{-6}$	136.723
$1. \times 10^{-7}$	136.723
$1. \times 10^{-8}$	136.723

This suggests that $\frac{\partial f}{\partial x} \approx 136.723$ when $(x, y) = (3, 1)$.

Here's how to find an expression for the derivative at an arbitrary point (x, y) using *Joy* and *Mathematica* (or you can find it by hand using logarithmic differentiation).

- Click in your *Mathematica* notebook beneath the last output.
- Type f[x, y] and drag across it, or click the **Select** button on the *Joy* palette several times until you have selected this expression.
- Click the **Partial Derivative** button on the *Joy* palette.

The field for the independent variable is selected (highlighted).

- Type x.
- Press $\boxed{\text{SHIFT}} - \boxed{\text{RET}}$.

Here is the result:

$$In[5]:= \quad \frac{\partial}{\partial x} \; (f[x, y])$$

$$Out[5]= \quad x \; (x + y)^{-1+x} + (x + y)^x \; \text{Log}[x + y]$$

⚠ In *Mathematica*, if you have created the function as f[x,y] then you must use the notation $\frac{\partial}{\partial x}$ (f[x, y]) for the partial derivative. You cannot use $\frac{\partial f}{\partial x}$ or the other notations that are accepted in written mathematics.

Now we evaluate the derivative at $(x, y) = (3, 1)$.

- From the *Joy* menus, choose **Algebra** ▷ **Substitute**.
- From the popup menu, choose **Last Output**.
- Click the radio button **Make the substitutions**.
- Fill in x->3,y->1.

You can type -> by typing - and then >.

- Click **OK**.

The result is:

▷ Making the substitutions {x -> 3, y -> 1} in the expression
   x (x + y)$^{-1+x}$ + (x + y)^x Log[x + y] gives

$$Out[6]= \quad 48 + 64 \, \text{Log}[4]$$

This is the exact value of the derivative, which we can express in numeric form:

- Drag across the last output, or click the **Select** button on the *Joy* palette repeatedly until you have selected the entire expression.
- Click the **Numeric** button and press $\boxed{\text{SHIFT}} - \boxed{\text{RET}}$.

Mathematica calculates the numerical value:

$$In[7]:= \quad \textbf{N[48 + 64 Log[4]]}$$

$$Out[7]= \quad 136.723$$

This is the same result we found in our estimate, to six significant digits.

Exercises

In Exercises 1–4:

a) Estimate the partial derivative at (x, y) by calculating the average rate of change for the given values of x, y, and h.

b) If you know how to calculate the exact value of the partial derivative by hand, do it. Otherwise, let *Mathematica* find it as in the example of this section (Example 24.2.1, `In[5]`).

1. $f(x, y) = x^3 y^2 + 2x - 3y$; $\frac{\partial f}{\partial x}$, $x = -2$, $y = 3$, $h = 0.001$.

2. Repeat, but estimate $\frac{\partial f}{\partial y}$.

3. $f(x, y) = \cos\left(\frac{x}{y}\right)$; $\frac{\partial f}{\partial x}$, $x = 5$, $y = -2$, $h = 10^{-n}$, $n = 0, 1, 2, ..., 8$.

4. Repeat, but estimate $\frac{\partial f}{\partial y}$.

■ 24.3 Differentiation Practice

24.3.1 Example: The Chain Rule

Let $u = xy - x^3$ and $v = \sin(1 + x \ln(y)^2)$. You can find the partial derivative $\frac{\partial}{\partial x}(u^2 v^2)$ by hand using the chain rule, or by substituting for u and v and then differentiating. Here's how to find it using *Joy*.

- Click in your *Mathematica* notebook beneath the last output, if any.
- Type u = x__y - x³ and press ⌈SHIFT⌉ − ⌈RET⌉.

    The symbol __ means to type a blank space to indicate multiplication.

Mathematica defines *u*:

In[2]:= **u = x y - x³**

Out[2]= $-x^3 + x y$

- Repeat with v = **Sin[1 + x Log[y]²]**.

Mathematica defines *v*:

In[3]:= **v = Sin[1 + x Log[y]²]**

Out[3]= $\text{Sin}[1 + x \text{Log}[y]^2]$

- Type $u^2 \; v^2$.
- Drag across $u^2 \; v^2$, or click the **Select** button on the *Joy* palette several times until you have selected the entire expression.
- Click the **Partial Derivative** button on the *Joy* palette.
- Type x for the independent variable.
- Press ⌷SHIFT⌷ − ⌷RET⌷.

Mathematica calculates the partial derivative:

In[4]:= $\dfrac{\partial}{\partial \text{ x}}$ (u^2 v^2)

Out[4]= 2 (-x^3 + x y)^2 Cos[1 + x Log[y]^2] Log[y]^2 Sin[1 + x Log[y]^2] +
          2 (-3 x^2 + y) (-x^3 + x y) Sin[1 + x Log[y]^2]^2

We can verify that the chain rule gives the same result:

In[5]:=   u^2 2 v $\left(\dfrac{\partial}{\partial \text{ x}} \text{ (v)} \right)$ + v^2 2 u $\left(\dfrac{\partial}{\partial \text{ x}} \text{ (u)} \right)$

Out[5]= 2 (-x^3 + x y)^2 Cos[1 + x Log[y]^2] Log[y]^2 Sin[1 + x Log[y]^2] +
          2 (-3 x^2 + y) (-x^3 + x y) Sin[1 + x Log[y]^2]^2

⚠ When a derivative is a factor in a product, as in this example, you must enclose it in parentheses ( ) for it to be interpreted correctly. Otherwise, *Mathematica* will give an error message. This applies to ordinary as well as partial derivatives.

Exercises

In these exercises:

a) Find the derivative by hand.

b) Check your answer using the **Partial Derivative** button on the *Joy* palette. If *Mathematica* gives the answer in a different form, make sure it is equivalent to yours.

1. $\frac{\partial}{\partial x} (x^5 \, y^{3/2} + x^{3/2} \, y^5)$.

2. $\frac{\partial}{\partial y} \left(\frac{x^2 + y^2}{x^2 - y^2} \right)$.

3. $\frac{\partial}{\partial x} \left(\cos\left(\sqrt{2x - 3y} \right) \right)$.

4. $\frac{\partial}{\partial x} (\ln(u + v^2))$, where $u = \sin(x\,y)$, $v = \sin\left(\frac{x}{y}\right)$.

5. $\frac{\partial}{\partial y} (\ln(u + v^2))$, where $u = \sin(x\,y)$, $v = \sin\left(\frac{x}{y}\right)$.

■ 24.4 Directional Derivatives and Gradients

24.4.1 Example: Visualizing Directional Derivatives

The directional derivative of a function measures its rate of change in a particular direction. Here is how to visualize this rate of change using level curves and the gradient field. We'll use the example $f(x, y) = 4\,x^2 + 6\,xy - 3\,y^2$.

- Create the function $f\,(x, y) = 4\,x^2 + 6\,xy - 3\,y^2$. (**Create** ▷ **Function**)

Mathematica creates the function:

▷ Define f[x, y] to be $4\,x^2 + 6\,x\,y - 3\,y^2$

- Graph the level curves of $f(x, y)$ for x and y between ±4. Check **Contour Shading**. (**Graph 3D** ▷ **Level Curves**)

Here is the result:

▷ Plot the level curves f[x, y] = c for $-4 \le x \le 4$ and $-4 \le y \le 4$

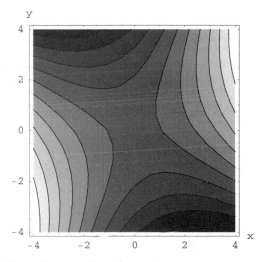

Out[3]= - ContourGraphics -

At any point (x, y), the directional derivative $D_u\, f(x, y)$ of f is positive in all directions u where the function increases and is negative in all directions where it decreases. For example, at $(x, y) = (1, 0)$ the function appears to increase toward $(2, 1)$, so the derivative in that direction should be positive. But toward $(2, -2)$ it appears to decrease, so that derivative should be negative.

We can verify this by calculating the directional derivatives using the relation $D_u f(x, y) = \nabla f(x, y) \cdot u$, where u is a unit vector.

> ♀ This example is easy enough to calculate by hand, but Section 12.5 shows how to use *Joy* to find directional derivatives.

The vector from $(1, 0)$ to $(2, 1)$ is $(2, 1) - (1, 0) = (1, 1)$, so a unit vector in this direction is $u = (1, 1)/\sqrt{2}$. The gradient of f is $\nabla f(x, y) = (8x + 6y, 6x - 6y)$, so $\nabla f(1, 0) = (8, 6)$. The directional derivative of f at $(1, 0)$ in the direction u is thus

$$D_u f(1, 0) = \nabla f(1, 0) \cdot u = (8, 6) \cdot \frac{(1,1)}{\sqrt{2}} = \frac{14}{\sqrt{2}} \approx 9.9$$

and is positive.

Similarly, you can verify that the directional derivative of f at $(1, 0)$ in the direction $v = (1, -2)/\sqrt{5}$ from $(1, 0)$ to $(2, -2)$ is $D_v f(1, 0) = -4/\sqrt{5} \approx -1.8$ and is negative.

The magnitude of the directional derivative is greater in directions where the function changes more rapidly and the level curves (which correspond to equally spaced values of the function) are closer together. For instance, starting at $(1, 0)$ the function appears to change more rapidly in the direction u than in the direction v. This is reflected in the magnitudes of the directional derivatives, 9.9 and 1.8.

24.4.2 Example: The Gradient Field

It can be proved that the gradient at (x, y) points in the direction where f increases most rapidly and away from the direction where it decreases most rapidly. Using *Joy*, you can graph the gradients of a function and superimpose them on the graph of the level curves. Especially when printing, it is easier to visualize the gradient field if we graph the level curves without contour shading. We do this first, using the function of Example 24.4.1.

- If you didn't create the function $f(x, y) = 4x^2 + 6xy - 3y^2$ in Example 24.4.1, do it now.
- Graph the level curves of $f(x, y)$ for x and y between ±4, being sure to *uncheck* **Contour Shading**. (**Graph 3D ▷ Level Curves**)

Mathematica plots the level curves without shading.

- From the *Joy* menus, choose **Graph ▷ Gradient Field**.
- Fill in the function **f[x,y]** and the intervals $-4 \leq x \leq 4, -4 \leq y \leq 4$.
- Check **Combine with the graph** and fill in the output name or number of the level curve graph.
- Click **OK**.

Mathematica superimposes the gradient field on the level curves. Next, we modify this graph so that it has equal scales on the x and y axes. This will cause angles to be displayed correctly.

- Choose **Graph ▷ Modify Graph** to modify the preceding graph.
- Make sure the correct output name or number appears and check **Equal scales**.

- Check **Assign name** and fill in the name `levels`.
- Click **OK**.

The result is (`%` is *Mathematica*'s abbreviation for `Out`) :

▷ Redisplay `%5` with equal scales on both axes.

`levels =`

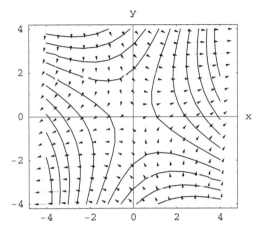

Out[6]= - Graphics -

Although there is no shading you can use the gradients to conclude, for example, that the values of *f* are greater near (4, 3) and less near (−4, −3) than they are near the origin.

⚠ The gradients are not drawn to scale, i.e., the lengths shown in the graph are correct only relative to one another. For instance, the true length of $\nabla f(0, -4) = (-24, 24)$ is $24\sqrt{2} \approx 34$, but it is drawn much shorter in the graph.

The gradients appear to cross the level curves at right angles. It can be proved that the gradient $\nabla f(x, y)$ is always perpendicular to the tangent to the level curve of *f* passing through (*x*, *y*).

♡ In order for angles to be displayed correctly, the two axes must have equal scales. In an example where this doesn't happen automatically, choose **Graph** ▷ **Modify Graph** and check **Equal scales**.

More generally, if θ is the angle between $\nabla f(x, y)$ and the unit vector ***u*** then

$$D_u f(x, y) = \nabla f(x, y) \cdot u = \| \nabla f(x, y) \| \cos \theta.$$

So the derivative is positive and the function increases when the angle is acute ($0 \le \theta < \pi/2$), and the derivative is negative and the function decreases when the angle is obtuse ($\pi/2 < \theta \le \pi$).

We can picture this in the graph by including the tangent to a point on one of the level curves. We take the point (*x*, *y*) = (1, 0), which looks quite close to one of the curves. Since $f(1, 0) = 4$, the level curve through this point is $4x^2 + 6xy - 3y^2 = 4$. We'll find the slope of the tangent by implicit differentiation.

⚲ Here we differentiate by hand, but Section 7.10 shows how to use *Joy* for implicit differentiation.

We have

$$8x + 6\left(x\,\frac{dy}{dx} + y\right) - 6y\,\frac{dy}{dx} = 0,$$

$$\frac{dy}{dx} = \frac{-8x - 6y}{6x - 6y} = -\frac{4x + 3y}{3x - 3y}.$$

So at $(1, 0)$ the tangent has slope $\frac{dy}{dx} = -4/3$ and its equation is

$$y = -\frac{4}{3}(x - 1).$$

We'll graph the line and combine it with the gradient field and level curves.

- Graph $-\frac{4}{3}(x - 1)$ for $-4 \le x \le 4$, assigning the name `tangent`. (**Graph** ▷ **Graph f(x)**)
- Combine the graphs *levels* and *tangent*. (**Graph** ▷ **Combine Graphs**)

The result is:

▷ Combining the graphs {levels, tangent} gives

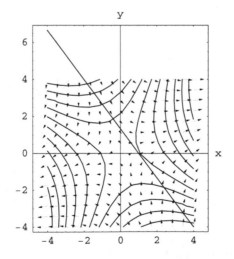

Out[8]=   - Graphics -

The graph illustrates the relation between the tangent and the gradient. (In general this is only approximate, because *Mathematica* may not have graphed the level curve through the exact point where you chose the tangent.) The directional derivative $D_u f(1, 0)$ is positive when u points to the same side of the tangent line as $\nabla f(1, 0)$, negative when it points to the opposite side, and zero when it points along the tangent. Example 24.4.1 showed that $\nabla f (1, 0) = (8, 6)$.

Exercises

Where necessary in these exercises, you can represent angles correctly by applying **Equal scales** to the axes. (**Graph ▷ Modify Graph**)

1. a) Graph the level curves of $z = x^3 - x^2 y + 4 y^2$ for x and y between ±2, showing contour shading. On a printout of the graph, draw the approximate directions in which the directional derivative of z at $(1, -1)$ will be positive and in which it will be negative.

   b) From the graph, predict the sign of $D_u f(1, -1)$ in the direction of the origin. Then calculate this derivative by hand and check your answer.

2. a) Graph the level curves of $z = x^3 - 2 y^2 + x - y$ for x and y between ±2. On a printout of the graph, draw the directions in which the directional derivative at $(-1, -1)$ will be largest, smallest, and zero.

   b) By hand, use the gradient to calculate unit vectors for the directions you found in a). There should be four vectors.

3. a) Graph the level curves, *without* contour shading, of $f(x, y) = 10 - 3 x^2 - 4 y^2$ for x, y between ±2. Then combine the graph with the gradient field of $f(x, y)$ for the same values of x, y.

   b) From the graph, predict the sign of the directional derivative $D_u f(1, 0)$ in the direction from $(1, 0)$ toward $(2, 2)$. Then calculate the derivative by hand and check your answer.

   c) In which directions u is $D_u f (1, 0)$ positive? Explain in terms of the graph.

   d) On a printout of the graph, draw the directions in which the directional derivative at $(\frac{-1}{2}, \frac{3}{2})$ will be largest, smallest, and zero. By hand, use the gradient to calculate unit vectors for the directions you found. There should be four vectors.

4. a) Find the equation of the tangent to the level curve of $f (x, y) = 10 - 3 x^2 + 4 y^2$ passing through $(-\frac{1}{2}, \frac{3}{2})$.

   b) Combine the graphs of the tangent in a) and the level curves of the function for x and y between ±2. If the line doesn't appear to be tangent to the level curve through $(-\frac{1}{2}, \frac{3}{2})$, review your reasoning.

   c) Verify that the tangent in a) is perpendicular to the gradient $\nabla f(-\frac{1}{2}, \frac{3}{2})$ by using their slopes.

5. a) Calculate by hand, for $z = x^2 - x y + 2 y^2$:

   (i) the gradient of z at $(3, 2)$;

(ii) the directional derivative of z in the direction of a unit vector $\boldsymbol{u} = (\cos\theta, \sin\theta)$ at the point (3, 2).

b) Graph the directional derivative in a) as a function of θ, for $0 \leq \theta \leq 2\pi$. Estimate to one decimal place its maximum M, minimum m, and the angles θ_M, θ_m where they occur. (**Graph** ▷ **Graph f(x)**)

> Enter θ using the **Greek** button on the *Joy* palette. When you are done, close the Greek palette by clicking its close box in the upper-left corner of the palette (Macintosh) or the close button **X** in the upper-right corner of the palette (Windows).

c) Calculate, by hand or directly in your *Mathematica* notebook, the vectors $M(\cos\theta_M, \sin\theta_M)$ and $M(\cos\theta_m, \sin\theta_m)$.

> If you do this in *Mathematica*, enter the vectors using curly brackets { } rather than parentheses ( ). If you do this by hand, remember that the values of θ are radians, not degrees.

d) Compare the results with the gradient in a). What would you expect to see if your estimates in b) were exactly correct, and why?

■ 24.5 Tangent Planes

24.5.1 Example: Visualizing the Tangent Plane Approximation

Here's how to visualize the tangent plane approximation to a surface at a given point. This example uses the function $f(x, y) = -10x^2 - 8y^2$ and the point (2, 1, −48) on its graph.

The equation of the tangent plane at this point is
$$
\begin{aligned}
z &= f(2, 1) + f_x(2, 1)(x - 2) + f_y(2, 1)(y - 1) \\
&= -48 - 40(x - 2) - 16(y - 1) \\
&= 48 - 40x - 16y .
\end{aligned}
$$

- Graph $-10x^2 - 8y^2$ for x and y between ±3, assigning the name `surface`. (**Graph 3D** ▷ **Surface**)

Mathematica graphs the surface:

```
▷ Plot  -10 x² - 8 y²  for  -3 ≤ x ≤ 3   and  -3 ≤ y ≤ 3

surface =
```

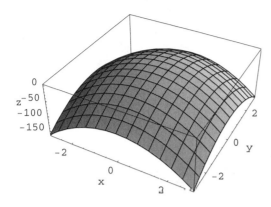

Out[2]= - SurfaceGraphics -

Now we graph the plane and combine it with the surface.

- Repeat for $48 - 40\,x - 16\,y$, assigning the name `plane`.
- Combine the graphs *surface* and *plane*. (**Graph ▷ Combine Graphs**)

Mathematica combines the graphs:

▷ Combining the graphs {surface, plane} gives

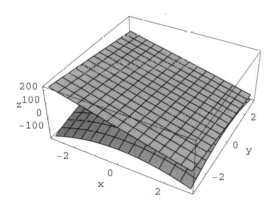

Out[4]= - Graphics3D -

Now rotate the combined graph to see the point of tangency more clearly:

- If you are using *Mathematica* Version 4 with interactive 3D graphics, drag the graph to rotate it upward to see between the plane and the surface.
- Otherwise, choose **Graph 3D ▷ 3D ViewPoint** and follow the on-screen instructions.

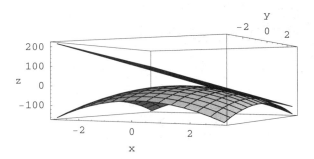

The tangent plane gives the linear (affine) approximation to the function: for all (x, y) near $(2, 1)$, we have $f(x, y) \approx 48 - 40\,x - 16\,y$. For example, when $(x, y) = (2.1, 0.9)$ the height of the tangent plane is:

$In[6]:=$ **48 − 40(2.1) − 16(0.9)**

$Out[6]=$ −50.4

The height of the surface is:

$In[7]:=$ **−10 (2.1)² − 8 (0.9)²**

$Out[7]=$ −50.58

This verifies that the tangent plane overestimates the function at $(2.1, 0.9)$, as suggested by the graph.

Tangent plane to a level surface. When the surface is a level surface for a function of three variables, the procedure is a bit different. For example, the equation of the tangent plane to the surface $h(x, y, z) = x + y^2 + z^2 = 8$ at $(3, 1, 2)$ is

$$\nabla h(3, 1, 2) \cdot (x - 3, y - 1, z - 2) = (1, 2, 4) \cdot (x - 3, y - 1, z - 2) = 0,$$

i.e., $x + 2\,y + 4\,z = 13$.

- Plot $x + y^2 + z^2 = 8$ for x, y, z between ± 4. (**Graph 3D ▷ Level Surface**)

▷ Graph of the level surface $x + y^2 + z^2 = 8$ for
$-4 \le x \le 4$, $-4 \le y \le 4$, $-4 \le z \le 4$.

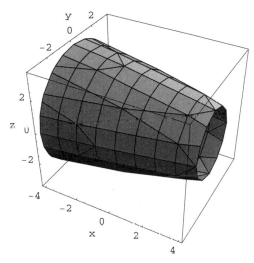

Out[8]= - Graphics3D -

Now we graph the plane and combine it with the level surface.

- Repeat for $x + 2y + 4z = 13$.

  You can graph the plane as a level surface, or you can solve for z and use **Graph 3D ▷ Surface**.

- Combine the graphs of the level surface and plane, using their output numbers or names. **(Graph ▷ Combine Graphs)**

The result is (the paraphrase contains %..., *Mathematica*'s shortcut for Out[...]):

▷ Combining the graphs {%8, %9} gives

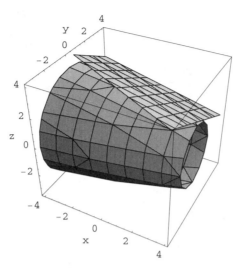

Out[10]= - Graphics3D -

Exercises

In Exercises 1 and 2:

    a) Graph the surface for x, y between ± 2 and the tangent plane at the given point.

    b) Combine the graphs to confirm that the plane appears tangent to the surface.

    c) Does the tangent plane approximation appear to overestimate or underestimate the function at the specified point? Verify this by calculating the exact and approximate values.

1. $z = x^2 + y^2$, $(1, -1, 2)$; approximate z when $(x, y) = (1.2, -0.95)$.

2. $z = \sin(x\,y)$, $(1, \frac{\pi}{6}, \frac{1}{2})$; approximate when $(x, y) = (1.2, \frac{\pi}{6} - 0.1)$.

In Exercises 3 and 4:

    a) Graph the level surface for the given values of w, for x, y, z between ± 2.

    b) Graph the tangent plane to the surface at the given point.

    c) Combine the graphs and confirm that the plane appears tangent to the surface, rotating the graph to change the viewpoint if necessary.

3. $w = x^2 - y + z^2$, $w = 1$, $(0, 0, -1)$.

4. $w = x^2 - x\,y + z^2$, $w = 1$, $(1, 1, 1)$.

■ 24.6 Optimization

Additional Prerequisites:

      Sec. 3.2   How to Graph Implicit Functions

      Secs. 5.1,2   Solving equations algebraically and numerically

24.6.1 Example: The Second Derivative Test

There is a test for functions $f(x, y)$ that is analogous to the second derivative test for functions of one variable. Suppose that (x, y) is a *critical point* of f, i.e., the partial derivatives $\frac{\partial f}{\partial x}$ and $\frac{\partial f}{\partial y}$ both equal 0. Let

$$\text{disc} = \frac{\partial^2 f}{\partial x^2} \frac{\partial^2 f}{\partial y^2} - \left(\frac{\partial^2 f}{\partial x \, \partial y} \right)^2,$$

the *discriminant* of f. Under appropriate smoothness conditions on f, it can be proved that

     if disc > 0 and $\frac{\partial^2 f}{\partial x^2} > 0$, then f has a local minimum at (x, y)

     if disc > 0 and $\frac{\partial^2 f}{\partial x^2} < 0$, then f has a local maximum at (x, y)

     if disc < 0, then f has neither a maximum nor a minimum at (x, y) (*saddle point*)

     if disc $= 0$, the test is indeterminate (anything can happen)

   ⚠ When the discriminant at a critical point is zero, the test doesn't provide any information. You need to examine the values of the function near the critical point to determine its behavior.

This example shows how to use *Joy* to visualize this result with the function $f(x, y) = x^3 - y^2 - 6\,x$.

- Create the function $f(x, y) = x^3 - y^2 - 6\,x$. (**Create ▷ Function**)

Mathematica creates the function:

  ▷ Define f[x, y] to be $x^3 - y^2 - 6\,x$

- Graph $f(x, y)$ for x and y between ± 2. (**Graph 3D ▷ Surface**)

Here is the surface:

  ▷ Plot f[x, y] for $-2 \le x \le 2$ and $-2 \le y \le 2$

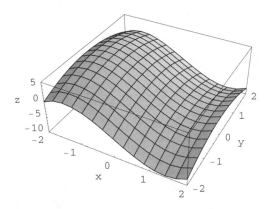

Out[3]= - SurfaceGraphics -

      ✧ If you are using *Mathematica* Version 4 with interactive 3D graphics, you can display the coordinates by choosing **Graph ▷ Modify Graph** and specifying **Axis Labels**.

Next, plot the level curves of the function.

    • Graph the level curves of $f(x, y)$ for x and y between ± 2, with **Contour Shading**. (**Graph 3D ▷ Level Curves**)

Here is the result:

▷ Plot the level curves f[x, y] = c for −2 ≤ x ≤ 2   and −2 ≤ y ≤ 2

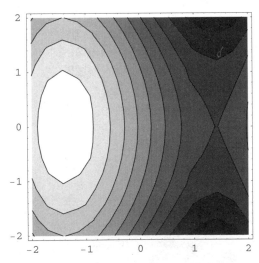

Out[4]= - ContourGraphics -

      ✧ In some cases, you may need to increase the number of contour lines that *Mathematica* draws in order to see the pattern more clearly.

The graphs of the surface and level curves both suggest that the function has a local maximum at $(x, y) \approx (-1.5, 0)$. At $(x, y) \approx (1.5, 0)$ the function seems to have a local minimum in the x-direction and a local maximum in the y-direction, so this appears to be a saddle point.

We can calculate the partial derivatives by hand:

$$\frac{\partial f}{\partial x} = 3\,x^2 - 6, \quad \frac{\partial^2 f}{\partial x^2} = 6\,x, \quad \frac{\partial^2 f}{\partial x\,\partial y} = 0$$

$$\frac{\partial f}{\partial y} = -2\,y, \quad \frac{\partial^2 f}{\partial y^2} = -2, \quad \text{disc} = -12\,x$$

So there are critical points at $(x, y) = \left(\sqrt{2}, 0\right)$ and $\left(-\sqrt{2}, 0\right)$. These are the two points we have identified. At $\left(\sqrt{2}, 0\right)$ we have disc $= -12\sqrt{2}$, so this is a saddle point. At $\left(-\sqrt{2}, 0\right)$ we have disc $= 12\sqrt{2}$ and $\frac{\partial^2 f}{\partial x^2} = -6\sqrt{2}$, so this is a local maximum.

24.6.2 Example: Lagrange Multipliers

This example shows how to use *Joy* to find the extreme values of an *objective function* $z = f(x, y)$ along a *constraint curve* $g(x, y) = c$. We'll take the example $z = f(x, y) = x^2 + y$, $g(x, y) = x^3 - y^3 = 4$.

- Graph the objective function $z = x^2 + y$ as a surface with x, y between ±4. (**Graph 3D ▷ Surface**)

The surface is:

▷ Plot x² + y for 4 ≤ x ≤ 4   and −4 ≤ y ≤ 4

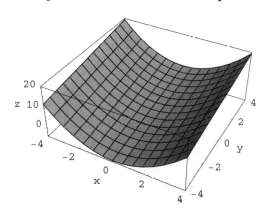

Out[2]=   - SurfaceGraphics -

We can think of the constraint $x^3 - y^3 = 4$ as a curve in space that is the intersection of a level surface with the graph of the objective function. We'll graph the level surface and combine the two graphs.

- Graph $x^3 - y^3 = 4$ as a level surface, with x, y between ±4 and $-10 \le z \le 25$. (**Graph 3D ▷ Level Surface**)

    The level surface is a *cylindrical surface* because z is arbitrary. We choose the z-interval large enough to include all the values shown in the preceding graph of the objective function.

- Combine the graphs of the surface and level surface using their output numbers or names. (**Graph ▷ Combine Graphs**)

The combined graph is:

▷ Combining the graphs {%2, %3} gives

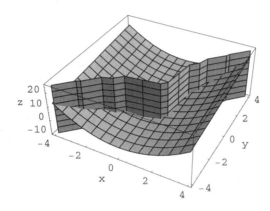

Out[4]=   - Graphics3D -

The intersection of the two surfaces is the curve on the graph of the objective function $z = x^2 + y$ that satisfies the constraint $x^3 - y^3 = 4$. The picture suggests that the objective function has a minimum on the constraint and no maximum (within the entire domain of the function, not just the part shown).

To find the minimum we look at the level curves, as follows.

- Graph the level curves of $x^2 + y$ for x and y between ±4 without shading, naming the graph `levels`. (**Graph 3D ▷ Level Curves, Assign name**)

Mathematica graphs the level curves:

▷ Plot the level curves $x^2 + y = c$ for $-4 \le x \le 4$ and $-4 \le y \le 4$

levels =

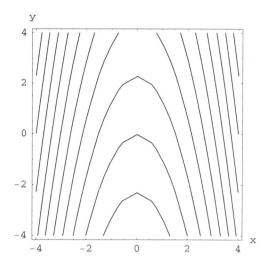

Out[5]= - ContourGraphics -

Now we'll graph the constraint curve and combine it with the level curves.

- Graph $x^3 - y^3 = 4$ for $-4 \le x \le 4$, naming it `constraint`. (**Graph ▷ Implicit Function, Assign name**)
- Combine the graphs *levels* and *constraint*. (**Graph ▷ Combine Graphs**)

The combined graph is:

▷ Combining the graphs {levels, constraint} gives

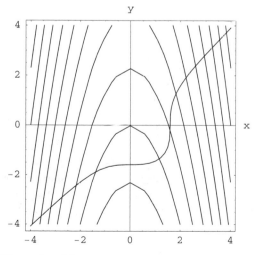

Out[7]= - Graphics -

We graphed the level curves of the objective function $z = f(x, y)$ without shading to make it easier to see the constraint curve $g(x, y) = c$ against them. We want to know where $f(x, y)$ takes its minimum and maximum values along the constraint. Now we'll use shading to help.

- Click **Recent ▷ Level Curves**.
- Click **Contour Shading** and **OK**.

Mathematica now adds shading to the level curves:

▷ Plot the level curves $x^2 + y = c$ for $-4 \le x \le 4$ and $-4 \le y \le 4$

levels =

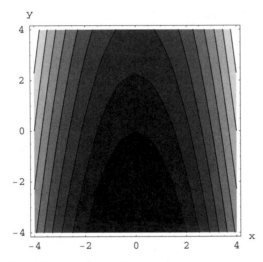

Out[8]= - ContourGraphics -

As we move along the constraint starting with $x = -4$, the shading shows that the value of the objective function $f(x, y)$ decreases until about $x = 0$ and then begins to increase. So there should be a minimum at or near $x = 0$ and no maximum on the constraint.

Suppose the constraint crosses a level curve at some point. Then the value of f must steadily increase or decrease as you move across the level curve, and f cannot have a maximum or minimum at that point.

Under suitable smoothness assumptions it can be proved that at any point along $g(x, y) = c$ where f has a minimum or maximum, the level curve and the constraint curve are tangent. Since the gradients of f and g are perpendicular to its level curves, this says that $\nabla f(x, y)$ is a multiple of $\nabla g(x, y)$ at such a point (or, as an exceptional case, $\nabla g(x, y) = \mathbf{0}$). So we are looking for points (x, y) where

$$\nabla f(x, y) = \lambda \nabla g(x, y) \text{ for some scalar } \lambda, \text{ or}$$
$$\nabla g(x, y) = \mathbf{0}.$$

A scalar λ that satisfies this condition is called a *Lagrange multiplier*. The sign of λ in a solution corresponds to whether the vectors $\nabla f(x, y)$ and $\nabla g(x, y)$ have the same or opposite directions. If $\lambda > 0$, then f and g increase in the same direction at a solution, and if $\lambda < 0$, then they increase in opposite directions. If $\lambda = 0$, then $\nabla f(x, y) = \mathbf{0}$.

In this example $\nabla g(x, y) = (3x^2, -3y^2)$, so $\nabla g(x, y) \neq \mathbf{0}$ on the constraint $x^3 - y^3 = 4$. Since $\nabla f(x, y) = (2x, 1)$, the Lagrange equations are

$$2x = 3\lambda x^2,$$
$$1 = -3\lambda y^2,$$
$$x^3 - y^3 = 4.$$

Here's how to solve the equations.

- From the *Joy* menus, choose **Algebra ▷ Solve**.
- Fill in $2 \text{ x} = 3 \, \lambda\_\text{x}^2$, $1 = -3 \, \lambda\_\text{y}^2$, $\text{x}^3 - \text{y}^3 = 4$.

  The symbol — means to leave a space to indicate multiplication.

- Fill in the variables **x, y,** λ.
- Make sure **Solve algebraically** is selected.
- Check **Assign name** and use the name **solns1** or a name of your choice.
- Click **OK**.

To save space, we show only part of the solution:

▷ Solve the equations $\{2\text{ x} = 3\,\lambda\,\text{x}^2, 1 = -3\,\lambda\,\text{y}^2, \text{x}^3 - \text{y}^3 = 4\}$ for $\{\text{x, y,}\ \lambda\}$ algebraically.

solns1 =

$Out[9]=$ $\left\{ \left\{ \lambda \to \frac{1}{6}\left(-\frac{1}{2}\right)^{1/3}, \text{x} \to 0, \text{y} \to (-1)^{1/3}\, 2^{2/3} \right\}, \left\{ \lambda \to -\frac{1}{6\,2^{1/3}}, \text{x} \to 0, \text{y} \to -2^{2/3} \right\} \right.$
$\left. \left\{ \lambda \to -\frac{(-1)^{2/3}}{6\,2^{1/3}}, \text{x} \to 0, \text{y} \to -(-2)^{2/3} \right\}, \ . \ . \ . \right\}$

It's easier to see what's happening if we express this in numeric form:

- Click the **Numeric** button on the *Joy* palette.
- Type **solns1**, or the name you assigned if it is different.
- Press SHIFT − RET.

The result is:

In[10]:= **N[solns1]**

Out[10]= {{$\lambda \to 0.0661417 + 0.114561\,I$, $x \to 0$, $y \to 0.793701 + 1.37473\,I$},
 {$\lambda \to -0.132283$, $x \to 0$, $y \to -1.5874$}, {$\lambda \to 0.0661417 - 0.114561\,I$,
 $x \to 0$, $y \to 0.793701 - 1.37473\,I$}, {$\lambda \to -0.188174 - 0.375458\,I$,
 $x \to -0.711254 + 1.41914\,I$, $y \to 0.758066 - 0.468014\,I$},
 {$\lambda \to -0.231069 + 0.350692\,I$, $x \to -0.873385 - 1.32553\,I$,
 $y \to -0.784345 - 0.422497\,I$}, {$\lambda \to 0.419243 + 0.0247651\,I$,
 $x \to 1.58464 - 0.0936064\,I$, $y \to 0.0262788 + 0.890511\,I$},
 {$\lambda \to -0.188174 + 0.375458\,I$, $x \to -0.711254 - 1.41914\,I$,
 $y \to 0.758066 + 0.468014\,I$}, {$\lambda \to 0.419243 - 0.0247651\,I$,
 $x \to 1.58464 + 0.0936064\,I$, $y \to 0.0262788 - 0.890511\,I$},
 {$\lambda \to -0.231069 - 0.350692\,I$, $x \to -0.873385 + 1.32553\,I$,
 $y \to -0.784345 + 0.422497\,I$}}

You can check that there is only one real solution, the second one in the list. This corresponds to

$$\lambda = -\frac{1}{6 \times 2^{1/3}},\ x = 0,\ y = -2^{2/3}\,.$$

For this solution,

$$\nabla f(0, -2^{2/3}) = (0, 1),\quad \nabla g(0, -2^{2/3}) = (0, -3 \times 2^{4/3})\,,$$

and $\lambda < 0$ because the gradients point in opposite directions (*f* increases upward at the solution and *g* increases downward). This is the minimum we saw in the graphs Out[7], Out[8] on pages 449–450. The value of the objective function here is $z = f(0, -2^{2/3}) = -2^{2/3}$. There is maximum.

> *Mathematica* treats roots of negative numbers in a special way. For example, it interprets $(-1)^{1/3}$ not as -1 but as $\frac{1}{2} + i\,\frac{\sqrt{3}}{2} \approx 0.5 + 0.866\,i$. There are three cube roots of -1, namely -1, $\frac{1}{2} \pm i\,\frac{\sqrt{3}}{2}$. Unless you are aware of this, you may think a solution is real when it is not.

> ⚡ Sometimes, neither you nor *Mathematica* will be able to solve the Lagrange equations exactly. In this case, choose **Algebra ▷ Solve** and click **Solve numerically**, finding an initial estimate from the graph.

Exercises

In Exercises 1–4:

a) Graph the surface and level curves for *x* and *y* between ±2.

If you are using *Mathematica* Version 4 with interactive 3D graphics, you can display coordinates by choosing **Graph ▷ Modify Graph** and specifying **Axis Labels**.

b) Use the graphs to identify possible local maxima, minima, and saddle points within this region. Mark these points on a printout of your graphs.

c) By hand, calculate the first- and second-order partial derivatives and confirm your answers to b).

1. $z = x^3 - y^3 - 3x + 3y$.

2. $z = x^2 + y^3 - y - x$ (use 20 contour lines).

3. $z = yx^2 - y^3 + x$.

4. $z = \sin(xy + x)$.

In Exercises 5–7:

a) As in Example 24.6.2, combine the graphs of the level curves of the objective function and the constraint curve, for x, y between ±4.

b) From the combined graph, estimate the coordinates (x, y) where the objective function has its maximum and minimum values, if any, along the constraint.

c) Solve the Lagrange equations by hand or using *Joy*, and compare your answer to b).

5. objective function $z = x + y$;

   constraint curve $x^2 + y^2 = 4$.

6. Find the points on the constraint curve $(x - 1)^2 + 2(y - 2)^2 = 4$ that are closest to and furthest from the origin.

   This will be easier if you use the *square* of the distance from the origin as the objective function. Use *Joy* to solve the Lagrange equations and use the **Numeric** button to express the solution in decimal form.

7. objective function $z = x^2 + y$;

   constraint curve $x^3 - y^3 + x = 4$.

   Solve the Lagrange equations numerically, beginning with an estimate taken from the graph. You will need initial estimates of x, y, λ. Explain why $\lambda < 0$, and use the initial estimate $\lambda = -1$.

8. a) Graph the level surface $(x - 1)^2 + 2(y - 2)^2 = 4$ for $-1 \le x \le 4$, $0 \le y \le 4$, $0 \le z \le 20$.

   b) Graph the surface $z = x^2 + y^2$ for $-1 \le x \le 4$, $0 \le y \le 4$ and combine it with the graph in a). (**Graph ▷ Combine Graphs**, listing the second surface first in the dialog to get a better picture)

   On a printout of the result, indicate the highest and lowest points where the surfaces meet.

   c) Explain why the two points in b) correspond to where $x^2 + y^2$ has its maximum and minimum values on the *curve* $(x - 1)^2 + 2(y - 2)^2 = 4$ in the plane.

d) Use *Joy* to solve the Lagrange multiplier equations for finding the exact maximum and minimum in c), and express the solutions in decimal form. Which solutions correspond to points on the constraint?

The solutions (x, y) are also the solutions to Exercise 6. These are the points on the ellipse $(x - 1)^2 + 2 (y - 2)^2 = 4$ that are nearest to and furthest from the origin.

Lab 24.1 Local Linearity

Prerequisites:	
Ch. 1	A Brief Tour of *Joy*
Secs. 10.1,2	Graphing surfaces, changing viewpoint

In this lab, you can explore the meanings of local linearity and differentiability for functions of two variables.

■ Before the Lab (Do by Hand)

1. Find the equation of the tangent plane to $f(x, y) = \sin(xy)$ where $(x, y) = (\frac{\pi}{3}, 1)$.

2. Let

$$g(x, y) = \begin{cases} \dfrac{xy}{x^2+y^2} & (x, y) \neq (0, 0) \\ 0 & (x, y) = (0, 0) \end{cases}.$$

   a) Find equations for $\frac{\partial g}{\partial x}$ and $\frac{\partial g}{\partial y}$, where $(x, y) \neq (0, 0)$.

   b) Find $\frac{\partial g}{\partial x}$ and $\frac{\partial g}{\partial y}$, when $(x, y) = (0, 0)$. You will need to do this by evaluating limits, since $\frac{xy}{x^2+y^2}$ is indeterminate at $(0, 0)$.

■ In the Lab

1. a) Graph $\sin(xy)$ for x between $\frac{\pi}{3} \pm 2$ and y between 1 ± 2, i.e., $\frac{\pi}{3} - 2 \leq x \leq \frac{\pi}{3} + 2$ and $-1 \leq y \leq 3$. (**Graph 3D ▷ Surface**)

   b) Repeat, replacing ±2 by smaller increments until the surface appears flat.

2. Graph the tangent plane you found in Question 1 (Before the Lab), using the most recent x- and y-intervals in your graph of the surface.

3. Combine the graphs of the tangent plane and the flat portion of the surface. If the two graphs are not virtually identical, review your reasoning. (**Graph 3D ▷ Combine Graphs**)

4. a) Graph $\frac{xy}{x^2+y^2}$ for x and y between ±1. For graphing purposes, you do not need to specify a value at the origin.

   b) Repeat, replacing ±1 by smaller intervals centered at the origin. Does the surface ever appear flat?

5. Graph $\frac{\partial g}{\partial x}$, which you found in Question 2 (Before the Lab), using the most recent x- and y-intervals in your graph of the surface. For graphing purposes, you do not need to specify a value at the origin.

6. Print your work.

■ After the Lab

If zooming in on the graph of $z = f(x, y)$ around some point $(x, y) = (a, b)$ makes it indistinguishable from a nonvertical plane, we say that f is *locally linear* at (a, b). The plane is tangent to the graph at the point $(x, y) = (a, b)$. The function f is differentiable at (a, b) if there is a nonvertical plane that is tangent to the surface where $x = a$, $y = b$.

It's possible to prove the following theorem, which you can use to help answer the questions below.

Theorem.

a) If the partial derivatives of f exist and are continuous in a region around (a, b), then f is differentiable at (a, b).

b) If f is differentiable at (a, b), then f is continuous there.

1. According to its graph, does $f(x, y) = \sin(xy)$ appear to be locally linear at $(x, y) = (\frac{\pi}{3}, 1)$? Why?

2. a) Find the partial derivatives of $f(x, y) = \sin(xy)$. Are they continuous at all points (x, y)?

   b) Does the theorem guarantee that f is differentiable at every point (a, b)?

3. Does
$$g(x, y) = \begin{cases} \dfrac{xy}{x^2+y^2} & (x, y) \neq (0, 0) \\ 0 & (x, y) = (0, 0) \end{cases}$$

   appear to be locally linear at the origin? Why?

4. a) Based on its graph, does g appear to be continuous at the origin? Why?

   b) Verify your answer to a) by using the equation for $g(x, y)$.

5. a) Based on its graph, does $\frac{\partial g}{\partial x}$ appear to be continuous at the origin? Why?

   b) You found the values of $\frac{\partial g}{\partial x}$ in Question 2 (Before the Lab). Use them to verify your answer to a).

6. According to the theorem, is g differentiable at the origin? Why?

Lab 24.2 Constrained Optimization: Lagrange Multipliers

Prerequisites:

Ch. 1	A Brief Tour of *Joy*
Secs. 3.2,4	Graphing implicit functions, combining graphs
Sec. 5.2	How to Solve Equations Numerically
Sec. 10.3	How to Plot the Level Curves of a Surface
Secs. 11.1,2	Creating and evaluating functions of several variables

In this lab, you can see how to visualize max/min problems in two variables that are subject to some constraint. You can also see how to estimate the solutions to the Lagrange multiplier equations numerically when you can't solve them algebraically.

■ Before the Lab (Do by Hand)

This picture shows the graph of $z = x - \cos(xy)$ being intersected with the cylindrical surface $x^4 - xy + y^4 = 2$. In this graph, x and y range between ± 2.

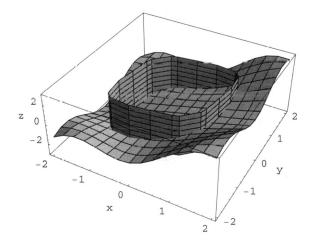

We want to find the maximum and minimum values of the *objective function* $z = x - \cos(xy)$ along the *constraint curve* where the two surfaces intersect. We can also visualize this in two dimensions, using level curves. Here are the level curves of $z = x - \cos(xy)$ together with $x^4 - xy + y^4 = 2$ in the plane, again for x and y between ± 2.

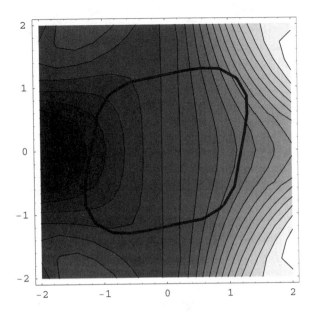

1. If you haven't already done so, read Example 24.6.2 on Lagrange multipliers.

2. On a copy of the graph showing the level curves, mark the approximate points (x, y) where you think $x - \cos(xy)$ will take its maximum and minimum values along $x^4 - xy + y^4 = 2$.

In order to find the coordinates of the max/min points you need to solve the Lagrange multiplier equations, which you will do in the lab.

3. Write down the Lagrange multiplier equations for this max/min problem.

The solutions to the Lagrange multiplier equations may include points where the objective function does not have a maximum or minimum. The solutions with real coordinates are the points where the constraint is tangent to the contour of the objective function. This is the same as saying that the gradients of the objective and constraint functions have the same or opposite direction (or are zero). There may also be solutions containing imaginary numbers. The Lagrange multiplier λ is positive when the two gradients have the same direction and negative when they have opposite directions.

4. Mark the points that you think will be the solutions to the Lagrange equations with real coordinates. It will help to sketch in some possible additional contours where the objective function changes slowly and the contours are far apart.

5. At each point that you marked, use the level curves to sketch the direction of the gradient of $x - \cos(xy)$. The gradient of $x^4 - xy + y^4$ always points away from the origin. Use that to predict the sign of λ at each of these points.

■ In the Lab

1. Create the function $f(x, y) = x - \cos(xy)$. (**Create ▷ Function**)

2. Repeat for $g(x, y) = x^4 - xy + y^4$.

3. a) Choose a point where you think the Lagrange equations might have a solution. Graph the level curves of $f(x, y)$ in a small region around that point. (**Graph 3D ▷ Level Curves**)

   > Choosing a small region helps ensure that there will be enough contours for you to visualize how the constraint and contour interact.

   b) Graph the constraint $g(x, y) = 2$ using the same x-values. Then redraw the graph to make the curve thicker. (**Graph ▷ Implicit Function, solving for one variable; Graph ▷ Modify Graph, Thick curves**)

   c) Combine the two graphs. Decide if the constraint appears to be tangent to the contour of the objective function through the point in question. If necessary, you can recreate the graphs using more contours or over a smaller region. (**Graph ▷ Combine Graphs**)

4. Repeat for any other points where you think the Lagrange equations might have a solution.

5. From the *Joy* menus:

   - Choose **Algebra ▷ Solve**.
   - Fill in the Lagrange multiplier equations and the three variables **x, y,** λ.
   - Click **Solve numerically**.
   - Pick one of the points you think is a solution and fill in its coordinates for the initial estimates of x and y.
   - If you predicted $\lambda > 0$ in Question 5 (Before the Lab), fill in 1 for the initial estimate of λ. If you predicted $\lambda < 0$, fill in -1. If you predicted $\lambda = 0$, then fill in 0.
   - Click **OK**.

6. If *Mathematica* finds a solution, check to see if it is one of the solutions you predicted.

   Mathematica may say that "Newton's method failed to converge to the prescribed accuracy after 15 iterations." It will then give you the latest x, y, λ values it found in trying to approximate a solution.

   You can try putting these values into the **Solve** dialog as the initial estimates for another attempt. *Mathematica* will find a solution if your initial estimate is close enough. You can also try your original initial x- and y-values but change the value of λ. This may help you find the solution you are looking for. On the other hand, there may not be a solution near that point after all.

7. Repeat for all the other points where you think there may be a solution to the Lagrange equations.

8. In your *Mathematica* notebook, evaluate $f(x, y)$ for all the points (x, y) you calculated as solutions to the Lagrange equations.

9. Print your work.

◾ After the Lab

1. If you found any solutions in Questions 5–7 (In the Lab) that you didn't anticipate, add them to the graph you prepared in Question 4 (Before the Lab). Use a different color for these points.

2. Compare the signs for λ that you calculated in the lab with the signs that you predicted before the lab and explain any differences.

3. According to your calculations in Question 8 (In the Lab), which solutions give the maximum and minimum values for the objective function along the constraint? Compare this with your prediction before the lab and explain any differences.

Lab 24.3 Directional Derivatives, Gradients, and the Cobb–Douglas Function

---

Prerequisites:

Ch. 1	A Brief Tour of *Joy*
Sec. 3.4	How to Combine Graphs
Sec. 10.3	How to Plot the Level Curves of a Surface
Sec. 11.1	How to Create a Function of Several Variables
Sec. 14.1	How to Plot a Gradient Field in 2D

The Cobb–Douglas function $q(x, y) = c\,x^a\,y^b$ plays an important role in the economic theory of production. In this lab, you can use the Cobb–Douglas function to model a production process having two inputs and see the economic consequences of the mathematical properties of the function.

■ Before the Lab (Do by Hand)

In the Cobb–Douglas function you can interpret x as the total cost of *labor* used in a production process by a firm, y as the total cost of *capital*, and $q(x, y)$ as the total value of the output. We call q the *production function*. Its level curves (contours) are called *isoquants*, since the output value is the same along each curve.

In this example, we'll take $q(x, y) = 2.2\,x^{0.7}\,y^{0.3}$. The actual units might be, for instance, thousands of dollars. Here are the isoquants:

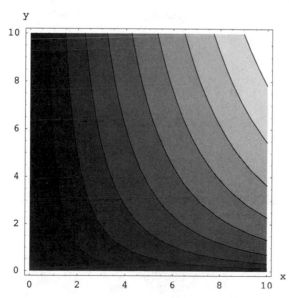

1. If either the labor or capital variable is increased but the output remains the same, what happens to the other variable?

2. If the firm increases both labor and capital by the same factor k, how will the output $q(x, y)$ change?

   Scaling up production means to increase the labor and capital inputs x and y without changing their proportions. The answer to this question illustrates the property called *constant returns to scale* in economics. The Cobb–Douglas function has this property as long as the exponents add up to 1.

3. Suppose the current costs of production are $x = 3$ (labor) and $y = 1$ (capital). The *expansion path* consists of the points (x, y) where capital and labor costs continue in the same proportion as production is scaled up. Find an equation for the expansion path that keeps costs in the same proportion as currently.

■ In the Lab

1. Create the function $q(x, y) = 2.2\, x^{0.7}\, y^{0.3}$. (**Create ▷ Function**)

2. Graph the level curves of $q(x, y)$ for $2 \le x \le 5,\ 0 \le y \le 3$, with 20 contours. Do *not* shade the contours. Assign the name `isoquants` to the graph. (**Graph 3D ▷ Level Curves**)

   Turning off shading will make the graphs you construct easier to read.

3. Graph the expansion path you found in Question 3 (Before the Lab) for $2 \le x \le 5,\ 0 \le y \le 3$. Assign the name `path`. (**Graph ▷ Graph f(x)**)

4. Plot the gradient field of $q(x, y)$ for $2 \le x \le 5,\ 0.001 \le y \le 3$, and combine it with the graph *isoquants*. Assign the name `field`. (**Graph ▷ Gradient Field**)

   Choosing y this way avoids the x-axis, where the gradient is undefined.

5. Combine the graphs *field* and *path*. (**Graph ▷ Combine Graphs**)

   If your graph doesn't have **Equal Scales**, apply **Graph ▷ Modify Graph** to the result. This is needed to display right angles correctly.

6. Print your work.

■ After the Lab

1. a) Find the gradient of $q(x, y)$ at $(3, 1)$. What is the equation of the line through $(3, 1)$ along the gradient?

   b) Sketch the line in a) on your graph from Question 5 (In the Lab). If it doesn't have the same direction as the gradient drawn by *Mathematica*, review your reasoning.

   ⚠ The vectors in the gradient field are not plotted to scale. Their lengths are correct relative to each other but are scaled to give a good overall picture.

2. Suppose you increase production starting at (3, 1) and follow the gradient rather than the expansion path. If you double the cost of labor, what will be the cost of capital? What will be the value of the output? Are labor, capital, and output scaled up proportionally along the gradient?

3. The directional derivative $D_u q(a, b)$ of q at (a, b) in the direction of a unit vector u measures the rate of change of output when the inputs change according to u.

    a) Calculate $D_u q(3, 1)$ in the direction of a unit vector u having the direction of the original expansion path.

    b) Calculate the directional derivative $D_u q(3, 1)$ in the direction of a unit vector u having the direction of the gradient.

    c) Which directional derivative is larger? What property of the gradient makes this happen?

    d) Explain how the graph in Question 1b) (After the Lab) gives the answer to 3c) without any calculations.

Chapter 25
Multiple Integrals

Examples and Exercises

Prerequisites:
Ch. 1	A Brief Tour of *Joy*
Sec. 10.1	How to Graph a Surface
Secs. 13.1,2	Symbolic and numeric multiple integrals

■ 25.1 Riemann Sums

Additional Prerequisites:
Sec. 9.1	How to Create a Sequence
Sec. 10.3	How to Plot the Level Curves of a Surface (Contours)
Sec. 11.1	How to Create a Function of Several Variables

25.1.1 Example: Calculating a Riemann Sum

Here's how to compute a Riemann sum approximation to

$$\int_0^4 \int_1^2 (1 + \sin(3x)\ln(2y))\, dy\ dx\,.$$

- From the *Joy* menus, choose **Create ▷ Function**.
- Fill in `f[x, y]` in the first input field and `1 + Sin[3 x] Log[2 y]` in the second.

    You can also use the **Functions** button on the *Joy* palette to enter the latter.

- Click **OK**.

Mathematica defines the function:

▷ `Define f[x, y] to be 1 + Sin[3 x] Log[2 y]`

The intervals $0 \le x \le 4$ and $1 \le y \le 2$ determine a rectangle in the xy-plane. The integral represents the volume between the surface $z = f(x, y)$ and the xy-plane over this rectangle.

⚠ For a function that takes negative values, the integral represents the net or signed volume between the surface and the xy-plane. This is the volume above the plane minus the volume below it.

• Graph $f(x, y)$ for $0 \le x \le 4$, $1 \le y \le 2$. (**Graph 3D ▷ Surface**)

Here is the graph:

▷ `Plot f[x, y] for 0 ≤ x ≤ 4  and 1 ≤ y ≤ 2`

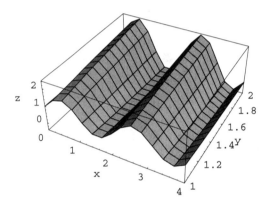

Out[3]= `- SurfaceGraphics -`

💡 If you are using *Mathematica* Version 4 with interactive 3D graphics, you can display the coordinates by choosing **Graph ▷ Modify Graph** and specifying **Axis Labels**.

We partition each interval into n parts, which partitions the rectangle into n^2 subrectangles. Then we choose a point in each subrectangle.

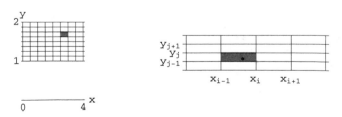

Partition the x- and y-intervals and choose a point in each subrectangle.

Each subrectangle is defined by intervals $x_{i-1} \le x \le x_i$, $y_{j-1} \le y \le y_j$ for some indices i and j between 1 and n. In this subrectangle, we choose a point (c_i, d_j).

For example, partition the intervals into $n = 2$ subintervals, each with widths Δx and Δy:

$$\Delta x = \frac{4-0}{2} = 2, \quad \Delta y = \frac{2-1}{2} = 0.5 .$$

The partition points are

$$x_0 = 0, \quad x_1 = 2, \quad x_2 = 4;$$
$$y_0 = 1, \quad y_1 = 1.5, \quad y_2 = 2.$$

In each of the four subrectangles let (c_i, d_j) be the point in the upper-right corner. That is, let $c_i = x_i$ and $d_j = y_j$ for $i, j = 1, 2$. This is an arbitrary choice, and you can choose other points in the exercises at the end of this section.

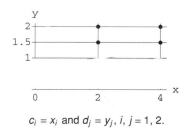

$c_i = x_i$ and $d_j = y_j$, $i, j = 1, 2$.

The Riemann sum for these points is

$$f(c_1, d_1)\,\Delta y\,\Delta x \; + \; f(c_1, d_2)\,\Delta y\,\Delta x \; + \; f(c_2, d_1)\,\Delta y\,\Delta x \; + \; f(c_2, d_2)\,\Delta y\,\Delta x$$

$$= ((1 + \sin(3 \times 2)\ln(2 \times 1.5)) + (1 + \sin(3 \times 2)\ln(2 \times 2)) +$$
$$(1 + \sin(3 \times 4)\ln(2 \times 1.5)) + (1 + \sin(3 \times 4)\ln(2 \times 2)))\,(0.5 \times 2)$$

$$= 1.97234 .$$

Using summation notation, we can write this sum as

$$\sum_{i=1}^{2} \sum_{j=1}^{2} f(c_i, d_j)\,\Delta y\,\Delta x \;=\; \sum_{i=1}^{2} \sum_{j=1}^{2} f(x_i, y_j)\,\Delta y\,\Delta x$$

$$= \sum_{i=1}^{2} \sum_{j=1}^{2} (1 + \sin(3\,x_i)\ln(2\,y_j))\,\Delta y\,\Delta x .$$

Each term approximates the volume over the corresponding subrectangle. For instance, $f(x_1, y_1)\,\Delta y\,\Delta x$ is the volume of a box whose height is $f(2, 1.5)$ and whose base corresponds to $0 \le x \le 2, 1 \le y \le 1.5$:

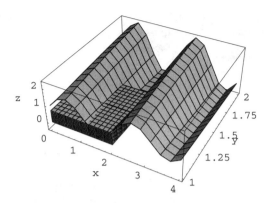

Approximating the volume over one subrectangle.

The picture suggests that we will get a better approximation with more subrectangles. Suppose we partition the intervals into 100 subintervals each and again use the right endpoint of each subinterval to choose x and y.

The Riemann sum for this partition is

$$\sum_{i=1}^{100}\sum_{j=1}^{100} f(c_i, d_j)\,\Delta y\,\Delta x = \sum_{i=1}^{100}\sum_{j=1}^{100} (1 + \sin(3\,x_i)\ln(2\,y_j))\,\Delta y\,\Delta x,$$

where

$$\Delta x = 0.04, \quad \Delta y = 0.01,$$

$$
\begin{aligned}
c_i &= x_i = 0.04\,i, & i &= 1, 2, 3, \ldots, 100 \\
d_j &= y_j = 1 + 0.01\,j, & j &= 1, 2, 3, \ldots, 100
\end{aligned}
$$

Here's how to use *Joy* to evaluate the corresponding Riemann sum approximation to the integral. First we'll define Δx, Δy, c_i, d_j:

- If necessary, click in your *Mathematica* notebook beneath the last output, so that a horizontal line appears.
- Type Δx = 0.04 and press ⌈SHIFT⌉ − ⌈RET⌉.

  We'll explain later in this example why it's important to use decimals here. Use the **Greek** button on the *Joy* palette to enter Δ. Do *not* leave a space between Δ and x because they are not individual quantities being multiplied together.

  ♡ Since you'll need to enter Δ several times, it's helpful to type the shortcut ⌈ESC⌉ − D − ⌈ESC⌉ instead (don't type the dashes –). If you use the Greek palette, you can drag it out of the way so you also use the *Joy* palette when you need it. When you are done, close the Greek palette by clicking the close box in its upper-left corner (Macintosh) or the close button **X** in the upper-right corner (Windows).

In[4]:= Δx = 0.04

Out[4]= 0.04

- Similarly, enter Δy = 0.01 and press SHIFT – RET.

    Mathematica may warn of a possible spelling error. You may ignore the message.

In[5]:= Δy = 0.01

Out[5]= 0.01

- Choose **Create ▷ Sequence**.
- Fill in c_i in the first field and (Δx) i in the second.
- Check **Define a second sequence**.
- Fill in d_j and 1 + (Δy) j.
- Click **OK**.

Mathematica defines the two sequences:

▷ Define c_i to be (Δx) i and define d_j to be $1 + \Delta y$ j

Now we'll evaluate the Riemann sum:

- Click in your *Mathematica* notebook beneath the last cells, so that a horizontal line appears.
- Type f[c_i, d_j] Δy␣Δx and drag across this expression to select it.

    The symbol ␣ means to leave a space to denote multiplication. You can also click in the expression and then click the **Select** button on the *Joy* palette several times until it is selected.

- Click the **Sum** button on the *Joy* palette.

    This creates the inner sum for the Riemann sum, with the input field for *j* selected. Since Δx is constant, it can be placed inside the inner sum as shown here or else entered with the outer sum.

$$\sum_{\square=\square}^{\square} f[c_i, d_j] \, \Delta y \, \Delta x$$

- Type j, TAB, 1, TAB, 100 Then select the whole expression.
- Click the **Sum** button again.
- Type i, TAB, 1, TAB, 100
- Press SHIFT – RET.

The Riemann sum is:

$In[7]:=$ $\displaystyle\sum_{i=1}^{100}\sum_{j=1}^{100} \mathtt{f[c_i, d_j]}\ \Delta y\ \Delta x$

$Out[7]=$ 4.04467

> ♡ Expressing Δx and Δy using decimals rather than fractions causes *Mathematica* to evaluate this sum numerically rather than give a symbolic answer in terms of sines and logarithms. Since there are $100^2 = 10,000$ terms, this is important to keep in mind. Symbolic evaluation could cause you to run short of memory.

In this case, you can find the exact value of the integral by hand or using the **Definite Integral** button on the *Joy* palette. Section 13.1 gives a step-by-step illustration of how to enter a multiple integral.

$In[8]:=$ $\displaystyle\int_0^4 \left(\int_1^2 \mathtt{f[x, y]}\ \mathtt{dy} \right) \mathtt{dx}$

$Out[8]=$ $\dfrac{11}{3} + \dfrac{\mathtt{Cos[12]}}{3} - \dfrac{\mathtt{Log[2]}}{3} + \dfrac{1}{3}\,\mathtt{Cos[12]\ Log[2]} + \dfrac{2\,\mathtt{Log[4]}}{3} - \dfrac{2}{3}\,\mathtt{Cos[12]\ Log[4]}$

You can express this as a decimal using the **Numeric** button on the *Joy* palette:

$In[9]:=$ $\mathtt{N}\Big[\dfrac{11}{3} + \dfrac{\mathtt{Cos[12]}}{3} - \dfrac{\mathtt{Log[2]}}{3} + \dfrac{1}{3}\,\mathtt{Cos[12]\ Log[2]} + \dfrac{2\,\mathtt{Log[4]}}{3} - \dfrac{2}{3}\,\mathtt{Cos[12]\ Log[4]}\Big]$

$Out[9]=$ 4.05618

So the Riemann sum 4.04467 was a fair approximation to the integral. You can improve the approximation by taking more subrectangles.

Exercises

In Exercises 1–3, use the **Sum** button to construct the Riemann sums, as shown in the example in this section, Example 25.1.1.

1. Estimate $\int_0^4 \int_1^2 (1 + \sin(3\,x)\ln(2\,y))\,dy\ dx$ with a Riemann sum, dividing the x- and y-intervals into 100 parts each and using the left-hand endpoint of each subinterval to choose a point in each subrectangle.

2. Estimate the integral in Exercise 1 using the midpoint of each subinterval to choose a point in each subrectangle.

3. a) Graph $\sin x \, \sin y$ for x and y between 0 and $\pi/2$. Use the graph to explain whether choosing left and right endpoints in a Riemann sum will overestimate or underestimate $\int_0^{\pi/2} \int_0^{\pi/2} \sin x \, \sin y\, dy\ dx$.

b) Find Riemann sums to overestimate and underestimate $\int_0^{\pi/2} \int_0^{\pi/2} \sin x \sin y \, dy \, dx$, using 100 subintervals each for the x- and y-intervals. Use decimals instead of fractions so *Mathematica* will evaluate the sums numerically.

c) Find the exact value of the integral and verify that the estimates in b) act the way you predicted (you can integrate this function by hand).

In Exercises 4–6, use the sign of the function and the idea that the integral is a limit of Riemann sums to help answer part b).

4. a) Graph $\sin(x + y)$ for x and y between ± 2.

    b) Using the graph, predict whether $\int_{-2}^2 \int_{-2}^2 \sin(x + y) \, dx \, dy$ will be positive, negative, or zero. Check your prediction by calculating the integral (you can do this by hand).

5. a) Graph the surface $z = \sin(x^2 + y^2)$ for x and y between ± 2. Then graph its level curves for the same intervals. **(Graph 3D ▷ Surface, Level Curves)**

    b) For which x, y between ± 2 is $\sin(x^2 + y^2)$ positive, zero, and negative?

    c) Using a) and b), predict whether $\int_{-2}^2 \int_{-2}^2 \sin(x^2 + y^2) \, dx \, dy$ will be positive, negative, or zero. Check your prediction by calculating the integral with *Joy* (integrate numerically).

6. a) Graph the level curves of $\ln(x^2 + y)$ for x and y between 0.1 and 2. Find the parts of the plot where the function is zero, positive, and negative. **(Graph 3D ▷ Level Curves)**

    b) Choose a rectangle within the plot on which the integral will be positive and another where it will be negative. Use *Joy* to integrate the function over these rectangles to confirm your prediction.

■ 25.2 Volume and Cross-Section Area

25.2.1 Example: The Cross-Section Area of a Solid

This example illustrates how to visualize the *volume of a solid as the integral of its cross-section area*.

If the parallel cross-sections of a solid are congruent to one another, then the volume of the solid is the cross-section area A times the length L of the solid in the perpendicular direction. For example, the volume of a right circular cylinder is the area of its cross-sections (πr^2) times its height (h).

When the cross-sections are not all congruent, A changes as you move along the length of the solid. For example, if the sections are perpendicular to the y-axis, then A becomes a function $A(y)$ and the volume is $\int A(y) \, dy$, integrated between the smallest and largest values of y in the solid. Each term $A(y_i) \, \Delta y$ in a Riemann sum approximates the volume of a thin slice of the solid with thickness Δy.

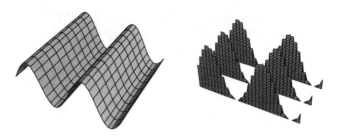

Volume = *AL* = area × length.

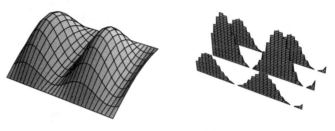

Volume = $\int A(y)\, dy$.

Here's an example showing how to calculate the volume using this method.

The double integral

$$\int_0^\pi \int_0^4 (1 + \sin 3\, x) \sin y \, dx \ dy$$

represents the volume of the solid whose top is the surface $z = (1 + \sin 3\, x) \sin y$ and whose base is the *xy*-plane, for $0 \le x \le 4, 0 \le y \le \pi$.

• Graph $(1 + \sin 3\, x) \sin y$ for $0 \le x \le 4, \ 0 \le y \le \pi$. (**Graph 3D ▷ Surface**)

▷ Plot (1 + Sin[3 x]) Sin[y] for 0 ≤ x ≤ 4   and 0 ≤ y ≤ π

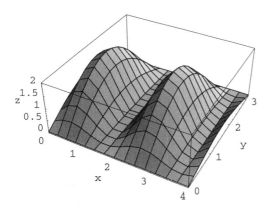

Out[2]= - SurfaceGraphics -

The cross-section perpendicular to the y-axis changes as you change the value of y. For each value of y in $[0, \pi]$, the area of the cross-section is the inner integral $A(y) = \int_0^4 (1 + \sin 3\,x) \sin y \, dx$. The volume of the solid is $\int_0^\pi A(y)\,dy$.

For example, here is what the cross-section looks like when $y = 1.5$. It is the shaded region on the front face of the solid.

Cross-section at $y = 1.5$ (shaded), area $= A(1.5)$.

You can construct the graphs in this section and similar graphs that show cross-sections in 3D with *Mathematica*'s If function. Express the cross-section using If, graph it as a parametric surface, and combine the graph with the top surface.

The cross-section area $A(1.5)$ is the area under the curve $z = (1 + \sin 3\,x) \sin 1.5$ for $0 \le x \le 4$. We'll graph the curve.

- Graph $(1 + \sin 3\,x) \sin 1.5$ for $0 \le x \le 4$ and $0 \le z \le 2$. (**Graph ▷ Graph f(x)**)

The result is:

▷ Graph $(1 + \text{Sin}[3\,x])\,\text{Sin}[1.5]$ on $0 \le x \le 4$ and $0 \le z \le 2$.

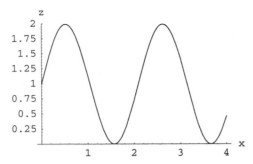

Out[3]= - Graphics -

You can integrate by hand or using the **Definite Integral** button to find the area. Section 13.1 gives a step-by-step illustration of how to use the button.

In[4]:= $\displaystyle\int_0^4 (1 + \text{Sin}[3\,x])\,\text{Sin}[1.5]\,d\mathbf{x}$

Out[4]= 4.0419

So the volume of the slice of the solid from $y = 1.5$ to $y = 1.6$ has $\Delta y = 0.1$ and volume $\approx (4.0419)(0.1) = 0.40419$.

For any y, the cross-section area is $A(y) = \int_0^4 (1 + \sin 3\,x)\sin y\,dx$:

In[5]:= **area** $= \displaystyle\int_0^4 (1 + \text{Sin}[3\,x])\,\text{Sin}[y]\,d\mathbf{x}$

Out[5]= $\dfrac{\text{Sin}[y]}{3} - \dfrac{1}{3}\,(-12 + \text{Cos}[12])\,\text{Sin}[y]$

The volume is approximated by Riemann sums $\sum_{i=1}^n A(y_i)\,\Delta y$ and given exactly by the integral $\int_0^\pi A(y)\,dy$:

In[6]:= **volume** $= \displaystyle\int_0^\pi \mathbf{area}\,d\mathbf{y}$

Out[6]= $\dfrac{2}{3}\,(13 - \text{Cos}[12])$

By using the **Numeric** button on the *Joy* palette, you can express the volume in decimal form:

In[7]:= **N[volume]**

Out[7]= 8.1041

You can also find the volume by integrating cross-section areas perpendicular to the x-axis (see Exercise 2). The final result must be the same in each case.

Exercises

1. a) Graph $z = e^x \sin y$ for $0 \le x \le 2$, $0 \le y \le \pi$. This is the top of a solid whose base is a rectangle in the xy-plane.

    If you are using *Mathematica* Version 4 with interactive 3D graphics, display the coordinates by choosing **Graph** ▷ **Modify Graph** and specifying **Axis Labels**.

    b) Predict which cross-section will have the larger area, the section where $x = 0.5$ or the one where $y = ?$. Confirm your answer by integration.

    c) Find the volume of the solid.

2. a) The graph of $z = (1 + \sin 3\,x) \sin y$ appears in the example of this section, Example 25.2.1. Interpret these integrals as cross-section areas and use the graph to predict which will be larger:

    $$\int_0^\pi (1 + \sin 3) \sin y \, dy \text{ and } \int_0^\pi (1 + \sin 10.5) \sin y \, dy.$$

    b) Confirm your answer to a) by integrating.

    c) Find the volume of the solid by integrating first with respect to y and then with respect to x. Compare your answer to the volume found in Example 25.2.1.

3. a) Graph the solid whose top is $z = x^2 - xy + 2\,y^2$ and whose base is the xy-plane, for x and y between ±2. Then graph its cross-section at $x = -1$.

    b) By hand, find the cross-section area at $x = -1$. Then find the area for arbitrary x.

    c) By hand, find the volume of the solid by integrating the area in b).

■ 25.3 Double Integrals on Nonrectangular Regions

Additional Prerequisite:
    Secs. 10.4,5   Graphing level surfaces, combining graphs

25.3.1 Example: Visualizing a Solid Whose Base Is Not a Rectangle

Here's how to graph the solid whose top is the plane $z = 5x - y + 5$ and whose base is the triangle bounded by the x- and y-axes, the parabola $y = x^2 + \frac{1}{2}$, and the line $x = 2$.

The base region (shaded) is contained in the rectangle with $0 \le x \le 2$, $0 \le y \le 4.5$.

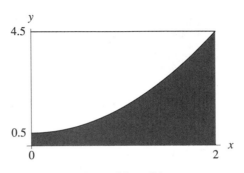

Base of the solid.

- Graph $5x - y + 5$ for $0 \le x \le 2$, $0 \le y \le 4.5$. (**Graph 3D ⊳ Surface**).

▷ Plot 5 x - y + 5 for 0 ≤ x ≤ 2 and 0 ≤ y ≤ 4.5`

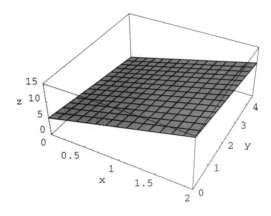

Out[2]= - SurfaceGraphics -

Next, we put in the points on and directly above the parabola, forming a *parabolic cylinder*. This is a level surface, $y - x^2 = \frac{1}{2}$ with z arbitrary. We'll graph the level surface and combine it with the plane.

- Choose **Graph 3D ⊳ Level Surface**.
- Fill in the expression $y - x^2$ and the value $1/2$.
- Fill in the intervals $0 \le x \le 2$, $0 \le y \le 4.5$, $0 \le z \le 15$. Click **OK**.

    We choose the z-values from the graph of the plane. If you are using *Mathematica* Version 4 with interactive 3D graphics, you can display the coordinates by choosing **Graph ⊳ Modify Graph** and specifying **Axis Labels**.

- Combine the graphs of the plane and the surface using their output names or numbers. (**Graph ⊳ Combine Graphs**)

The result is:

▷ Combining the graphs {%2, %3} gives

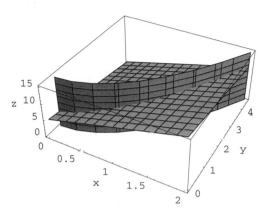

Out[4]= - Graphics3D -

The solid we want is in the front of the picture, facing the viewer. Its top is the slanted plane and its base is the region in the *xy*-plane shown earlier in this example. Its rear side is the parabolic cylinder, and its remaining sides are the vertical planes $y = 0$, $x = 0$, and $x = 2$.

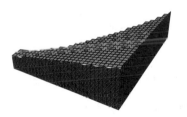 

Solid with nonrectangular base (two views).

The volume of our solid is the *integral of the cross-section area*. We can take cross-sections parallel to the *yz*-plane at each *x*, $0 \leq x \leq 2$. The area $A(x)$ of each cross-section can be found by hand or using the **Definite Integral** button (Section 13.1):

In[5]:= **area** $= \int_0^{x^2 + \frac{1}{2}} (5\,x - y + 5) \; d y$

Out[5]= $5 \; (1 + x) \; \left(\frac{1}{2} + x^2 \right) - \frac{1}{2} \; \left(\frac{1}{2} + x^2 \right)^2$

The volume of the solid is $\int_0^2 A(x)\, dx$:

In[6]:=   **volume =** $\int_0^2$ **area d x**

Out[6]= $\dfrac{771}{20}$

or, in decimal form,

In[7]:=   **N[volume]**

Out[7]=   38.55

That is, the volume is

$$\int_0^2 \int_0^{x^2+\frac{1}{2}} (5\,x - y + 5)\,dy\ dx = 38.55\,.$$

Taking cross-sections at each y instead gives the same result (see Exercise 3).

Exercises

1. a) Graph the curve $y^2 - 2\,y - x + 3 = 0$ for x between ± 6. (**Graph ▷ Implicit Function**)

    b) Use a double integral to find the area between the curve and the line $x = 6$.

2. a) By hand, sketch the region in the xy-plane determined by the limits of integration in

    $$\int_1^3 \int_0^{\ln x} (x + y)^3\ dy\ dx\,.$$

    b) Use *Joy* to graph the solid whose volume is the integral in a). Enter $\ln x$ as **Log[x]** or use the **Functions** button on the *Joy* palette. (**Graph 3D ▷ Surface, Level Surface**)

    c) Print the graph and sketch in some typical cross-sections whose areas are being integrated in a).

3. This exercise refers to the example of this section, Example 25.3.1.

    a) Take a cross-section at an arbitrary value of y instead of x and find the cross-section area of the solid. You will need to consider the cases $y < 1/2$ and $y \geq 1/2$ separately.

    b) Find the volume of the solid using the cross-sections in a) and compare it to the value found in Example 25.3.1.

4. a) Calculate (by hand or with *Joy*)

    $$\int_{-1}^1 \int_{x^3}^x (x^2 + 2\,xy + y^2)\,dy\ dx\,.$$

b) Why can't the integral in a) be the volume of a solid? Graph the solid the integral appears to represent and explain why the integral is not the volume. (**Graph 3D** ▷ **Surface, Level Surface**)

c) Find the volume of the solid.

■ 25.4 Triple Integrals

Additional Prerequisite:
Secs. 10.4,5   Graphing level surfaces, combining graphs

25.4.1 Example: Intersecting Two Cylinders

Here's an illustration of how to use *Joy* to help visualize a solid and to find its volume by triple integration. In this example, the solid is formed by intersecting the two cylinders $x^2 + y^2 = 4$ and $z^2 + y^2 = 4$.

- From the *Joy* menus, choose **Graph 3D** ▷ **Level Surface**.
- Fill in $x^2 + y^2$ and the value 4.
- Fill in the intervals $-3 \le x \le 3$, $-3 \le y \le 3$, $-3 \le z \le 3$. Click **OK**.

  The equations of the two cylinders show that x, y, and z will each vary between ± 2. Taking slightly larger intervals than is necessary often gives a better picture.

▷ Graph of the level surface $x^2 + y^2 = 4$ for
 $-3 \le x \le 3$, $-3 \le y \le 3$, $-3 \le z \le 3$.

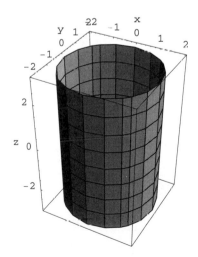

Out[2]= - Graphics 3D -

- From the *Joy* palette, click **Last Dialog**.
- Replace the function by $z^2 + y^2$.
- Click **OK**.

▷ Graph of the level surface $z^2 + y^2 = 4$ for
  $-3 \leq x \leq 3$, $-3 \leq y \leq 3$, $-3 \leq z \leq 3$.

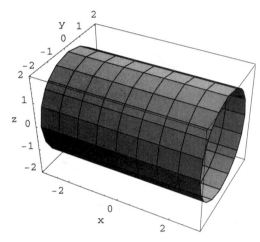

Out[3]= - Graphics3D -

- Combine `Out[2]`, `Out[3]`, using your output numbers if they are different. (**Graph** ▷ **Combine Graphs**)

The result is (the paraphrase shows *Mathematica*'s abbreviation %... for `Out[...]`):

▷ Combining the graphs {%2, %3} gives

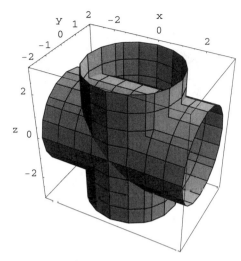

Out[4]= - Graphics3D -

The solid consists of the points that lie within both cylinders. Because of the symmetry of the solid, we can focus on the points with coordinates ≥ 0, which make up 1/8 of the whole. We can visualize them better by graphing and combining the level surfaces for x, y, z between 0 and 2:

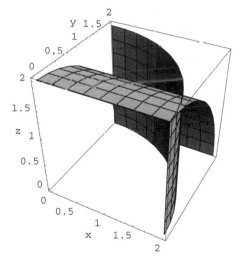

1/8 of the intersection.

For these points,

for each x between 0 and 2, y varies between 0 and $\sqrt{4 - x^2}$;
for each such x and y, z varies between 0 and $\sqrt{4 - y^2}$.

So the volume of the solid is

$$8 \int_0^2 \int_0^{\sqrt{4-x^2}} \int_0^{\sqrt{4-y^2}} dz \ dy \ dx \ .$$

We can calculate this by hand or using the **Definite Integral** button (Sec. 13.1):

$$In[78]:= \quad 8 \int_0^2 \int_0^{\sqrt{4-x^2}} \int_0^{\sqrt{4-y^2}} dz \ dy \ dx$$

$$Out[78]= \quad \frac{128}{3}$$

Exercises

If you are using *Mathematica* Version 4 with interactive 3D graphics, you can display the coordinates of the surfaces in these exercises by choosing **Graph ▷ Modify Graph** and specifying **Axis Labels**.

1. a) Graph $z = x^2 + 2y^2$ and $z = 10 - x^2 - y^2$, over x- and y-intervals large enough to see where these surfaces intersect.

   b) Find the volume of the solid contained between the surfaces, using triple integration.

2. Find the volume in the example of this section, Example 25.4.1, using a triple integral of the form $\int\int\int dz \ dx \ dy$ and compare your result with what we found in that example. Does switching the order of integration make a difference?

3. a) Graph the level surface $z^2 - x^2 - y^2 = 1$ for x, y, z between ±5. The surface will consist of two symmetric "bowls" facing away from each other. (**Graph 3D ▷ Level Surface**)

   b) Find the total volume contained by the two bowls for z between ±5.

■ 25.5 Integrating in Polar Coordinates

> Additional Prerequisite:
>   Sec. 3.8   How to Graph in Polar Coordinates

In these exercises, recall that the area of a region in polar coordinates is given by an integral of the form $\int\int f(r, \theta) \, r \ dr \ d\theta$. To graph curves in polar coordinates, choose **Graph ▷ Polar Plot**.

Exercises

1. a) Graph the curve $r = \cos(2\,\theta)$ in polar coordinates for $0 \le \theta \le 2\pi$.

    b) Find the values of θ that determine one "leaf" of the graph and find its area.

2. a) Graph the two cardioids $r = 1 - \cos\theta$ and $r = 1 + \cos\theta$.

    b) Find the area between the cardioids.

3. a) Graph the circle $r = \cos\theta$ in polar coordinates for $0 \le \theta \le 2\pi$,

    b) Evaluate $\int_0^{2\pi} \int_0^{\cos\theta} r\, dr\, d\theta$.

    c) Why does this integral *not* give the area within the circle? Answer by tracing the motion of a point around the circle for $0 \le \theta \le 2\pi$.

    d) What integral *does* give the area?

4. a) The standard equation for an ellipse in Cartesian coordinates is

    $$\frac{x^2}{a^2} + \frac{y^2}{b^2} = 1,$$

    where $a > 0$, $b > 0$. Find the area contained by the ellipse by converting the equation to polar coordinates and calculating an appropriate integral.

    Mathematica's output may contain `Sign[a]`. This function is defined by

    $$\text{Sign}(a) = \begin{cases} 1, & a > 0 \\ -1, & a < 0 \\ 0, & a = 0 \end{cases}.$$

    b) When $a = b$, your answer should turn out to be the area of a circle of radius a. Does it?

5. The *centroid* of a region R in the plane is the point $(\bar{x}, \bar{y})$ given by

    $$\bar{x} = \frac{\iint x\, dy\, dx}{\text{area}(R)}, \quad \bar{y} = \frac{\iint y\, dy\, dx}{\text{area}(R)}.$$

    The centroid is the balancing point for a thin plate in the shape of R when the density of the plate is constant. In this case, it is the same as the center of mass.

    Find the centroid of the region contained by the cardioid $r = 1 - \cos\theta$.

Lab 25.1 Double Integrals and Riemann Sums

Prerequisites:
Ch. 1	A Brief Tour of *Joy*
Sec. 4.2	How to Tabulate an Expression
Sec. 11.1	How to Create a Function of Several Variables
Sec. 13.2	How to Calculate Multiple Integrals Numerically
Sec. 25.1	Riemann Sums

This lab will help you understand how Riemann sums approximate a double integral. It also generalizes results about left- and right-endpoint Riemann sums from one variable to two.

■ Before the Lab (Do by Hand)

Let $f(x, y) = 1 + \ln(x + y) \sin(x^2)$. *Mathematica* cannot find the exact value of

$$\int_0^1 \int_1^3 f(x, y)\, dy\ dx$$

but can only approximate it. Here is a graph of this function over the rectangle R defined by $0 \le x \le 1, 1 \le y \le 3$.

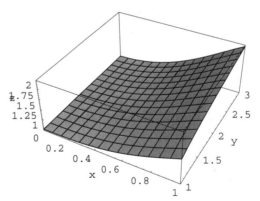

1. If you haven't already read Example 25.1.1 on Riemann sums, do that now.

2. a) Use a partial derivative to show that $f(x, y)$ is increasing in the x-direction at all points in the rectangle R.

    b) Repeat for the y-direction.

3. At which points in the highlighted subrectangle of R on page 485 does $f(x, y)$ reach its maximum and minimum?

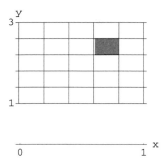

4. Partition each interval $0 \le x \le 1$ and $1 \le y \le 3$ into two equal parts.

    a) Use your answer to Question 3 to find a Riemann sum that is guaranteed to be an overestimate of $\int_0^1 \int_1^3 f(x, y)\, dy\, dx$.

    b) Repeat to find an underestimate of this integral.

5. Suppose you use your overestimate as an approximation to the integral. Explain why the error in your approximation is at most the difference between the two estimates. This is called an *error bound* for the integral.

6. We divide each interval $0 \le x \le 1$, $1 \le y \le 3$ into n parts, where n is arbitrary. Using the notation of Example 25.1.1, let (c_i, d_j) be the point we choose in the subrectangle defined by intervals $x_{i-1} \le x \le x_i$, $y_{j-1} \le y \le y_j$.

    a) Express Δx, Δy in terms of n

    b) Suppose you want to construct a Riemann sum that overestimates the integral. Express c_i in terms of Δx and i. Express d_j in terms of Δy and j.

    c) Repeat b) for an underestimate, but to distinguish this from b) call the point (a_i, b_j).

■ In the Lab

1. Create the function $f(x, y) = 1 + \ln(x + y)\sin(x^2)$. (**Create ▷ Function**)

    Enter $\ln(x + y)$ as `Log[x + y]` or use the **Functions** button on the *Joy* palette.

Now you'll construct a Riemann sum to estimate $\int_0^1 \int_1^3 f(x, y)\, dy\, dx$. Example 25.1.1 gives detailed instructions for entering expressions such as Δx and Σ. We will proceed somewhat differently so that you can evaluate the Riemann sum for different values of n and also find overestimates and underestimates.

2. First, you'll define Δx, Δy:

    - Click in your *Mathematica* notebook beneath the last cells, so that a horizontal line appears.

- Type $\triangle\mathbf{x}$ = and fill in the value you used in Question 6a) (Before the Lab), using decimals so that *Mathematica* will evaluate the sums numerically (e.g., 1.0 instead of 1).
- Press [SHIFT] − [RET].

Mathematica's output should be in terms of *n* and not have a specific numeric value.

- Repeat for $\triangle\mathbf{y}$. *Mathematica* may warn of a possible spelling error.

3. Next you'll define c_i, d_j for overestimating the integral.

- Choose **Create** ▷ **Sequence**.
- Fill in c_i and then fill in the expression for c_i you found in Question 6b) (Before the Lab).
- Check **Define a second sequence**.
- Fill in d_j and then fill in the expression for d_j in Question 6b).
- Click **OK**.

4. The next step is to construct a Riemann sum for overestimating the integral.

- If necessary, click in your notebook beneath the last output to make a horizontal line appear.
- Type **over** = $\sum_{i=1}^{n} \sum_{j=1}^{n} \mathbf{f[c_i, d_j]} \, \triangle\mathbf{y} \, \triangle\mathbf{x}$.
- Press [SHIFT] − [RET].

5. Next you'll define a_i, b_j for underestimating the integral.

- Choose **Last Dialog** to reopen the **Sequence** dialog.
- Define a_i, b_j using your answers to Question 6c) (Before the Lab).
- Press [SHIFT] − [RET].

6. Now you'll construct the Riemann sum for underestimating the integral.

- In your notebook, type **under** = $\sum_{i=1}^{n} \sum_{j=1}^{n} \mathbf{f[a_i, b_j]} \, \triangle\mathbf{y} \, \triangle\mathbf{x}$.

  You can copy and paste the summation from **over** and make the necessary changes.

- Press [SHIFT] − [RET].

7. Next, you'll create a table comparing the under- and overestimates for several values of *n*.

- Choose **Create** ▷ **List**.
- Fill in **n, over, under, over – under**.
- Make sure the popup menu is set to **x Equally Spaced**.
- Fill in $1 \le \mathbf{n} \le 4$ in increments of 1.

  Be sure to use **n** as the variable.

- Check **Display in decimal form** and **Display as a table**.
- Press [SHIFT] − [RET].

8. Check the values of *over* and *under* in the table for $n = 2$ against your answers to Question 4 (Before the Lab). If these are not the same (except for rounding off), review your reasoning and correct any mistakes. When you are ready, proceed.

- Choose **Last Dialog** to reopen the **List** dialog.
- Fill in $10 \leq n \leq 100$ in increments of 10.
- Press $\boxed{\text{SHIFT}} - \boxed{\text{RET}}$.

   Depending on your computer, this may take a little while.

9. Although *Mathematica* cannot evaluate $\int_0^1 \int_1^3 f(x, y)\, dy\ dx$ exactly, you can use its built-in algorithm to approximate it numerically. Section 13.2 illustrates the process in more detail.

- In your notebook, type f[x, y] and select it by dragging or by clicking the **Select** button several times.
- Click the **Numeric Integral** button on the *Joy* palette and fill in the inner integral $N \int_1^3 f[x, y]\, dy$.
- Type parentheses ( ) around the inner integral and select the entire expression. Then click the **Numeric Integral** button again.
- Fill in the outer integral to obtain $N \int_0^1 \left(N \int_1^3 f[x, y]\, dy \right) dx$ and press $\boxed{\text{SHIFT}} - \boxed{\text{RET}}$.

10. Print your work.

■ After the Lab

1. a) What general pattern did the overestimates follow in the tables in Questions 7 and 8 (In the Lab)? How did they compare to the estimate of the integral in Question 9?

   b) Repeat for the underestimates.

2. a) The difference *over* − *under* is an error bound for approximating the integral with these Riemann sums. By about what factor does the error bound appear to change when n is doubled, e.g., $n = 2$, 4 and $n = 10$, 20, 40, 80?

3. By about what factor does the error bound appear to change when n is multiplied by 10, e.g., $n = 1$, 10, 100, $n = 2$, 20; 3, 30; 4, 40?

4. The actual error in an approximation is unknown unless you know the true value of the integral. However, the error bound tells you the most the error can possibly be and is usually a conservative estimate of the error. Based on your results so far, predict the general pattern for this method of approximation by filling in the blanks below:

   Error($2\,n$) $\approx$ ___ Error (n) (express in terms of 2)
   Error($10\,n$) $\approx$ ___ Error (n) (express in terms of 10)
   Error ($c\,n$) $\approx$ ___ Error (n) (express in terms of c)

   With 10 times as many subintervals, you gain $\approx$ ___ decimal places of accuracy.

5. Based on your results so far, about how large would n have to be to make the error bound < 0.001 using this method of approximation?

Lab 25.2 Double Integrals and Housing Prices

General Prerequisites:

Ch. 1	A Brief Tour of *Joy*	
Sec. 10.3	How to Plot the Level Curves of a Surface (Contours)	
Sec. 11.1	How to Create a Function of Several Variables	
Sec. 13.1	How to Calculate a Multiple Integral	

In this lab, you can explore using double integrals to calculate the average value of a function of two variables as a way of estimating average housing prices.

■ Before the Lab (Do by Hand)

The members of the Town Council in Seaside Bluffs are concerned about real estate prices in their ritzy seashore community. They have hired you as a consultant to get a better understanding of how prices vary with location and how this determines the average house price in the town.

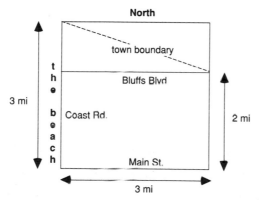

After some extended research, you determine that real estate prices in the area increase as you move away from Main Street. You estimate that where Main Street meets the beach houses cost about $400,000, and that along the beach prices increase by $100,000 per mile as you go north.

1. About how much does a house cost at the corner of Bluffs Boulevard and Coast Road?

You also determine that prices decrease away from the beach, and estimate that they drop 20% per mile as you move inland.

2. About how much does a house cost 1 mile inland on Main Street? 3 miles inland?

3. About how much does a house cost 1 mile inland on Bluffs Boulevard? 3 miles inland?

4. Find a formula that estimates the price $p(x, y)$ of a house x miles inland from the beach and y miles north of Main Street, in units of $100,000.

5. About how much do the most expensive and least expensive houses in Seaside Bluffs cost, and where are they located?

■ In the Lab

1. Create the price function $p(x, y)$ whose formula you found before the lab. Enter the function as p[x, y]. (**Create** ▷ **Function**)

2. Graph the level curves (contours) of $p(x, y)$ over the 3×3 square in which the town lies. (**Graph 3D** ▷ **Level Curves**)

3. The level curves in the graph correspond to house prices that are equally spaced between the cheapest house and the most expensive house. How much more do the houses cost when you go from one level curve to the next?

4. The town forms a trapezoid within the square. Find the integral of $p(x, y)$ over the trapezoid.

5. Print your work.

■ After the Lab

1. On your printout of the contours, label the contours with the prices to which they correspond, to the nearest $1000.

The average house price in the town can be thought of as

$$\frac{\int\int p(x, y)\, dy\, dx}{A},$$

where A is the area of the town and the integration is over the trapezoid defining the town.

> Here is why this integral expresses average value. Imagine dividing the town into n equal rectangles, each with area $\Delta y\, \Delta x = A/n$. If $\Delta y\, \Delta x$ is small, then for each individual rectangle the price function is nearly constant and equal to its value at the center (x_i^*, y_i^*). So the average price over the town is approximately the average of the n numbers $p(x_i^*, y_i^*)$,
>
> $$\frac{1}{n}\sum p(x_i^*, y_i^*) = \frac{\Delta y \Delta x}{A} \sum p(x_i^*, y_i^*) = \frac{1}{A}\sum p(x_i^*, y_i^*)\,\Delta y\, \Delta x.$$
>
> As $\Delta y, \Delta x \to 0$, this approximation approaches the limit $\frac{1}{A}\int\int p(x, y)\, dy\, dx$.

2. Find the average house price in Seaside Bluffs.

3. On your printout of the contours, sketch the approximate contour where the price of a house equals the town average.

4. Now sketch the approximate contours where the price is halfway between the average and the maximum and halfway between the average and the minimum.

Chapter 26
Differential Equations

Examples and Exercises

Prerequisites:
     Ch. 1   A Brief Tour of *Joy*
     Secs. 15.1–4 Slope fields, solving differential equations

■ 26.1 Slope Fields and Visualizing Solutions

26.1.1 Example: Estimating a Solution from a Slope Field

The slope field of a differential equation allows you to predict the qualitative and quantitative behavior of its solutions. For example, consider the equation

$$y'(t) = t - y.$$

We can plot the slope field for t and y between ± 3.

- Plot the slope field of $y'(t) = t - y$ for $-3 \le t \le 3, \; -3 \le y \le 3$. (**Calculus** ▷ **Slope Field**)

    Enter the derivative as $\texttt{y'[t]}$, not as $\frac{dy}{dt}$.

The result is:

▷ The slope field for y′[t] = t - y for -3 ≤ t ≤ 3 and -3 ≤ y ≤ 3

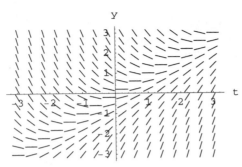

Out[2]=   - Graphics -

The solutions to the equation are curves that follow the tangents shown in the slope field. For example, if $y(-1) = 2$, then starting at $(t, y) = (-1, 2)$ and following tangents should produce a curve like this:

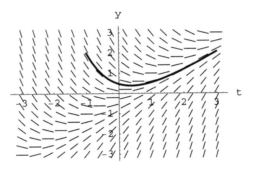

It appears that as $t \to \infty$ all solution curves approach a line that is their asymptote and also seems to be a solution. Its equation has the form $y = mt + b$ and must satisfy $y'(t) = t - y$, i.e.,

$$m = t - (mt + b),$$
$$m + b = (1 - m)t.$$

In order for this to be true for *all* values of t, we must have $m = 1$ and $b = -1$.

If it is true for $t = 0$, then $b = -m$. If it is also true for $t = 1$, then $m = 1$.

So the only possible line that is a solution is $y = t - 1$, and you can check that it satisfies $y'(t) = t - y$.

To show that all solutions are asymptotic to this one, we'll solve the equation. (If you know how, you can also solve it by hand.)

- Solve $y'(t) = t - y$ with no initial condition. (**Calculus ▷ Differential Equations**)

▷ Solve   y'[t] = t - y.

Define y[t] to be the solution; then y[t] =

Out[3]= $-1 + t + E^{-t} C[1]$

This shows that $y(t) \to t - 1$ as $t \to \infty$. (In Version 4, *Mathematica*'s output displays e instead of E.)

Exercises

In Exercises 1 and 2:

a) Plot the slope field for the given values of t and y.

b) As a point (t, y) moves to the right, how does the slope change? How could you have predicted this from the differential equation?

c) Repeat when point (t, y) moves upward.

1. $y'(t) = 1.1\,y + 1$, for t and y between ± 2.

2. $\frac{dy}{dt} = 3\,y - 2\,y^2$, for $-2 \le t < 2$ and $-1 \le y \le 3$.

In Exercises 3–5:

a) Plot the slope field for the given values of t and y and use it to predict any solutions to the differential equation that are constant functions, called *equilibrium solutions*.

b) Substitute your answers to a) into the differential equation to check that they are solutions.

3. $y'(t) = y - 2$ for $-3 \le t \le 3$ and $-1 \le y \le 4$.

4. $\frac{dy}{dt} = y\,t - 1$, for t and y between ± 2.

5. $y'(t) = 2\,y - 4\,y^2$, for t and y between ± 2.

In Exercises 6 and 7:

a) Print the slope field for the given values of t and y. Sketch the solution satisfying the given initial condition and use your sketch to estimate its value at the specified point.

b) Solve the equation over the interval between the given initial and final points, graphing the solution combined with the slope field in a). (**Calculus ▷ Differential Equations**)

c) Check your estimate in a) against the graph in b) and also by evaluating $y(t)$ at the specified value of t. (**Graph ▷ Combine Graphs**)

6. $y'(t) = t^2 - y$, for t and y between ± 3. Given $y(-2) = 2$, estimate $y(2)$.

7. $\frac{dy}{dt} = y\,t - 1$, for t and y between ± 2. Given $y(0) = 1$, estimate $y(1.5)$. Solve numerically or use *Mathematica*'s symbolic solution (the symbolic solution contains the special function Erf[x], the Gaussian error function).

8. a) Plot the slope field of $y'(t) = 2t - 3y + 1$ for t and y between ±2.

b) By hand, find the equation of a straight line that is a solution to the differential equation. Check that it appears to be an asymptote of all the solutions.

c) Use *Joy* to solve the differential equation without an initial condition. Find the limit of the solution as $t \to \infty$ and check your answer to b).

■ 26.2 Euler's Method

26.2.1 Example: How Euler's Method Works

Euler's method for estimating the solution to a differential equation is based on the idea of linear approximation to a differentiable function. It is generally less efficient than *Mathematica*'s built-in method for approximating solutions, but it is easy to undertand and makes clear the principles that underlie numerical solutions. Here is a brief explanation of Euler's method for an equation of the form $y'(t) = f(t, y)$.

The algorithm pieces together linear approximations to the solution that follow the slope field for the equation. At each point, the slope field indicates the direction of the tangent line to the solution curve. We start at a point (t_0, y_0) corresponding to an initial condition $y(t_0) = y_0$. We follow the slope field at the initial point for a short distance (the step size), arriving at a second point (t_1, y_1). We then change direction, following the slope field at (t_1, y_1) to a third point, etc. We can then connect the points with straight lines, i.e., create an interpolating function.

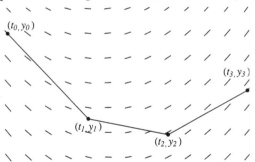

Euler's method follows the slope field and interpolates between points.

By doing this, we can estimate the solution $y = y(t)$ at any t. The estimate generally improves as we make the steps shorter. So if we want to approximate y at a particular value of t, increasing the number of steps generally increases the accuracy of the result.

Here is an illustration with the equation $y'(t) = \ln(ty)$, where we are given the initial condition $y(0.5) = 1.5$ and want to find $y(2)$. This is a differential equation that *Mathematica* cannot solve exactly.

First, we'll graph the slope field for this equation.

- Plot the slope field of $y'(t) = \ln(ty)$. Use the interval $0.01 \le t \le 2$, $0.01 \le y \le 2$ and fill in the name `slopes`. (**Calculus ▷ Slope Field, Assign name**)

  Enter $\ln(ty)$ as `Log[t⌣y]`. Here ⌣ means to insert a space to indicate multiplication. Since $\ln(ty)$ is only defined when $ty > 0$, we choose positive values of t, y.

Mathematica plots the slope field:

▷ The slope field for `y'[t] = Log[t y]` for `0.01 ≤ t ≤ 2` and `0.01 ≤ y ≤ 2`

slopes =

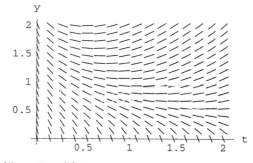

Out[2]= - Graphics -

Now we apply Euler's method.

- Solve $y'(t) = \ln(ty)$ by Euler's method, for $0.5 \le t \le 2$ with $y(0.5) = 1.5$. Use 10 steps, set the popup menu to **All**, and fill in the name `approx`. (**Calculus ▷ Euler's Method, Assign name**).

The result is:

Equations:	`y'[t] = Log[t y]`
Initial Values:	`y = 1.5`
Method of Solution:	`Euler`
Name of Solutions:	`y[t]`
Number of Steps:	`10`
Output:	`All`

approx =

t	y
0.5	1.5
0.65	1.45685
0.8	1.44867
0.95	1.4708
1.1	1.52097
1.25	1.59817
1.4	1.70197
1.55	1.83221
1.7	1.98878
1.85	2.1715
2.	2.38009

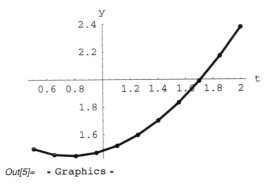

Out[5]=   - Graphics -

The table shows that Euler's method moves in 10 steps of length 0.15 each from $t = 0.5$ to $t = 2$. We can read the approximation $y(2) \approx 2.38009$ from the table.

We can combine the Euler graph and the slope field:

- Choose **Graph ▷ Combine Graphs**.
- If necessary, fill in the names `slopes,approx`.
- Click **OK**.

Mathematica combines the graphs:

▷ Combining the graphs {slopes, approx} gives

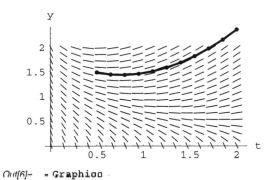

Out[6]=   **- Graphics -**

The slope field suggests that the graph of the solution beginning at $(t, y) = (0.5, 1.5)$ will be concave up. Since Euler's method follows the tangents, it will tend to underestimate the solution. You can confirm this by using *Mathematica*'s built-in method for solving differential equations numerically.

- Solve $y'(t) = \ln(ty)$ numerically, for $0.5 \le t \le 2$ with $y(0.5) = 1.5$. You don't need to graph the solution. (**Calculus ▷ Differential Equations**)

Mathematica creates an interpolating function:

▷ Solve y'[t]=Log[t y] numerically
  with y=1.5`.

Define y[t] to be the solution; then y[t] =

Out[7]=   InterpolatingFunction[{{0.5, 2.}}, <>][t]

Now we can evaluate this function at $t = 2$:

- Click in your *Mathematica* notebook beneath the last output, so that a horizontal line appears.
- Type y[2].
- Press SHIFT — RET.

In[8]:= **y[2]**

Out[8]= 2.60106

So the Euler estimate is in fact less than *Mathematica*'s internal estimate because the solution is concave up. In the exercises, you can see that increasing the number of steps brings the estimates closer together.

Exercises

1. a) Plot the slope field of $y'(t) = y + t + 1$ for $-2 \le t \le 2$ and $-2 \le y \le 2$.

b) Use the slope field to predict the graph of the solution with $y(-2) = \frac{1}{10}$. Apply Euler's method to approximate the solution on $-2 \leq t \leq 2$ with 100 steps (choose **Graph** from the popup menu).

c) Solve the equation exactly (not numerically) and graph the solution for $-2 \leq t \leq 2$, combining it with the graph of the Euler approximation. (**Calculus ▷ Differential Equations**)

d) How well do the approximation and the solution agree when t is near -2 compared to when t is far from -2?

2. a) Apply Euler's method to $y'(t) = 2t$ for $1 \leq t \leq 4$, with the initial condition $y(1) = 0$. Find the **Final value** $y(4)$ with 100 steps and 1000 steps, simultaneously.

   b) Solve the equation in a) by hand and find the exact value of $y(4)$.

Exercises 3 and 4 refer to the differential equation in Example 26.2.1, $y'(t) = \ln(ty)$ for $0.5 \leq t \leq 2$ with $y(0.5) = 1.5$.

3. a) Apply Euler's method and find the **Final Value** for 10, 100, 1000, and 10,000 steps, simultaneously (enter 10,000 *without* a comma). How do the estimates change and how do they compare to *Mathematica*'s built-in estimate $y(2) \approx 2.60106$? Are they all underestimates?

   b) Find the difference between 2.60106 and the estimates for 10, 100, 1000, and 10,000 steps this way:

   • Click in your *Mathematica* notebook beneath the last output, to make a horizontal line appear.
   • Type 2.60106 - Out [...], filling in the output number in a).
   • Press ⌷SHIFT⌷ − ⌷RET⌷.

   The difference is approximately the error in approximating $y(2)$ by Euler's method (only approximately, because 2.60106 is not the exact value of $y(2)$). As you multiply the number of steps by 10, by about what factor does this difference change?

4. a) Apply Euler's method and find the **Graph** for 10, 20, 40, and 80 steps simultaneously.

   b) Solve the differential equation numerically for this initial condition, and combine the graph with the graph of the Euler approximations. (**Calculus ▷ Differential Equations**)

   c) How do the graphs change as the number n of steps increases? In particular, how does the vertical distance between the approximations and the graph of *Mathematica*'s numerical solution appear to change as you double n?

Lab 26.1 Least Squares, Population, and the Logistic Equation

---

Prerequisites:

Ch. 1	A Brief Tour of *Joy*
Secs. 6.3,4	Making and plotting a table of values
Sec. 15.2	How to Solve a DE Symbolically

How quickly is the population of the United States growing? In this lab, you can explore a model that describes how the population has changed over time and use it to predict how it will change in the future. The model is based on the logistic differential equation and estimating its parameters with a least squares fit to census data. This approach is suggested by a similar example in Deborah Hughes-Hallett, Andrew Gleason, et al., *Calculus*, 2nd ed. (John Wiley & Sons, New York, 1998).

■ Before the Lab (Do by Hand)

Here is a table of U.S. Census data, gathered from the *Statistical Abstract of the United States* (U.S. Department of Commerce, 1992). The population is in millions of people. The results of the 2000 census will be available in 2002. In May 1999, the Census Bureau estimated the population at 272.5 million.

Year	Population	Year	Population
1790	3.9	1900	76.0
1800	5.3	1910	92.0
1810	7.2	1920	105.7
1820	9.6	1930	122.8
1830	12.9	1940	131.7
1840	17.1	1950	151.3
1850	23.2	1960	179.3
1860	31.4	1970	203.3
1870	39.8	1980	226.5
1880	50.2	1990	248.7
1890	62.9	2000	??

The *logistic equation*, $P'(t) = a P(M - P)$, models the rate of change $P'(t)$ of a function $P(t)$ as being proportional to $P(t)$ and also to the difference between $P(t)$ and some number M. Here a is a proportionality constant, and we take $a > 0$ so that the population grows when $0 < P < M$.

1. What are the equilibrium values of P for a population that is changing according to the logistic equation? (Hint: what is the value of P' at equilibrium?)

2. The *exponential* model for population growth is $P'(t) = kP$. What are the equilibrium values of P for this model?

The *relative* rate of population change is

$$\frac{P'(t)}{P(t)},$$

i.e., the rate of change as a proportion of the population.

3. How is the relative rate of change different for the logistic and exponential models?

4. Graph $P'(t)/P(t)$ as a function of P for the exponential model, i.e., with P on the horizontal axis and $P'(t)/P(t)$ on the vertical axis.

5. Graph $P'(t)/P(t)$ as a function of P for the logistic model. The graph should be a straight line. What are its slope and intercept?

In the lab, you'll use *Joy* to estimate the relative rate of change of population from the data in the census table. Then you'll graph this rate against the population data and see whether a pattern emerges that is closer to the exponential or to the logistic model.

How can we approximate $P'(t)/P(t)$ from the data at hand? The derivative $P'(t)$ represents the instantaneous rate of change of population, in millions of people per year. We can estimate it by taking the average rate of change over an interval of time. For example,

$$P'(1800) \approx \frac{P(1810) - P(1790)}{20}$$

$$\frac{P'(1800)}{P(1800)} \approx \frac{1}{P(1800)} \left(\frac{P(1810) - P(1790)}{20} \right).$$

6. Estimate $P'(1800) / P(1800)$ from the population table.

■ In the Lab

1. Enter the population data (this is tedious, but it makes the rest of the analysis go faster):

    *   In your *Mathematica* notebook, type P[1790] = 3.9 and press SHIFT − RET.
    *   Repeat for all the data through 1990.

2. Create a table *popdata* of the population values so they will be easy to work with:

    *   Choose **Create ▷ List**.
    *   Fill in t, P[t].
    *   Specify the interval 1790 ≤ t ≤ 1990, in intervals of 10.
    *   Check **Display as a table** and **Assign name**, using the name popdata.
    *   Click **OK**.

3. Plot the points in *popdata*:

   - Choose **Graph ▷ Points**.
   - Fill in `popdata`.
   - Click **OK**.

4. Create a function $R(t)$ to approximate the relative rate of change of the population:

   - Choose **Create ▷ Function**.
   - Fill in `R[t]` in the first input field.
   - In the second field, fill in

   $$\frac{1}{P[t]} \left(\frac{P[t + 10] - P[t - 10]}{20} \right).$$

   - Click **OK**.

5. Create a list *rate* showing the population and relative rate of change:

   - Choose **Create ▷ List**.
   - Fill in `P[t], R[t]`.
   - Type the interval $1800 \le t \le 1980$, in intervals of 10.

     You need to take 1800 and 1980 for the beginning and end of the list because the census data do not give you enough information to calculate the relative rate of change for 1790 or 1990.

   - Check **Display as a table** and **Assign name**, using the name `rate`.
   - Click **OK**.

6. Plot the points in *rate*. (**Graph ▷ Points**) Then add a **Grid** to the plot to make it easier to answer the questions below. (**Graph ▷ Modify Graph**)

Real data usually do not fit any model exactly, so you'll need to estimate the line that fits the plot best.

7. a) By eye, estimate the slope and vertical intercept for the line that seems to fit the graph of *rate* best.

   b) Use your answer to Question 5 (Before the Lab) to estimate values of a and M in the logistic model for the census data.

8. a) Solve the logistic equation Pop$'(t) = a$ Pop$(M -$ Pop$)$, using the values of a and M that you just estimated. Graph the solution for $1790 \le t \le 2200$, taking the population in 1790 as the initial condition. (**Calculus ▷ Differential Equations**)

   Since you used P for population when entering the actual data, you should use a different name such as *Pop* when solving the differential equation. This will prevent overwriting the data values you entered earlier if you need to use them again.

b) Combine the graph in a) with your graph of *popdata*. Try modifying your values of *a* and *M* to get a curve that seems to give a reasonable fit to the data.

9. Print your work. Save your notebook if you plan to do the **For Further Exploration** section later.

■ After the Lab

Models often do a better job of fitting the data over a short period of time than over a long one. These questions concern the success of the exponential and logistic models in describing the census data and predicting future levels of population.

1. a) The graph in Question 6 (In the Lab) shows the relative rate of change as a function of population. Over the period 1800–1980, does the exponential or logistic model fit this graph better? Why?

    b) Over which time period is the exponential model a good fit to the data? Why?

2. Explain why the logistic model predicts that if the initial population is less than *M* then the population can never grow beyond *M*. What happens to P' as *P* approaches *M*?

3. a) What is the maximum population predicted by your logistic model in Question 8 (In the Lab)? How does it compare to the U.S. Census Bureau estimate of 272.5 million for 1999? You can also find the current estimate at the Census Bureau website, `http://www.census.gov/`

    b) What population does your logistic model predict for the U.S. in 2000? in 2010?

4. In a logistic model, the graph of the relative rate of change as a function of population is a straight line. The graph in Question 6 (In the Lab) should appear roughly straight except for a big jump. When did this jump take place and what are some historical reasons for it?

■ For Further Exploration

Rather than estimate by eye the slope and intercept of the relative rate of change line, you can use calculus to find the values that fit the data best. The *least squares estimate* is the line $R = mP + b$ that minimizes the sum of the squared differences between the vertical coordinates $R(t)$ and the points on the line.

1. Minimize the sum:

    - In your *Mathematica* notebook, type
        ```
        ssq = ∑¹⁸ₖ₌₀ (R[1800 + 10 k] - (-m P[1800 + 10 k] + b))².
        ```

    - Press SHIFT – ⏎CT⏎.

2. a) Solve $\frac{\partial}{\partial m}$ (ssq) = 0 and $\frac{\partial}{\partial b}$ (ssq) = 0 for m and b. These are *partial* derivatives, where only m or b is changing and the other is constant. (**Algebra ▷ Solve, Partial Derivative button**)

    b) Compare these values for the slope and intercept to the ones you estimated by eye.

    ♡ *Mathematica* has a built-in command for finding the least squares line in one step. It is
      `Fit[rate,{1,P},P]`

3. Find the values of a and M corresponding to the slope and intercept.

Lab 26.2 Numerical Solution of Differential Equations: Designing a Waterslide

---

Prerequisites:
      Ch. 1   A Brief Tour of *Joy*
      Secs. 5.1,2   Solving equations algebraically and numerically
      Sec. 15.3   How to Solve a DE Numerically

In this lab, you can design a waterslide that will meet certain specifications and then explore what a ride down the slide would be like. How long would it last? How fast would you go?

■ Before the Lab (Do by Hand)

The Shape of the Waterslide

Suppose the waterslide is to have the following shape, with a pool of water at the bottom:

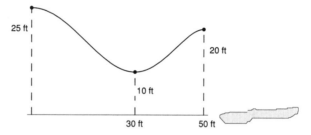

There are various ways to design a curve with this shape. One way is to find a polynomial that has the correct height and slope at key points along the curve. For example, if $y = y(x)$ is a function whose graph is the slide, then $y(0) = 25$. The slide is flat at the start, so $y'(0) = 0$.

1. The slide is also flat at its lowest point and at the far end. Find the equations for y and y' that describe the slide at these points.

In the lab, you will use these equations to find a function whose graph acts as the waterslide.

Traveling on the Waterslide

As you travel along the slide your total energy is divided in two parts,

$$\text{total energy} \;=\; \text{potential energy} + \text{kinetic energy}$$
$$=\; mgy + \frac{mv^2}{2},$$

where m = mass, $g = 32 \, \text{ft} / \text{sec}^2$, y = height, and v = speed along the slide. We assume that the water in the slide makes friction negligible. Suppose you push off at the beginning of the ride with a speed of 5 ft/sec. Your total energy at the beginning is

$$mgy + \frac{mv^2}{2} = m(32)(25) + \frac{m(5^2)}{2}$$
$$= 812.5 \, m.$$

The principle of conservation of energy states that the total energy in the system *at any time t* equals $812.5 \, m$, although the way it is divided into potential and kinetic energy changes as your speed and height change.

2. Find the value of v when the height is y. Your answer should be given in terms of y and constants.

3. Use your answer to Question 2 to determine when you speed up and when you slow down as you travel along the slide.

To determine how long the ride lasts, you will also need to find $\frac{dx}{dt}$, your velocity in the x-direction along the slide. We can find it using the equation

$$v^2 = \left(\frac{dx}{dt} \right)^2 + \left(\frac{dy}{dt} \right)^2, \tag{*}$$

which is justified by the accompanying sketch.

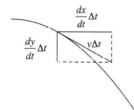

At a point (x, y) on the slide, during a short time interval $\Delta t > 0$ you move approximately a distance of $v \Delta t$ in the direction of the tangent. You move approximately $\frac{dx}{dt} \Delta t$ horizontally and $\frac{dy}{dt} \Delta t$ vertically.

4. a) By hand, substitute $\frac{dy}{dt} = \frac{dy}{dx} \frac{dx}{dt}$ (the chain rule) into equation (*) and solve for $\frac{dx}{dt}$.

 b) Use a) to show that

$$\frac{dx}{dt} = \sqrt{ \frac{1625 - 64 \, y(x)}{1 + \left(\frac{dy}{dx} \right)^2} }, \tag{**}$$

where we write $y(x)$ instead of just y to indicate that y depends on x.

■ In the Lab

Creating the Waterslide

First, we'll find a curve whose graph is the waterslide we want. The six equations you found to describe the slide suggest looking for a polynomial of degree 5. Here's how to do that.

1. Create a function $y(x) = a + bx + cx^2 + dx^3 + ex^4 + fx^5$. (**Create** ▷ **Function**)

   Remember to enter this function as $y[x]$.

2. Solve the six equations in Question 1 (Before the Lab) for the unknowns a, b, c, d, e, f, and name the solution `coeffs`. (**Algebra** ▷ **Solve, Assign name**)

   Enter the equations as $y[0] = 25$, $y'[0] = 0$, filling in the other equations you found before the lab. *Joy* will warn you that the equations don't explicitly contain the unknowns. Click **Continue** to override the warning.

3. Enter the solutions in your *Mathematica* notebook:

   - Click in the notebook beneath the last output, so that a horizontal line appears.
   - Type $a = 25$ and press $\boxed{\text{SHIFT}} - \boxed{\text{RET}}$.
   - Repeat for the other five coefficients.

4. Graph $y(x)$ for $0 \le x \le 50$. If the graph doesn't have the shape of the waterslide shown above, review your reasoning.

How Long Are You on the Waterslide?

Next, we'll find how long it takes to reach the end of the slide. To do this, we must first find $x(t)$.

5. Here we'll solve the differential equation in Question 4 (Before the Lab) *numerically* for $x(t)$, since *Mathematica* cannot solve it symbolically.

   - Create a function slope$(x) = y'(x)$. (**Create** ▷ **Function**)

   - Choose **Calculus** ▷ **Differential Equations**.
   - Fill in the equation

   $$x'[t] = \sqrt{\frac{1625 - 64\, y[x]}{1 + slope[x]^2}} \; .$$

   - Fill in the interval $0 \le t \le 10$.
   - Fill $x = 0$.

     ♡ There should be only one initial condition since we are only trying to solve for $x(t)$. We cannot refer to $y'(x)$ directly because y is not an unknown in this differential equation.

- Click **Solve numerically** and **Graph solution**.
- Click **OK**.

6. Here's how to find the time you spend on the slide. First you'll estimate this time from the graph and then use *Mathematica* to solve for it.

   - From the graph, estimate the time t_s that it takes to reach the end of the slide.

     This is the value of t that makes $x = 50$.

   - Choose **Algebra** ▷ **Solve**.
   - Fill in the equation **x[t] = 50**.
   - Fill in **t** as the variable for which to solve.
   - Check **Solve numerically** and fill in the initial estimate **t = ...** you found for t_s from the graph.

     Mathematica's solver needs a rough estimate of the solution to an equation to get started in finding a more accurate solution.

   - Click **OK**.

7. Print your work. Save your notebook if you plan to do the **For Further Exploration** section later.

■ After the Lab

1. In the lab, you found a polynomial of degree 5 that satisfied the conditions for the shape of the waterslide. Why was it reasonable to look for a polynomial of degree 5?

Landing in the Water

If you're designing a waterslide, you need to find how far beyond the end of the slide you travel and how long it takes before you land. You need to locate a pool of water at the bottom for a safe landing, set safety standards for the time between riders, etc.

Let xair(t) and yair(t) denote your x- and y-coordinates while sailing through the air after you fly off the slide. Let $t = 0$ when you leave the slide. Resetting the clock in this way makes it easier to find the coordinates.

2. After you reach the end of the slide, your height above the ground is subject only to the force of gravity.

   a) Find an equation for $\frac{d^2 \text{yair}}{dt^2}$, your acceleration in the vertical direction, and solve it to find $\frac{d\text{yair}}{dt}$, your velocity in the vertical direction. What is $\frac{d\text{yair}}{dt}$ when you leave the slide? (Hint: you're moving horizontally at the end of the slide.)

   b) Solve the equation for $\frac{d\text{yair}}{dt}$ to find yair(t). What is the initial value of yair?

c) When do you land in the water? How long does the whole ride last?

3. a) Equation (**) applies to x, y. Explain why it is valid for xair, yair at the moment you fly off the slide.

   b) Use equation (**) to find $\frac{d\text{xair}}{dt}$ at the moment you fly off the slide.

4. a) What is $\frac{d\text{xair}}{dt}$ when you are flying through the air?

   b) Find xair(t) for any value of t.

   c) What horizontal distance do you travel from the beginning of the slide until splashdown?

■ For Further Exploration

You can animate your trip along the slide and into the water in two stages, first when you are on the slide and then when you are in the air. Section 15.7 shows how to animate solutions of systems of differential equations. Here you'll use the same idea to animate a moving point.

Here is the first stage.

1. You'll divide your ride into segments of 0.3 sec, and plot your location at
   $t = 0, 0.3, 0.6, \ldots, 3$. This will give a good approximation of this part of the ride.

   - Choose **Graph ▷ Animation**.
   - Choose **Points** as the kind of graph to animate.
   - Fill in `{x[0.3 k], y[x[0.3 k]]}`.

     This plots the points on the curve every 0.3 sec.

   - Click in the **Animation Info** window to make it active.
   - Fill in 0 ≤ k ≤ 10.
   - Fill in the common plot frame, 0 ≤ x ≤ 75, −5 ≤ y ≤ 25.

     This will give a frame large enough to show the whole ride, including the splash landing.

   - Click **Combine with** and fill in `Out[...]`, using the output number of your graph of the waterslide in Question 4 (In the Lab).
   - Click in the **Plot Points** window and click **OK**.

2. Create the functions xair(t) and yair(t), using the expressions you found in Questions 4b), 2b) (In the Lab). (**Create ▷ Function**)

3. Animate $x = $ xair(t), $y = $ yair(t) every 0.3 sec for the time you are in the air. Using 0 ≤ k ≤ 4 will carry you into the water. Use the same common plot frame as before.

   Be sure to use yair(0.3 k), not yair(xair(0.3 k)). Why?

4. Select the cell brackets for the two animations (see Section 1.4.2) and press ⌘-Y (Macintosh) or CTRL-Y (Windows). This animates your whole ride.

Chapter 27
Systems of Differential Equations

Examples and Exercises

Prerequisites:

Ch. 1	A Brief Tour of *Joy*
Secs. 15.5–7	Solving and animating systems of differential equations
Secs. 16.4,15	Creating a matrix, finding eigenvalues

■ 27.1 Linear Systems of First-Order Equations

27.1.1 Example: A Linear System $v'(t) = A\,v$

We can study the linear system

$$
\begin{aligned}
x'(t) &= x - 3\,y \\
y'(t) &= -6\,x - 2\,y
\end{aligned}
$$

by comparing the behavior of the system for different initial conditions and by looking for certain special solutions.

First, we'll graph the direction field for the system with x and y between ± 15.

- From the *Joy* menus, choose **Graph ▷ Vector Field**.
- Fill in $\{x - 3\ y, -6\ x - 2\ y\}$.
- Fill in $-15 \le x \le 15, -15 \le y \le 15$.
- Check **Normalize lengths** and **Assign name**, filling in `field`.
- Click **OK**.

The graph is:

▷ `Graph of the vector field {x - 3 y, -6 x - 2 y} for -15 ≤ x ≤ 15, -15 ≤ y ≤ 15.`

`field =`

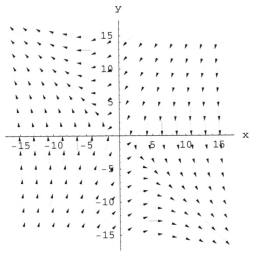

Out[3]= `- Graphics -`

Now we'll find the solution corresponding to the initial conditions $x(0) = 1$, $y(0) = 3$ and plot its trajectory.

- From the *Joy* menus, choose **Calculus ▷ Differential Equations**.
- Fill in `x'[t] = x - 3 y,  y'[t] = -6 x - 2 y`.
- Fill in `0 ≤ t ≤ 10` and the initial conditions `x = 1,  y = 3`.
- Check **Graph solution** and click **OK**.

Mathematica solves the system:

▷ `Solve  {x'[t]=x-3 y,y'[t]=-6 x-2 y}`
  `with  {x=1,y=3}.`

  `Graph styles:`

  `x      Black     Solid`
  `y      Red       .....`

  `Define {x[t],y[t]} to be the solution; then {x[t],y[t]} =`

$$\left\{ -E^{-5t}\left(-\frac{4}{3}+\frac{E^{9t}}{3}\right),\ E^{-5t}\left(\frac{8}{3}+\frac{E^{9t}}{3}\right)\right\}$$

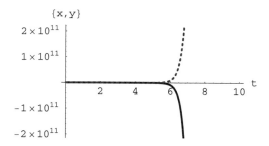

Out[5]= - Graphics -

(In Version 4, *Mathematica*'s output denotes the constant e by e instead of E.) Because e^{9t} grows so quickly, $0 \le t \le 10$ is too big to show the behavior of x, y when t is near 0. We'll graph the trajectory of the solution in the phase plane and keep the value of t small, letting $0 \le t \le 1$.

- Choose **Graph ▷ Parametric Plot**.
- Fill in x = x[t], y = y[t].
- Fill in $0 \le t \le 1$.
- Click **OK**.

The graph is:

▷ Graph the curve (x,y) = (x[t],y[t]) for 0 ≤ t ≤ 1.

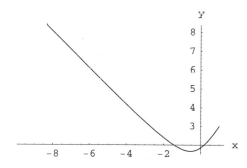

Out[6]= - Graphics -

The direction field suggests that (x, y) begins at the initial point $(1, 3)$ and follows a curve that quickly approaches a straight line. To confirm this, we'll make the graph of the trajectory thicker for better visibility and then combine it with the direction field.

- Choose **Graph ▷ Modify Graph**.
- Make sure the output number of the preceding graph appears, in this case Out[6].
- Check **Assign name** and fill in trajectory.

- Click **OK**.

Now we combine the two graphs.

- Choose **Graph ▷ Combine Graphs**.
- Fill in `field,trajectory`.
- Click **OK**.

The result is:

▷ Combining the graphs {field, trajectory} gives

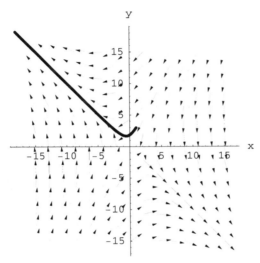

Out[9]= ⁃ Graphics ⁃

Two important kinds of solutions are equilibrium and straight-line solutions. An *equilibrium solution* is one where $x(t)$ and $y(t)$ are constant, in which case $x'(t)$ and $y'(t)$ are always zero. You can show by hand or using *Joy* (**Algebra ▷ Solve**) that the only equilibrium solution for this system is the origin $x(t) = 0$, $y(t) = 0$. (Solve $x - 3y = 0$, $-6x - 2y = 0$.)

A *straight-line solution* is a solution to the system whose trajectory follows a line through the equilibrium point. The direction field shows two lines through the equilibrium where we might find straight-line solutions. One appears to be an asymptote for the solution we found earlier.

We can think of the system in matrix form,

$$v'(t) = Av, \text{ where } A = \begin{pmatrix} 1 & -3 \\ -6 & -2 \end{pmatrix}, \ v = \begin{pmatrix} x \\ y \end{pmatrix}.$$

If v is a straight-line solution, then $v \neq \mathbf{0}$ and $v'(t)$ points along the line through the origin with the same or opposite direction as v. So for such a solution

$$Av = \lambda v,$$

where λ is a real number. We call λ an *eigenvalue* and v an *eigenvector* of A. The eigenvectors indicate the directions of any straight-line solutions of the system. Here's how to find them.

- From the *Joy* menus, choose **Matrices ▷ Eigenvalues**.
- The default matrix is selected automatically. Click the **Matrix** button on the *Joy* palette.
- In the dialog that appears, fill in 2 for the number of rows and the number of columns, and click **Paste**.
- Fill in the matrix, pressing TAB to move to each successive entry:

$$\begin{pmatrix} 1 & -3 \\ -6 & -2 \end{pmatrix}$$

- Make sure the boxes for **eigenvalues** and **eigenvectors** are both checked.
- Click **Algebraically**.

   The other choice finds numeric estimates and is best for more complicated matrices.

- Click **OK**.

Mathematica finds the eigenvalues and eigenvectors:

▷ The eigenvalues and eigenvectors of $\begin{pmatrix} 1 & -3 \\ -6 & -2 \end{pmatrix}$ are

Out[6]//MatrixForm=
$$\begin{pmatrix} -5 & \{1, 2\} \\ 4 & \{-1, 1\} \end{pmatrix}$$

This says there are two solutions to $A v = \lambda v$: when $\lambda = -5$, $v = (1, 2)$ and when $\lambda = 4$, $v = (-1, 1)$. Any multiple of each v would also be a solution. So any straight-line solutions of the system must be along the lines $y = 2x$ and $y = -x$.

We can relate the solution we found earlier for the initial conditions $x(0) = 1$, $y(0) = 3$ to the eigenvector $(-1, 1)$. That solution (on p. 511 just before the graphs in Out[5]) shows that

$$x(t) = \frac{4}{3} e^{-5t} - \frac{1}{3} e^{4t}, \quad y(t) = \frac{8}{3} e^{-5t} + \frac{1}{3} e^{4t}.$$

Notice that the coefficients in the exponential functions are the eigenvalues $\lambda = -5, 4$. As t grows, we have $y(t) \approx -x(t)$. That is, the trajectory in Out[9] on page 512 approaches the line through the origin with the direction of the eigenvector $(-1, 1)$ for $\lambda = 4$. As the trajectory approaches the line we have $v'(t) \approx 4v$, so we expect it to move away from the origin as t increases.

In the exercises for this section, you can experiment with other initial points to see how the solution depends on the initial conditions, to see that an initial point on an eigenvector produces a straight-line solution, and to see how the behavior differs for the two eigenvalues. All solutions to this system have the form

$$\begin{pmatrix} x(t) \\ y(t) \end{pmatrix} = c_1 e^{-5t} \begin{pmatrix} 1 \\ 2 \end{pmatrix} + c_2 e^{4t} \begin{pmatrix} -1 \\ 1 \end{pmatrix},$$

where the constants c_1, c_2 depend on the initial conditions. That is, all solutions are linear combinations of the straight-line solutions given by the eigenvalues.

It can be proved in general that if λ is an eigenvalue with eigenvector v, then $e^{\lambda t} v$ is a solution to the system. When λ is real, the trajectory corresponding to this solution is a straight line in the phase plane. When λ contains an imaginary term, the trajectory of $e^{\lambda t} v$ takes other geometric forms, such as periodic or spiraling motion. The eigenvalues and eigenvectors determine all solutions, although when λ is repeated or zero the situation is more complicated than shown here.

Exercises

Exercises 1–3 refer to the system in Example 27.1.1,

$$\begin{aligned} x'(t) &= x - 3y \\ y'(t) &= -6x - 2y \end{aligned}.$$

1. a) Solve the system for the initial condition $x(0) = 2$, $y(0) = 2$. Then graph the trajectory in the phase plane for $0 \le t \le 1$. (**Calculus** $\triangleright$ **Differential Equations, Graph** $\triangleright$ **Parametric Plot**)

   b) How does this solution compare with the one shown in the example? Use the equations of $x(t)$ and $y(t)$ to see what happens as t increases.

2. a) Choose any point $(x, y) \ne (0, 0)$ on the line $y = -x$ as the initial condition. Solve the system, and then graph the trajectory in the phase plane for $0 \le t \le 1$.

   b) Animate the motion of the solution along the trajectory to see in which direction the solution moves. Use five frames ($1 \le k \le 5$), graph for $0 \le t \le k/5$, and choose the common plot frame from the x, y values in the phase portrait in a). Section 15.7 shows how to animate solutions to systems of differential equations. (**Graph** $\triangleright$ **Animation**)

   c) Show algebraically that this is a straight-line solution and find the direction in which it moves along the trajectory. Does this agree with what the animation in b) shows?

3. Repeat Exercise 2 for the line through the origin with the direction of the eigenvector $(1, 2)$. How does the trajectory for this line differ from the one in a), and how does the eigenvalue cause this behavior?

Exercises 4 and 5 deal with a *simple harmonic oscillator*. These are systems, such as mechanical systems and electronic circuits, that can be modeled by the differential equation $y''(t) = -ky$. For example, this equation illustrates the motion of a mass attached to a spring, where $y(t)$ is the position of the mass at time t and the constant $k > 0$ depends on the mass and on the stiffness of the spring. One way to solve the equation is to solve the linear system of two first-order equations $v' = -ky$, $y' = v$. Pulling the mass from its rest position at $y = 0$ to $y = a$ and releasing it with initial velocity $y'(0) = b$ corresponds to the conditions $y(0) = a$, $v(0) = b$.

4. a) Consider the system $y''(t) = -4y$, where you pull the mass to $y = 3$ and just let go. Find the exact solution (not a numeric one) and graph for $0 \le t \le 10$. Then graph (y, v) in the phase plane. In which direction does (y, v) follow the trajectory?

☿ In **Graph** ▷ **Parametric Plot**, enter y = y[t] as the first equation since you want *y* on the horizontal axis and *v* on the vertical one.

b) Repeat for the initial position $y = 2$. How do the two trajectories compare?

c) According to the graphs, as the spring contracts and expands, where is the mass when it is moving fastest?

d) Find the eigenvalues of the system. How do they predict the behavior of the system?

5. a) Consider the system $y''(t) = -2y$, where you pull the mass to $y = 2$ and let go. Find the exact solution (not a numeric one) and graph for $0 \le t \le 10$. Do not graph the phase plane trajectory.

b) Repeat for $y''(t) = -8y$. Do the amplitude and frequency of $v(t)$ and $y(t)$ increase, decrease, or remain the same when you change the constant *k* from 2 to 8?

c) What is the effect of the change in *k* on the path and speed of the mass as it oscillates? If the spring remains the same, would this correspond to making the mass heavier or lighter? If the mass remains the same, would this correspond to making the spring stiffer (stronger) or slacker (weaker)?

■ 27.2 Nonlinear Systems

27.2.1 Example: Linearizing a Nonlinear System

A nonlinear system of differential equations such as

$$x'(t) = 2x - x^2 - xy$$
$$y'(t) = y - y^2 + xy$$

may describe the rates of change with respect to time of two competing populations $x(t)$ and $y(t)$. The system is nonlinear because the right-hand sides do not take the linear form $ax + by$.

We want to know how the populations interact over time and how this depends on the initial population levels.

We start by plotting the direction field for *x* and *y* between ±3.

- From the *Joy* menus, choose **Graph** ▷ **Vector Field**.
- Fill in { 2 x - x² - x‿y, y - y² + x‿y}.

  The symbol ‿ means that you should insert a space to indicate multiplication.

- Fill in − 3 ≤ x ≤ 3, − 3 ≤ y ≤ 3.
- Check **Normalize lengths** and **Assign name**, and fill in field.
- Click **OK**.

Mathematica plots the field:

▷ Graph of the vector field $\{2\,x - x^2 - x\,y,\ y - y^2 + x\,y\}$ for $-3 \le x \le 3,\ -3 \le y \le 3$.

field =

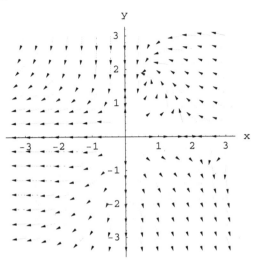

Out[2]=   - Graphics -

We'll take as initial values $x(0) = 2.5$, $y(0) = 0.5$. *Mathematica* cannot solve this system exactly, so we will estimate the solution numerically.

- From the *Joy* menus, choose **Calculus ▷ Differential Equations**.
- Fill in the equations x'[t] = 2 x - x² - x_y, y'[t] = y - y² + x_y.
- Fill in the **initial conditions** x = 2.5, y = 0.5, checking the box first if necessary.
- Fill in the interval 0 ≤ t ≤ 100.
- Check **Solve numerically** and **Graph solution**.
- Click **OK**.

Mathematica solves the system numerically:

▷ Solve {x'[t]=2 x-x²-x y,y'[t]=y-y²+x y} numerically
  with {x=2.5`,y=0.5`}.

Graph styles:

x	Black	Solid
y	Red	.....

Define {x[t],y[t]} to be the solution; then {x[t],y[t]} =

```
{InterpolatingFunction[{{0., 100.}}, <>][t],
 InterpolatingFunction[{{0., 100.}}, <>][t]}
```

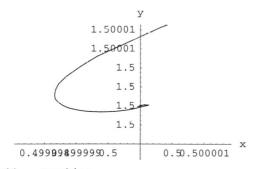

Out[3]= - Graphics -

The solutions are expressed as functions using interpolation and seem to approach limits rapidly as *t* grows, $x(t) \to 0.5$ and $y(t) \to 1.5$. Here's how to see this in the phase plane.

- Choose **Graph ▷ Parametric Plot**.
- Fill in x = x[t], y - y[t].
- Fill in 0 < t ≤ 100.
- Click **OK**.

The result is:

▷ Graph the curve (x,y) = (x[t],y[t]) for 0 ≤ t ≤ 100.

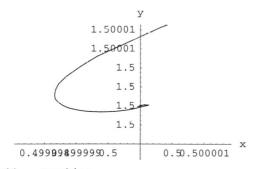

Out[4]= - Graphics -

The trajectory seems to spiral in toward (0.5, 1.5).

This graph doesn't start at $(x, y) = (2.5, 0.5)$ because *Mathematica* has automatically zoomed in around the limiting point. *Mathematica* displays only the first six significant digits on the axes.

Next, we'll thicken the trajectory to make it more visible and then combine it with the direction field.

- Choose **Graph ▷ Modify Graph**.
- Make sure the output number of the preceding graph appears, in this case `Out[4]`.
- Check **Assign name** and fill in `trajectory`.
- Click **OK**.

Now we combine the two graphs.

- Choose **Graph ▷ Combine Graphs**.
- Fill in `field,trajectory`.
- Click **OK**.

The result is:

▷ Combining the graphs {field, trajectory} gives

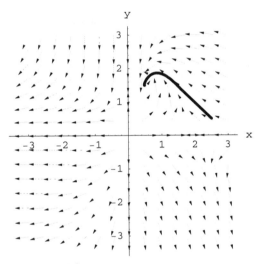

Out[6]= - Graphics -

This view shows how the trajectory follows the direction field beginning at the initial point $(x, y) = (2.5, 0.5)$, but shows less detail near the point $(0.5, 1.5)$ that the solution seems to approach.

Here's how we can understand why the system behaves this way. We first find the equilibrium solutions, i.e., those solutions that are constant functions of t and thus have $x'(t) = 0$, $y'(t) = 0$. You can solve

$$0 = 2x - x^2 - xy$$
$$0 = y - y^2 + xy$$

by hand or with **Algebra ▷ Solve** and find four solutions: $(0, 0)$, $(0, 1)$, $(2, 0)$, and $(\frac{1}{2}, \frac{3}{2})$. Each is an equilibrium solution to the system. The equilibrium $(\frac{1}{2}, \frac{3}{2})$ is the point that our trajectory seems to approach.

We can study the behavior of a nonlinear system near an equilibrium point by approximating it with a linear system. We then find the eigenvalues and eigenvectors of the new system as in Example 27.1.1. This method is called *linearization*. We apply it to an *autonomous* system, that is, a system

$$x'(t) = u(x, y)$$
$$y'(t) = v(x, y)$$

where the derivatives depend on x, y but not on t. We define the *Jacobian* matrix of the system at an equilibrium point (a, b) to be the matrix

$$A = \left. \begin{pmatrix} \frac{\partial u}{\partial x} & \frac{\partial u}{\partial y} \\ \frac{\partial v}{\partial x} & \frac{\partial v}{\partial y} \end{pmatrix} \right|_{(x,y)=(a,b)}.$$

In Example 27.1.1 the system was already in the form $v'(t) = A\,v$ where A contained only constants, and in that case the Jacobian matrix is the same as the original matrix.

The importance of the Jacobian is that, under appropriate smoothness assumptions, we have the linear approximation

$$\begin{pmatrix} x'(t) \\ y'(t) \end{pmatrix} \approx \begin{pmatrix} u(a, b) \\ v(a, b) \end{pmatrix} + A \begin{pmatrix} x - a \\ y - b \end{pmatrix} = A \begin{pmatrix} x - a \\ y - b \end{pmatrix}.$$

If we let $v(t) = \begin{pmatrix} x - a \\ y - b \end{pmatrix}$, then the original system is approximated by $v'(t) = A\,v$, and the eigenvalues and eigenvectors of A help describe the behavior of the system near the equilibrium (a, b).

For this example, at the equilibrium $(\frac{1}{2}, \frac{3}{2})$ the Jacobian matrix is

$$A = \begin{pmatrix} \frac{\partial u}{\partial x} & \frac{\partial u}{\partial y} \\ \frac{\partial v}{\partial x} & \frac{\partial v}{\partial y} \end{pmatrix} = \begin{pmatrix} 2 - 2x - y & -x \\ y & 1 - 2y + x \end{pmatrix} = \begin{pmatrix} -\frac{1}{2} & -\frac{1}{2} \\ \frac{3}{2} & -\frac{3}{2} \end{pmatrix}.$$

- Choose **Matrices ▷ Eigenvalues**.
- The default matrix is selected automatically. Click the **Matrix** button on the *Joy* palette.
- In the dialog that appears, fill in 2 for the number of rows and the number of columns, and click **Paste**.
- Fill in the matrix, pressing TAB to move to each successive entry:

$$\begin{pmatrix} -\frac{1}{2} & -\frac{1}{2} \\ \frac{3}{2} & -\frac{3}{2} \end{pmatrix}$$

- Make sure the boxes for **eigenvalues** and for **eigenvectors** are both checked.
- Click **Algebraically**.
- Click **OK**.

Mathematica finds:

▷ The eigenvalues and eigenvectors of $\begin{pmatrix} -\frac{1}{2} & -\frac{1}{2} \\ \frac{3}{2} & -\frac{3}{2} \end{pmatrix}$ are

Out[6]//MatrixForm=
$$\begin{pmatrix} \frac{1}{2}\,(-2 - \mathrm{I}\,\sqrt{2}\,) & \{-\frac{1}{3}\,\mathrm{I}\,(\mathrm{I} + \sqrt{2}\,)\,,\ 1\} \\ \frac{1}{2}\,(-2 + \mathrm{I}\,\sqrt{2}\,) & \{\frac{1}{3}\,\mathrm{I}\,(-\mathrm{I} + \sqrt{2}\,)\,,\ 1\} \end{pmatrix}$$

Mathematica uses I to denote $i = \sqrt{-1}$. In Version 4, this appears as i. Nonreal eigenvalues always come in conjugate pairs $r \pm is$, as in this case. Each eigenvalue λ contributes a solution $e^{\lambda t}\,v$ to the *linearized* system, where v is an eigenvector whose entries are complex numbers. These are not straight-line solutions as they are when λ is real. To see this, when $\lambda = r + is$ we begin by writing

$$e^{\lambda t} = e^{rt}\,e^{ist} = e^{rt}\,\cos(st) + i\,e^{rt}\,\sin(st).$$

For example, for the first eigenvalue in the solution we have

$$\begin{aligned} \lambda &= \tfrac{1}{2}\left(-2 - i\sqrt{2}\right) = -1 - \tfrac{1}{\sqrt{2}}\,i, \\ e^{\lambda t} &= e^{-t}\,\cos\left(-\tfrac{t}{\sqrt{2}}\right) + i\,e^{-t}\,\sin\left(-\tfrac{t}{\sqrt{2}}\right) \\ &= e^{-t}\,\cos\left(\tfrac{t}{\sqrt{2}}\right) - i\,e^{-t}\,\sin\left(\tfrac{t}{\sqrt{2}}\right). \end{aligned}$$

The corresponding eigenvector is

$$v = \begin{pmatrix} -\frac{1}{3}\,i\left(i + \sqrt{2}\right) \\ 1 \end{pmatrix} = \begin{pmatrix} \frac{1}{3} \\ 1 \end{pmatrix} + i\begin{pmatrix} -\frac{\sqrt{2}}{3} \\ 0 \end{pmatrix}.$$

Multiplication gives $e^{\lambda t}\,v = e^{-t}\,v_1 + i\,e^{-t}\,v_2$, where

$$v_1 = \begin{pmatrix} \frac{1}{3}\,\cos\!\left(\tfrac{t}{\sqrt{2}}\right) - \frac{1}{3}\,\sqrt{2}\,\sin\!\left(\tfrac{t}{\sqrt{2}}\right) \\ \cos\!\left(\tfrac{t}{\sqrt{2}}\right) \end{pmatrix}, \quad v_2 = \begin{pmatrix} -\frac{1}{3}\,\sqrt{2}\,\cos\!\left(\tfrac{t}{\sqrt{2}}\right) - \frac{1}{3}\,\sin\!\left(\tfrac{t}{\sqrt{2}}\right) \\ -\sin\!\left(\tfrac{t}{\sqrt{2}}\right) \end{pmatrix}.$$

It can be shown that we only need to use one of the two conjugate eigenvalues and that every solution to the linearized system has the form $c_1\,e^{-t}\,v_1 + c_2\,e^{-t}\,v_2$, where c_1, c_2 depend on the initial conditions. (When λ is repeated or zero, the situation is more complicated.) This expresses the solutions in terms of real numbers only. As t increases in such a solution, the sines and cosines in $e^{-t}\,v_1,\, e^{-t}\,v_2$ cause the trajectory to revolve around the equilibrium point and the exponential term causes it to spiral inward toward the equilibrium. The spiraling behavior carries over to the trajectories for the original system that the linearized system approximates. This kind of equilibrium point is called a *spiral sink*. It can be shown that the solutions to the linearized and original systems act the same way in the long run, except perhaps when $\lambda = 0$ or $\lambda = 0 \pm is$.

The exercises below let you experiment with other initial conditions and equilibrium points, to illustrate different kinds of behavior that can occur. You can find a systematic description of all the possibilities in a standard differential equations text.

Exercises

Exercises 1 and 2 refer to the system in Example 27.2.1:

$$x'(t) = 2x - x^2 - xy$$
$$y'(t) = y - y^2 + xy$$

1. Solve the system numerically for the initial condition $x(0) = 0.5$, $y(0) = 0.5$ and graph the solutions for $0 \le t \le 50$. Graph the trajectory in the phase plane for $0 \le t \le 50$. How does the trajectory compare to the one in the example?

2. a) Graph the normalized direction field for this system with $0 \le x \le 1$, $1 \le y \le 2$.

    b) Find the Jacobian matrix, eigenvalues, and eigenvectors for the equilibrium at $(0, 1)$. (**Matrices ▷ Eigenvalues**)

    c) Solve the system numerically for the initial condition $x(0) = 0.02$, $y(0) = 1.01$, for $0 \le t \le 100$. Graph the trajectory in the phase plane for $0 \le t \le 100$. Then use **Graph ▷ Modify Graph** to draw a thicker curve.

    d) Combine the graphs of the direction field and the trajectory. If the axes do not have equal scales, use **Graph ▷ Modify Graph** to make them equal. This makes the result easier to visualize.

    e) The initial point in c) is in the direction of the eigenvector $(2, 1)$ at the equilibrium $(0, 1)$ and is very close to that equilibrium. In which direction does $(x(t), y(t))$ move along the trajectory? Use b) to explain the direction of motion near the initial point. Use Example 27.2.1 to explain it at the other end of the trajectory.

3. a) Solve numerically and graph the solution for $0 \le t \le 10$:

    $$x'(t) = 2x - xy, \quad x(0) = 0.5$$
    $$y'(t) = -y + xy, \quad y(0) = 0.5$$

    b) Section 15.7 shows how to animate solutions to differential equations. Animate the solution in the phase plane using five frames. Choose the common plot frame from the x, y values shown in a). (**Graph ▷ Animation**)

    c) Describe its trajectory and the direction in which it moves. About which equilibrium solution does it move? Find the Jacobian matrix A of the system for this equilibrium.

d) Find the eigenvalues of A, by hand or using *Joy* (you don't need the eigenvectors). What is the function $e^{\lambda t}$ for these eigenvalues? (This shows that the trajectory for the linearized system is periodic, but without further analysis you can't be sure that it doesn't spiral in or out for the original system.) (**Matrices** ▷ **Eigenvalues**)

Lab 27.1 Sharks vs Fish: A Predator–Prey Model

Prerequisites:
Ch. 1	A Brief Tour of *Joy*
Secs. 15.1,4	Slope fields and Euler's method

During World War I, there was less fishing in the Adriatic Sea and so the supply of fish available in the Italian markets dropped. Not only was the total size of the catch down, but so was the proportion made up of fish that were in demand by consumers. The percentage of the catch made up of sharks and other predators not favored for eating tripled, jumping from about 12% when fighting broke out in 1914 to 36% when the war ended in 1918. The rest of the catch was made up of food fish destined for market. These were the prey of the predator fish. The fishing changes seemed to affect the shark and food fish populations differently. After the war, the percentage of sharks caught returned to its prewar level.

An investigation of this phenomenon led to the Lotka–Volterra differential equations for modeling predator–prey interactions. You can find the historical and mathematical background in Martin Braun, *Differential Equations and Their Applications*, 4th ed. (Springer-Verlag, New York, 1993). In this lab, you can use *Joy* to explore why this occurred. *Mathematica* cannot find symbolic solutions for some of the differential equations in the chapter. Instead, you can analyze them in a qualitative way, and then use *Joy* to illustrate the analysis by solving the system numerically for a particular example.

■ Before the Lab (Do by Hand)

The simplest type of interaction between predator and prey populations can be described by a pair of differential equations, as follows. Let $F(t)$ be the number of food fish (the prey) at time t and $S(t)$ be the number of sharks (the predators). Suppose that the fish eat food that is available in abundance, while the sharks feed mostly on the fish. The Lotka–Volterra equations for modeling this situation are

$$F'(t) = aF - bSF$$
$$S'(t) = -pS + qSF$$

where $a, b, p, q > 0$ are constants. Here a is the relative rate F'/F at which the fish population would increase in the absence of sharks ($S = 0$), and p is the relative rate $-S'/S$ at which the shark population would decrease without fish to eat ($F = 0$). The constants b and q represent the effect of the sharks on the fish population, and vice versa.

> The effect of each population on the other is taken to be proportional to the number of interactions of the two populations, and thus to the product SF. The proportionality constants are b and q.

Fishing reduces the overall population of sharks and fish by some fraction c, $0 < c < 1$. One way to incorporate the effect of fishing is to modify the equations in this way:

$$F'(t) = aF - bSF - cF = (a - c)F - bSF$$
$$S'(t) = -pS + qSF - cS = -(p + c)S + qSF \qquad (*)$$

1. What assumption about the fishing process would make it reasonable to use the same fraction c in each equation?

We say the two populations are in *equilibrium* if they don't change over time, i.e., $F(t) = F_e$ and $S(t) = S_e$ are constant functions. This corresponds to saying that $F'(t) = S'(t) = 0$ for all t.

2. Find the equilibrium values F_e, S_e for equations (*).

The *phase plane* picture of a solution graphs $(F(t), S(t))$ in the plane as t varies, i.e., $F = F(t)$, $S = S(t)$. The solution $F = F_e$, $S = S_e$ is just a single point, i.e., a solution at the equilibrium point remains there for all values of t. For those familiar with vectors, a nonequilibrium solution follows a curve in the direction of the tangent vector $(F'(t), S'(t))$.

3. Show that $F'(t) = (S_e - S)bF$, $S'(t) = (F - F_e)qS$.

Think of the phase plane as composed of four quadrants around the equilibrium point (F_e, S_e), according to whether F and S are greater than or less than their equilibrium values.

4. a) In each of the four quadrants about the equilibrium, find the signs of $F'(t)$, $S'(t)$.

   b) As the trajectory of a solution passes through a point (F, S) in the first quadrant around the equilibrium, in which direction will it tend to go? Repeat for the other quadrants.

   c) Sketch some possible trajectories $F = F(t)$, $S = S(t)$ around an equilibrium (F_e, S_e). Indicate the direction in which the solution follows the trajectory.

For simplicity, let $a = p = 1$. This amounts to choosing a time unit for which the relative growth rates of fish and sharks in the absence of the other is about $\pm 100\%$ per unit time. It makes the assumption that both rates are equal, which you can change later if you wish. Similarly, assume $F_e = 1$, $S_e = 1$. This means that the units chosen to count fish and sharks are just their equilibrium values, and that you are measuring the populations as multiples of their equilibrium sizes.

Take $c = 0.1$, i.e., assume that the catch rate is 10% per unit time, much less than the natural population growth rates.

5. a) Find the values of q and b that are implied by these choices of the parameters.

   b) Rewrite equations (*) using these parameters.

■ In the Lab

1. Plot the direction field for the system in Question 5b) (Before the Lab), for $0.01 \leq F \leq 3$, $0.01 \leq S \leq 3$. (**Graph ▷ Vector Field, Normalize lengths**)

2. a) Use Euler's method to approximate a solution to the system for $0 \leq t \leq 10$, with the initial conditions $F = 1.1$, $S = 1.1$. Choose 50 steps and **Graph** from the popup menu. (**Calculus ▷ Euler's Method**)

    The initial conditions mean that the fish and shark populations are initially 10% above their equilibrium values.

    b) Graph the curves $F = F(t)$, $S = S(t)$ for $0 \leq t \leq 10$. (**Graph ▷ Parametric Plot**)

3. Repeat the last question using 200 steps in Euler's method.

    ⚠ You cannot do this question and the next one together. While **Calculus ▷ Euler's Method** will let you choose 200, 800, 3200 steps, it will only define the functions $F(t)$, $S(t)$ for 3200 steps. This won't let you plot the trajectory for all three in **Graph ▷ Parametric Plot**.

4. Repeat using 800 steps in Euler's method. If your computer seems fast enough, repeat again with 3200 steps.

5. Print your work. If you will be doing the section **For Further Exploration**, save your notebook.

■ After the Lab

1. How does the graph of the direction field in Question 1 (In the Lab) compare to your predictions in Question 4 (Before the Lab)? Review your reasoning if there are any inconsistencies.

2. Where is the equilibrium point (F_e, S_e) in the trajectory graphs you created?

3. a) In Question 2 (In the Lab), explain how the graphs of $F(t)$ and $S(t)$ in a) show that the trajectory in b) will spiral outward.

    b) What happens to the trajectory of the Euler approximation as you increase the number of steps? What do you think the actual trajectory is?

4. Euler's method for two differential equations follows the tangents to the direction field to create an approximation to the trajectory. For the shape of the field in this example, would you expect Euler's method to produce a trajectory that was closer to or further from the origin than the actual one?

It can be proved that when you increase the number of steps in Euler's method by a factor of 4, as we did each time in the lab, you reduce the maximum possible error in the approximation by a factor of $1/4$ (i.e., the worst-case error will be 25% of its previous value). So in going from 50 to 3200 steps, we reduce the maximum error by a factor of $\frac{1}{4^3} \approx 0.016$.

5. Use the trajectories you graphed in the lab to explain how the fish and shark populations seem to interact with each other. For instance, what happens to the sharks and to the fish if there are many fish? What happens if there are few fish? Why doesn't the fish population seem to die out or to explode?

■ For Further Exploration

It can be shown that the average values of $F(t)$ and $S(t)$ over a long period of time are equal to the equilibrium values F_e and S_e. You can visualize this average behavior in the graphs of $F = F(t)$ and $S = S(t)$.

1. How does the percentage of the catch made up of fish change as the solution moves around its trajectory? How does the percentage along the trajectory compare to the percentage at equilibrium?

2. What happens to the equilibrium values as the level of fishing, expressed by the parameter c, decreases? Will the percentage of sharks in the total catch rise or fall?

3. Repeat the preceding question if the level of fishing increases. How do these questions help explain what was seen in the Italian fish markets during and after World War I?

Chapter 28
Matrices and Linear Equations

Examples and Exercises

Prerequisites:
    Ch. 1   A Brief Tour of *Joy*
    Secs. 16.4–8   Creating matrices, row reduction, multiplication

In many of the exercises in this chapter, you will need to transform a matrix to reduced row-echelon form. You can do this directly with **Matrices ▷ Row Reduce** or step by step with **Matrices ▷ Row Operations**. Some questions can also be answered with **Algebra ▷ Solve** or **Matrices ▷ Inverse** instead of reducing the matrix.

■ 28.1 Existence and Uniqueness of Solutions

Exercises

1. a) Convert to reduced row-echelon form:
$$\begin{pmatrix} 1 & 2 & 0 & 1 \\ 5 & 1 & 3 & 3 \\ 6 & 2 & 1 & 3 \\ 0 & 3 & 3 & 2 \end{pmatrix}.$$

   b) Use a) to determine how many solutions this system has:
$$x + 2y = 1$$
$$5x + y + 3z = 3$$
$$6x + 2y + z = 3$$
$$3y + 3z = 2$$

2. a) Solve using row reduction:

$$x + 3y = 1$$
$$3x + y + z = 0$$
$$2y + z = 3$$
$$2x + 3y + 3z = 2$$

b) Prove that if an augmented matrix $[A \mid b]$ reduces to the identity matrix, then $A u = b$ has no solution. (Compare with Question 1.)

3. The four equations in Question 2 represent four planes.

a) Find the intersection of planes 2, 3, 4.

b) Graph planes 2, 3, 4 together for x, y between ± 5 and mark the intersection on a printout of the graph. (**Graph 3D ▷ Graph Surface**)

c) Does the intersection in b) satisfy the equation of plane 1?

d) By hand, add a sketch of plane 1 to the graph in b).

In Questions 4–6:

a) For which values of a is the system consistent?

b) If the system is consistent, is the solution unique?

4.
$$x + 2y = a$$
$$2x + y = 1$$
$$3x + 2y = 3$$

5.
$$x + 2y = 1$$
$$2x + ay = 1$$
$$3x + 2y = 3$$

6.
$$x + y = a$$
$$x + y + 2z = 0$$
$$ax + y + z = 1$$

7. You inherit $25,000 from a generous uncle and decide to invest it in mutual funds. A higher rate of return carries more risk, so you plan to divide the money among different kinds of funds. You choose a money market fund that returns 5% per year, a bond fund that returns 8% per year, and a stock fund that returns 20% per year. You would like your return for the first year to be $3200 so you can take a nice vacation, but don't want to take the risk necessary to achieve a larger return.

a) How can you divide the $25,000 investment to achieve this goal? Is there more than one way to divide it up?

b) In a), what is the least amount of money that you must invest in stocks to achieve the goal? What is the most you can invest in stocks?

c) Suppose you decide to invest twice as much in stocks as in the money market. Can you get a return of $3200, and in how many ways?

■ 28.2 Homogeneous and Nonhomogeneous Systems

Additional Prerequisites:
          Secs. 10.5,6   Combining 3D graphs, graphing space curves

28.2.1 Example: The Geometry of Solutions

In this example, we'll compare the solutions to two related systems of linear equations,

$$
\begin{aligned}
x + y + z &= 5 \\
2x + y + 2z &= 3 \\
x + 2y + z &= 12
\end{aligned}
\quad \text{and} \quad
\begin{aligned}
x + y + z &= 0 \\
2x + y + 2z &= 0 \\
x + 2y + z &= 0
\end{aligned}.
$$

The second system is *homogeneous*, i.e., the right side is zero, but the first is not. We'll express these systems in matrix form and reduce the matrices, beginning with the nonhomogeneous system.

- From the *Joy* menus, choose **Matrices ▷ Row Reduce**.
- The default matrix is selected automatically. Click the **Matrices** button on the *Joy* palette.
- Fill in 3 rows and 4 columns, and click **Paste**.

    A 3×4 matrix replaces the default matrix in the **Row Reduce** dialog.

- Fill in the matrix $\begin{pmatrix} 1 & 1 & 1 & 5 \\ 2 & 1 & 2 & 3 \\ 1 & 2 & 1 & 12 \end{pmatrix}$ and click **OK**.

The result is:

▷ Row reduce the matrix $\begin{pmatrix} 1 & 1 & 1 & 5 \\ 2 & 1 & 2 & 3 \\ 1 & 2 & 1 & 12 \end{pmatrix}$

$Out[2]=$ $\begin{pmatrix} 1 & 0 & 1 & -2 \\ 0 & 1 & 0 & 7 \\ 0 & 0 & 0 & 0 \end{pmatrix}$

The reduced matrix corresponds to the equations

$$x + z = -2 \atop y = 7 \quad , \text{ i.e., } \quad \begin{matrix} x & = & -2 - z \\ y & = & 7 \\ z & = & z \end{matrix} .$$

The solution is not unique, since z can be set arbitrarily. We can think of z as a parameter and the solution as a line in R^3:

$$\boldsymbol{u} = \begin{pmatrix} x \\ y \\ z \end{pmatrix} = \begin{pmatrix} -2 - z \\ 7 \\ z \end{pmatrix} = \begin{pmatrix} -2 \\ 7 \\ 0 \end{pmatrix} + z \begin{pmatrix} -1 \\ 0 \\ 1 \end{pmatrix} .$$

The vectors $z(-1, 0, 1)$ describe a line through the origin. Adding $(-2, 7, 0)$ shifts this line so that it passes through $(-2, 7, 0)$ (its *displacement* vector). The vector $(-1, 0, 1)$ gives the *direction* of these two parallel lines.

Now we solve the homogeneous system. This will use the same row operations, so the reduced row-echelon form is

$$\begin{pmatrix} 1 & 0 & 1 & 0 \\ 0 & 1 & 0 & 0 \\ 0 & 0 & 0 & 0 \end{pmatrix} .$$

This shows that solution is now $x = -z$, $y = 0$, or in parametric vector form $\boldsymbol{u} = z(-1, 0, 1)$. This is just the direction vector for the nonhomogeneous system. The two systems have parallel solutions, and the one for the homogeneous system passes through the origin.

It can be proved in general that the solution to a homogeneous system always passes through the origin and changing the right side of a system just changes the displacement vector (if there is a solution at all).

 ♡ When a solution with three unknowns has one parameter it corresponds to a line in R^3 and can be plotted using **Graph 3D ▷ Space Curve**. When it has two parameters it is a plane and can be plotted using either **Graph 3D ▷ Surface** or **Parametric Surface**.

Here's how to picture the two lines. First we graph the solution to the nonhomogeneous system.

- Choose **Graph 3D ▷ Space Curve**.
- Fill in the equations x = - 2 - t, y = 7, z = t.
- Fill in the interval – 5 ≤ t ≤ 5.
- Click **OK**.

Now we graph the solution to the homogeneous system.

- Click **Last Dialog** on the *Joy* palette to reopen the **Space Curve** dialog.
- Change the equations to x = - t, y = 0, z = t and click **OK**.

Then we combine the two.

- Combine `Out[3]`,`Out[4]`, using your output numbers if they are different. (**Graph ▷ Combine Graphs**)

The result is (the paraphrase shows the abbreviation **%3** for `Out[3]`):

▷ Combining the graphs {%3, %4} gives

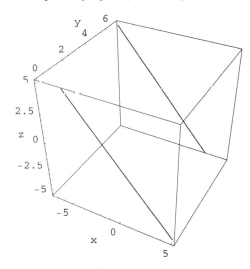

Out[5]= - Graphics3D -

The line that appears closest to the viewer passes through the origin. This is the solution to the homogeneous system. The solution to the nonhomogeneous system is parallel to it, passing through the point $(-2, 7, 0)$.

Exercises

1. a) Express the solution parametrically:
    $$x + 3y + 2z = 5$$
    $$2x - 2y + z = 1$$

    b) What are the direction and displacement of the solution?

    c) Graph the solution along with the solution to the corresponding homogeneous system for z between ±5.

2. a) Express the solution parametrically:
    $$x + 3y + 2z = 5k$$
    $$2x - 2y + z = k$$

    b) Describe how the direction and displacement of the solution change as k varies from 0 to 1.

c) Graph the solutions for $k = 0$, $\frac{1}{2}$, 1 and combine the graphs.

3. a) Express the solution parametrically:

$$x + 3\,y + 2\,z = a$$
$$2\,x - 2\,y + z = b$$

b) What is the solution when the system is homogeneous?

c) What combinations of a, b make the displacement vector **0**?

4. a) Express the intersection of these planes parametrically as a line:

$$2\,x - 3\,y - 4\,z = 1$$
$$2\,x + 3\,y - 4\,z = 3$$
$$x - 2\,z = 1$$

b) Graph the three planes and combine the graphs. Mark the intersection on a printout of the combined graph. You can graph the planes by solving for z and choosing **Graph 3D ▷ Surface**.

If you are using *Mathematica* Version 4 with interactive 3D graphics, display the coordinates by choosing **Graph ▷ Modify Graph** and specifying **Axis Labels**.

c) Observe from the graph that the intersection doesn't pass through the origin, and verify that in your solution to a). How can you change the equations so the intersection passes through the origin?

5. Express the solution to this system parametrically as a line in R^4:

$$x - 3\,y + z - w = 0$$
$$x - 2\,y + 3\,z - 4\,w = 4$$
$$2\,x - y + 2\,z - w = 2$$
$$4\,x + 4\,y + z + 4\,w = 1$$

6. a) Express the solution to this system parametrically:

$$x - 3\,y + z - w = 0$$
$$x - 2\,y + 3\,z - 4\,w = 4$$
$$4\,x - 9\,y + 10\,z - 13\,w = 12$$
$$2\,x - 5\,y + 4\,z - 5\,w = 4$$

b) How many parameters are needed for the solution? Is the solution a line in R^4, and if not, then how would you describe it?

c) Which terms in the solution to a) correspond to the solution to the associated homogeneous system?

■ 28.3 Matrix Operations

Additional Prerequisites:
Sec. 4.9	How to Simplify Expressions in General
Sec. 16.12	How to Transpose a Matrix

28.3.1 Example: The Distributive Law

The distributive law for matrix multiplication states that $P(Q+R) = PQ + PR$ and $(Q+R)P = QP + RP$ whenever the sizes of P, Q, R make the sums and products well-defined. Here is how to use *Joy* to prove $P(Q+R) = PQ + PR$ for the case where P is 4×3 and Q, R are 3×2.

Since a proof requires that the matrices be arbitrary, we let

$$P = \begin{pmatrix} a & b & c \\ d & e & f \\ g & h & i \\ j & k & l \end{pmatrix}, \quad Q = \begin{pmatrix} m & n \\ o & p \\ q & r \end{pmatrix}, \quad R = \begin{pmatrix} s & t \\ u & v \\ w & x \end{pmatrix}.$$

- Click in your *Mathematica* notebook beneath the last output, if any, so that a horizontal line appears.
- Type P = and click the **Matrices** button on the *Joy* palette.
- Fill in 4 rows and 3 columns and click **Paste.**
- Fill in

$$\begin{pmatrix} a & b & c \\ d & e & f \\ g & h & i \\ j & k & l \end{pmatrix} \cdot$$

- Press SHIFT − RET .

The result is:

$$In[2]:= \quad P = \begin{pmatrix} a & b & c \\ d & e & f \\ g & h & i \\ j & k & l \end{pmatrix}$$

$$Out[2]= \begin{pmatrix} a & b & c \\ d & e & f \\ g & h & i \\ j & k & l \end{pmatrix}$$

- Similarly, enter $Q = \begin{pmatrix} m & n \\ o & p \\ q & r \end{pmatrix}$ and $R = \begin{pmatrix} s & t \\ u & v \\ w & x \end{pmatrix}$.

In[3]:= $Q = \begin{pmatrix} m & n \\ o & p \\ q & r \end{pmatrix}$

Out[3]= $\begin{pmatrix} m & n \\ o & p \\ q & r \end{pmatrix}$

In[4]:= $R = \begin{pmatrix} s & t \\ u & v \\ w & x \end{pmatrix}$

Out[4]= $\begin{pmatrix} s & t \\ u & v \\ w & x \end{pmatrix}$

There are two ways to multiply matrices. We'll illustrate both of them.

- From the *Joy* menus, choose **Matrices ▷ Multiply**.
- Fill in P . (Q + R).
- Check **Assign name** and use the default name **matrix1** or a name of your choice.
- Click **OK**.

▷ Evaluating P . (Q + R) gives

matrix1 =

Out[5]= $\begin{pmatrix} a\,(m+s)+b\,(o+u)+c\,(q+w) & a\,(n+t)+b\,(p+v)+c\,(r+x) \\ d\,(m+s)+e\,(o+u)+f\,(q+w) & d\,(n+t)+e\,(p+v)+f\,(r+x) \\ g\,(m+s)+h\,(o+u)+i\,(q+w) & g\,(n+t)+h\,(p+v)+i\,(r+x) \\ j\,(m+s)+k\,(o+u)+l\,(q+w) & j\,(n+t)+k\,(p+v)+l\,(r+x) \end{pmatrix}$

We'll calculate *P Q+P R* directly in the notebook.

- If necessary, click below the last output so that a horizontal line appears.
- Type matrix2 = P . Q + P . R.
- Press ⎡SHIFT⎤ − ⎡RET⎤.

In[6]:= **matrix2 = P . Q + P . R**

Out[6]= $\begin{pmatrix} am+bo+cq+as+bu+cw & an+bp+cr+at+bv+cx \\ dm+eo+fq+ds+eu+fw & dn+ep+fr+dt+ev+fx \\ gm+ho+iq+gs+hu+iw & gn+hp+ir+gt+hv+ix \\ jm+ko+lq+js+ku+lw & jn+kp+lr+jt+kv+lx \end{pmatrix}$

You can check that the two matrices are equal, but here's how to let *Mathematica* do it for you.

- From the *Joy* menus, choose **Algebra** ▷ **Simplify**.
- Fill in `matrix1` – `matrix2`.
- Click **Simplify** and **OK**.

▷ Simplifying matrix1 – matrix2 gives

$$Out[7]= \begin{pmatrix} 0 & 0 \\ 0 & 0 \\ 0 & 0 \\ 0 & 0 \end{pmatrix}$$

Since the difference is the zero matrix, we have shown that for these matrices $P(Q+R) = PQ + PR$.

Exercises

1. Do the matrices $A = \begin{pmatrix} 5 & 2 & 1 \\ 2 & 5 & 7 \\ 1 & 7 & -3 \end{pmatrix}$ and $B = \begin{pmatrix} 1 & 4 & 2 \\ 3 & -2 & 1 \\ 2 & 1 & 0 \end{pmatrix}$ commute, i.e., does $AB = BA$?

2. Is there a way to choose scalars a, b, c so that $\begin{pmatrix} 5 & 2 & 1 \\ 2 & 5 & 7 \\ 1 & 7 & -3 \end{pmatrix}$ and $\begin{pmatrix} a & b & c \\ a & b & c \\ a & b & c \end{pmatrix}$ commute? What are all the ways to do this?

3. The associative law for matrix multiplication states that $P(QR) = (PQ)R$ whenever the sizes of P, Q, R are compatible. Use *Joy* to prove this for the case of arbitrary matrices P, Q, R where P is 2×3, Q is 3×4, and R is 4×2.

4. a) One property of transpose matrices is that $(AB)^T = B^T A^T$ whenever the sizes of A, B are compatible. Use *Joy* to prove this for the case of arbitrary matrices A, B where A is 3×4 and B is 4×2.

   b) Suppose A is $m \times n$ and B is $n \times p$. By hand, write an expression for the entry in row i, column j of $(AB)^T$. Do the same for $B^T A^T$ and prove these entries are the same.

5. a) Let P be an arbitrary 3×4 matrix. Use *Joy* to show that PP^T and $P^T P$ are square symmetric matrices.

   b) Use the general property $(AB)^T = B^T A^T$ of Exercise 4 to prove, by hand, that for every $m \times n$ matrix P, the matrices PP^T and $P^T P$ are square and symmetric.

■ 28.4 Inverse Matrices

Additional Prerequisite:
      Sec. 16.9   How to Invert a Matrix

Exercises

Solve the equations in Questions 1–4 by inverting an appropriate matrix.

1. $\begin{aligned} x + 2\,y &= 3 \\ 4\,x + 5\,y + 6\,z &= 4 \\ 8\,y + 9\,z &= 5 \end{aligned}$

2. $\begin{pmatrix} 1 & 4 & 2 \\ 3 & -2 & 1 \\ 2 & 1 & 0 \end{pmatrix} X = \begin{pmatrix} 5 & 2 & 1 \\ 2 & 5 & 7 \\ 1 & 7 & -3 \end{pmatrix}.$

3. $X \begin{pmatrix} 1 & 4 & 2 \\ 3 & -2 & 1 \\ 2 & 1 & 0 \end{pmatrix} = \begin{pmatrix} 5 & 2 & 1 \\ 2 & 5 & 7 \\ 1 & 7 & -3 \end{pmatrix}.$

4. $x = A\,x + b$, where

$$A = \begin{pmatrix} 1 & 2 & 3 \\ 4 & 0 & 6 \\ 7 & 8 & 9 \end{pmatrix}, \quad b = \begin{pmatrix} 1 \\ 2 \\ 3 \end{pmatrix}.$$

5. a) Which values of a make $\begin{pmatrix} 5 & 2 & 1 \\ 2 & a & 7 \\ 1 & 7 & -3 \end{pmatrix}$ an invertible matrix?

    ⚠ You can answer a) and b) using determinants or step-by-step row reduction (or by choosing **Matrices** ▷ **Inverse**). But **Matrices** ▷ **Row Reduce** will give a misleading answer since *Mathematica* will divide by a quantity that might be zero, depending on the value of a.

    b) Repeat for $\begin{pmatrix} 5 & 2 & 1 \\ 2 & a & 7 \\ 1 & 7 & a \end{pmatrix}.$

■ 28.5 Determinants

Additional Prerequisite:

      Sec. 16.11   How to Find a Determinant

Exercises

In Questions 1–3, use a determinant to help decide if the equations have a unique solution, multiple solutions, or no solution.

1.
$$\begin{aligned} x + 2y &= 3 \\ 4x + 5y + 6z &= 4 \\ 8y + 9z &= 5 \end{aligned}$$

2. $\begin{pmatrix} -24 & 2 & 0 \\ 4 & 5 & 6 \\ 0 & 8 & 9 \end{pmatrix} x = \begin{pmatrix} 3 \\ -1 \\ 2 \end{pmatrix}.$

3. a) $\begin{pmatrix} 2 & 2 & 2 \\ 2 & 2 & 2 \\ 2 & 2 & 2 \end{pmatrix} x = \mathbf{0}.$

   b) $\begin{pmatrix} 2 & 2 & 2 \\ 2 & 2 & 2 \\ 2 & 2 & 2 \end{pmatrix} x = 3x.$

   c) $\begin{pmatrix} 2 & 2 & 2 \\ 2 & 2 & 2 \\ 2 & 2 & 2 \end{pmatrix} x = 6x.$

4. An important property of determinants is that $\det(AB) = (\det A)(\det B)$, whenever A, B are square matrices of the same size. Use *Joy* to prove this for the case when A, B are arbitrary 3×3 matrices (see Example 28.3.1 for how to work with arbitrary matrices).

5. This question illustrates one way to define determinants. Let

$$A = \begin{pmatrix} a & b & c \\ d & e & f \\ g & h & i \end{pmatrix}$$

be an arbitrary 3×3 matrix.

a) Use *Joy* to calculate $\det A$. How many terms are in the determinant? How many factors are in each term (not counting the factor of ± 1)?

b) From how many rows and columns do the factors in each term come? How does this explain the number of terms?

c) How many terms carry a factor of $+1$ and how many carry -1? Show that the sign for each term is given by the following procedure, to be done by hand and illustrated here with the term $-c\,e\,g$:

> if necessary, arrange the factors in order of their *rows* ($c\,e\,g$)
> write the *column* numbers of those factors (*321*)
>> (this is called a *permutation* of the integers *123*)
> rearrange *123* into the column numbers by *transposing* adjacent integers
>> ($123 \to 132 \to 312 \to 321$)
> choose $+1$ for an *even* number of transpositions and -1 for an *odd* number
>> (there are three transpositions so the sign of $c\,e\,g$ is -1)

It can be proved that no matter how you choose the transpositions for a given permutation, the number is either always even or always odd. So there is no ambiguity in determining the sign associated with a permutation. The value of the determinant is the sum of all the permutation terms, multiplied by the corresponding sign factors.

6. Repeat Question 5 for an arbitrary 4×4 matrix.

■ 28.6 Elementary Matrices

Exercises

An *elementary matrix* is the matrix obtained by applying an elementary row operation to an identity matrix. For example,

$$\begin{pmatrix} 1 & 0 & 0 & 0 \\ 0 & 1 & 0 & 0 \\ 0 & 0 & 1 & 0 \\ 0 & -3 & 0 & 1 \end{pmatrix}$$

is the elementary matrix obtained by multiplying row 2 of the 4×4 identity matrix by -3 and adding it to row 4.

In Questions 1–3, let

$$A = \begin{pmatrix} a & b & c & d \\ e & f & g & h \\ i & j & k & l \end{pmatrix}.$$

1. a) Calculate MA, where M is the elementary matrix obtained by multiplying row 1 of the 3×3 identity matrix by r.

b) Repeat, multiplying row 3 by *r* instead of row 1.

c) Explain how *MA* appears to be related to *A* for this type of elementary row operation.

2. a) Calculate *MA*, where *M* is the elementary matrix obtained by interchanging rows 1 and 3 of the 3×3 identity matrix.

    b) Repeat, interchanging rows 2 and 3 instead.

    c) Explain how *MA* appears to be related to *A* for this type of elementary row operation.

3. a) Calculate *MA*, where *M* is the elementary matrix obtained by adding *r* times row 2 of the 3×3 identity matrix to row 3.

    b) Repeat, adding *r* times row 3 to row 1 instead.

    c) Explain how *MA* appears to be related to *A* for this type of elementary row operation.

4. a) Find the reduced row-echelon form of

$$A = \begin{pmatrix} 1 & 2 & 0 & 1 \\ 6 & 13 & 1 & 3 \\ 0 & 3 & 4 & 2 \end{pmatrix}$$

by performing elementary row operations step by step.

    b) For each row operation in a), create the elementary matrix to which it corresponds.

    c) Multiply *A* on the left by the first elementary matrix in b) and then multiply the result on the left by the others in turn.

    d) The final matrix in c) should be the same as the final matrix in a). Explain why.

Lab 28.1 Row Reduction and Solving Equations

Prerequisites:
Ch. 1	A Brief Tour of *Joy*
Secs. 16.4,6	Creating matrices, elementary row operations

In this lab, you will solve a system of linear equations by applying elementary row operations. These enable you to read off the solution and determine its geometric form. When you apply an elementary row operation, the equations change but the solutions don't. So transforming a system to one whose solutions are easier to find is a good way to solve the original system.

■ Before the Lab (Do by Hand)

1. What are all the possible kinds of intersections of three planes in R^3? Sketch an example of each kind.

The equations

$$\begin{aligned} x + 2y + 3z &= 1 \\ 4x + 5y + 6z &= 2 \\ 7x + 8y + 9z &= 3 \end{aligned} \qquad (*)$$

represent three planes in R^3. In matrix form, this system can be written as $A\,u = b$, with augmented matrix

$$[A \,|\, b] = \begin{pmatrix} 1 & 2 & 3 & 1 \\ 4 & 5 & 6 & 2 \\ 7 & 8 & 9 & 3 \end{pmatrix}.$$

2. What elementary row operation will create a 0 in the second row, first column of $[A \,|\, b]$?

3. What are the equations of the three planes corresponding to the new augmented matrix?

■ In the Lab

1. Apply the elementary row operation you chose in Question 2 (Before the Lab) to the matrix

$$\begin{pmatrix} 1 & 2 & 3 & 1 \\ 4 & 5 & 6 & 2 \\ 7 & 8 & 9 & 3 \end{pmatrix}$$

and verify that it works the way you predicted. (**Matrices ▷ Row Operations**)

2. Now apply row operations one after the other to transform the matrix so that it is in *row-echelon form (REF)*, that is:

    rows that are all 0 are at the bottom
    the first nonzero entry of each nonzero row is 1 (this is called a *pivot*)
    the entries below a pivot are all 0
    each pivot is to the right of the pivots above it

    Choose the operations so that you make steady progress toward the goal, e.g., don't change an entry that you want to be 0 to something else. If you change your mind about any operations you chose, you can backtrack by choosing **Last Dialog** from the *Joy* palette.

3. Apply additional operations so that the entries above the pivots are also 0. This is called *reduced row-echelon form (RREF)*.

4. Repeat Question 2 for the system
    $$x + 2y + 3z = a$$
    $$4x + 5y + 6z = b \tag{**}$$
    $$7x + 8y + 9z = c$$

5. Print your work.

■ After the Lab

1. How many solutions are there to system (*), and what are they?

2. Before the lab, you found all the ways three planes can intersect. What geometric shape is the intersection for system (*), and why?

3. a) What are the equations of the planes corresponding to the *REF* you found for system (*) ?

    b) What are the equations of the planes corresponding to the *RREF*?

    c) Prove that if (x, y, z) is in the intersection of the *REF* planes for this system then it is also in the intersection of the *RREF* planes, and vice versa.

4. a) How must a, b, c be related for system (**) to be consistent?

    b) When (**) is consistent, what shape is the intersection of the three planes, and why?

5. What geometric shape is the set of all (a, b, c) for which system (**) is consistent, and why?

Lab 28.2 Elementary Matrices and Inverses

Prerequisites:
Ch. 1	A Brief Tour of *Joy*
Secs. 16.6,8	Row operations, multiplication

This lab shows the connection between elementary matrices and inverses. An *elementary matrix* is the matrix obtained by applying an elementary row operation to an identity matrix.

■ Before the Lab (Do by Hand)

1. Find the reduced row-echelon form of $A = \begin{pmatrix} 1 & 2 \\ 3 & 4 \end{pmatrix}$. Keep track of the elementary row operations you use.

2. For each elementary row operation you used, form the elementary matrix obtained by applying that operation to $\begin{pmatrix} 1 & 0 \\ 0 & 1 \end{pmatrix}$. Call these matrices $M_1, M_2, ..., M_n$, where M_1 corresponds to the first operation, M_2 to the second, etc.

3. Calculate the product $P = M_n M_{n-1} ... M_2 M_1$.

4. Calculate PA and AP. Your result should be the identity matrix. If it is not, find your error and redo your work.

■ In the Lab

1. In your *Mathematica* notebook, enter $A = \begin{pmatrix} 1 & 2 & 0 \\ 6 & 13 & -3 \\ 0 & 3 & -8 \end{pmatrix}$. That is,

   - Click in your notebook below the last output, if any, so that a horizontal line appears.
   - Type A = and click the **Matrix** button on the *Joy* palette.
   - Set the number of rows and columns to 3 in the dialog, and click **Paste**.
   - Fill in the matrix and press SHIFT − RET.

2. Find the reduced row-echelon form of A, by applying elementary row operations step by step. Keep track of the elementary row operations you use. (**Matrices ▷ Row Operations**)

3. You can create elementary matrices efficiently by checking the boxes in the **Matrix Paste** dialog to create an identity matrix. After the identity matrix is pasted in, you can change one entry to turn it into the elementary matrix you want. Then press SHIFT   RET to create the matrix.

a) Create an elementary matrix M_1 in your notebook for the first row operation, just as you did in Question 1 (In the Lab). Use the **Subscript** button on the *Joy* palette to enter the subscript.

b) Repeat for the other elementary matrices, naming them M_2, M_3, etc.

4. Multiply the elementary matrices in order, as in Question 3 (Before the Lab), and assign the name P to the product. (**Matrices** ▷ **Multiply, Assign name**, or directly in your notebook)

5. Calculate *PA* and *AP*. Your result should be the identity matrix. If it is not, find your error and redo your work.

6. Print your work.

■ After the Lab

1. What is A^{-1}, where A is the 2×2 matrix you worked with before the lab? What is P^{-1}?

2. What is A^{-1}, where A is the 3×3 matrix you worked with in the lab? What is P^{-1}?

3. Explain why the inverse of every invertible matrix A is the product of elementary matrices.

4. Explain why every invertible matrix A is the product of elementary matrices.

5. If A is any square matrix, does the procedure in this lab produce A^{-1}? Is the matrix P always invertible?

Chapter 29
Vector Spaces and Linear Transformations

Examples and Exercises

General Prerequisites:

Ch. 1	A Brief Tour of *Joy*
Sec. 4.9	How to Simplify Expressions in General
Secs.16.4,7	Creating and reducing matrices

■ 29.1 Spanning and Independence

29.1.1 Example: Determining the Span

The vectors

$$\begin{pmatrix} 1 \\ -1 \\ 2 \\ 7 \end{pmatrix}, \begin{pmatrix} 2 \\ 0 \\ 5 \\ -1 \end{pmatrix}, \begin{pmatrix} 3 \\ 1 \\ 0 \\ 1 \end{pmatrix}$$

span a subspace of R^4 consisting of the linear combinations of these three vectors. The span contains those vectors $w \in R^4$ for which $A v = w$ has a solution $v \in R^3$, where

$$A = \begin{pmatrix} 1 & 2 & 3 \\ -1 & 0 & 1 \\ 2 & 5 & 0 \\ 7 & -1 & 1 \end{pmatrix}, v = \begin{pmatrix} x \\ y \\ z \end{pmatrix}, w = \begin{pmatrix} a \\ b \\ c \\ d \end{pmatrix}.$$

This is true because a linear combination of the columns of A with coefficients x, y, z is a product of the form $A v$. Here's how to determine the vectors w in the span.

- From the *Joy* menus, choose **Matrices** ▷ **Row Operations**.
- Click the **Matrix** button on the *Joy* palette.
- Fill in the augmented matrix

$$\begin{pmatrix} 1 & 2 & 3 & a \\ -1 & 0 & 1 & b \\ 2 & 5 & 0 & c \\ 7 & -1 & 1 & d \end{pmatrix}.$$

- Apply successive elementary row operations until the matrix is *upper-triangular*, i.e., all entries below the main diagonal are zero. The result is not unique, and one of the possibilities is:

$$Out[8]= \begin{pmatrix} 1 & 2 & 3 & a \\ 0 & 1 & -6 & -2\,a+c \\ 0 & 0 & 16 & a+b-2\,(-2\,a+c) \\ 0 & 0 & 0 & -7\,a+15\,(-2\,a+c)+\frac{55}{8}\,(a+b-2\,(-2\,a+c))+d \end{pmatrix}$$

There is a solution if and only if the last row is entirely zero. Here's how to determine which values of *a*, *b*, *c*, *d* make this happen.

- Choose **Algebra** ▷ **Simplify**.
- Copy the entry in row 4, column 4 of the last matrix and paste it into the **Simplify** dialog.
- Click **Expand**.
- Click **OK**.

Mathematica expands the expression:

▷ Expanding $-7\,a+15\,(-2\,a+c)+\dfrac{55}{8}\,(a+b-2\,(-2\,a+c))+d$ gives

$$Out[9]= \ -\frac{21\,a}{8}+\frac{55\,b}{8}+\frac{5\,c}{4}+d$$

So

$$-\frac{21\,a}{8}+\frac{55\,b}{8}+\frac{5\,c}{4}+d=0,$$

or $-21\,a+55\,b+10\,c+8\,d=0$, is an equation for the span. Since three of the four coordinates of *w* can be set arbitrarily, the span is a three-dimensional subspace of R^4, called a *hyperplane*.

- ⚠ You cannot use **Matrices** ▷ **Row Reduce** here because *Mathematica* will say that the reduced row-echelon form is the identity matrix. This is true except for the values of *a*, *b*, *c*, *d* that make the row 4, column 4 entry equal zero. *Mathematica* doesn't take this possibility into account.

Exercises

1. Are the vectors $\begin{pmatrix} 1 \\ 3 \\ 1 \end{pmatrix}, \begin{pmatrix} -1 \\ 9 \\ 2 \end{pmatrix}, \begin{pmatrix} 2 \\ 2 \\ 1 \end{pmatrix}$ linearly independent in R^3?

2. For which values of a are these vectors linearly independent?

$$\begin{pmatrix} a \\ 3 \\ 4 \\ 1 \end{pmatrix}, \begin{pmatrix} -1 \\ a \\ 6 \\ 2 \end{pmatrix}, \begin{pmatrix} 2 \\ ? \\ a \\ 1 \end{pmatrix}, \begin{pmatrix} 1 \\ 9 \\ -2 \\ 2 \end{pmatrix}.$$

3. a) Write b as a linear combination of u, v, w, where

$$b = \begin{pmatrix} 3 \\ 4 \\ -1 \\ 4 \end{pmatrix}, u = \begin{pmatrix} 1 \\ 0 \\ 1 \\ 0 \end{pmatrix}, v = \begin{pmatrix} 0 \\ 1 \\ -1 \\ 1 \end{pmatrix}, w = \begin{pmatrix} 1 \\ 1 \\ 0 \\ 1 \end{pmatrix}.$$

   b) What property of u, v, w makes the solution in a) not unique?

4. a) Show that the span of these vectors is a plane in R^3 and find its equation.

$$\begin{pmatrix} 1 \\ 5 \\ 9 \end{pmatrix}, \begin{pmatrix} 2 \\ 1 \\ 0 \end{pmatrix}, \begin{pmatrix} 3 \\ 1 \\ -1 \end{pmatrix}.$$

   b) Find a vector that is *not* in the span.

In Questions 5–11,

   a) Are the vectors linearly independent in the space V?

   b) Do the vectors span the space V?

5. $\begin{pmatrix} 3 \\ 8 \\ 4 \end{pmatrix}, \begin{pmatrix} -1 \\ 1 \\ 0 \end{pmatrix}, \begin{pmatrix} 2 \\ 0 \\ 1 \end{pmatrix}, V = R^3.$

6. $\begin{pmatrix} 3 \\ 5 \\ 4 \end{pmatrix}, \begin{pmatrix} -1 \\ 1 \\ 2 \end{pmatrix}, \begin{pmatrix} 2 \\ 2 \\ 1 \end{pmatrix}, V = R^3.$

7. $\begin{pmatrix} 3 \\ 3 \\ 4 \end{pmatrix}, \begin{pmatrix} -1 \\ 1 \\ 6 \end{pmatrix}, \begin{pmatrix} 2 \\ 2 \\ 1 \end{pmatrix}, \begin{pmatrix} 1 \\ 9 \\ -2 \end{pmatrix}, V = R^3$.

8. $2x^2 + 3x + 3, \ 6x^2 - 1, \ 26x^2 + 12x + 9, \ 14x^2 + 3x + 1, V = $ all polynomials of degree ≤ 2.

9. $\begin{pmatrix} 3 \\ 3 \\ 4 \\ 1 \end{pmatrix}, \begin{pmatrix} -1 \\ 1 \\ 6 \\ 2 \end{pmatrix}, \begin{pmatrix} 2 \\ 2 \\ 1 \\ 1 \end{pmatrix}, \begin{pmatrix} 1 \\ 9 \\ -2 \\ 2 \end{pmatrix}, V = R^4$.

10. $\begin{pmatrix} 1 & 2 \\ 3 & 4 \end{pmatrix}, \begin{pmatrix} 4 & 3 \\ 2 & 1 \end{pmatrix}, \begin{pmatrix} 1 & 3 \\ 2 & 4 \end{pmatrix}, V = $ all 2×2 matrices.

11. $\begin{pmatrix} 1 & 2 \\ 3 & 4 \end{pmatrix}, \begin{pmatrix} 1 & 2 \\ -3 & -4 \end{pmatrix}, \begin{pmatrix} -1 & -2 \\ 3 & 4 \end{pmatrix}, V = $ all 2×2 matrices.

■ 29.2 Bases

29.2.1 Example: Extending a Linearly Independent Set to a Basis

The vectors

$$v_1 = \begin{pmatrix} 1 \\ 2 \\ 3 \\ 4 \end{pmatrix}, v_2 = \begin{pmatrix} 3 \\ -1 \\ 1 \\ 2 \end{pmatrix}$$

are linearly independent in R^4. It can be proved that every linearly independent set in a vector space is a subset of some basis. One way to find two additional vectors to create a basis for R^4 is by choosing them from the standard basis $\{e_1, e_2, e_3, e_4\}$. Here's how to do that by reducing a matrix whose columns are v_1, v_2 together with the standard basis vectors.

- Find the reduced row-echelon form of the matrix: (**Matrices ▷ Row Reduce**)

$$\begin{pmatrix} 1 & 3 & 1 & 0 & 0 & 0 \\ 2 & -1 & 0 & 1 & 0 & 0 \\ 3 & 1 & 0 & 0 & 1 & 0 \\ 4 & 2 & 0 & 0 & 0 & 1 \end{pmatrix}.$$

▷ Row reduce the matrix $\begin{pmatrix} 1 & 3 & 1 & 0 & 0 & 0 \\ 2 & -1 & 0 & 1 & 0 & 0 \\ 3 & 1 & 0 & 0 & 1 & 0 \\ 4 & 2 & 0 & 0 & 0 & 1 \end{pmatrix}$

$$Out[2]= \begin{pmatrix} 1 & 0 & 0 & 0 & 1 & -\frac{1}{2} \\ 0 & 1 & 0 & 0 & -2 & \frac{3}{2} \\ 0 & 0 & 1 & 0 & 5 & -4 \\ 0 & 0 & 0 & 1 & -4 & \frac{5}{2} \end{pmatrix}$$

So the first four columns of the original matrix are linearly independent and, since $\dim R^4 = 4$, $\{v_1, v_2, e_1, e_2\}$ is a basis for R^4.

Another way. By changing the order of columns corresponding to the standard basis, you can find another basis for R^4 that contains v_1, v_2. For example, row reducing

$$\begin{pmatrix} 1 & 3 & 0 & 0 & 0 & 1 \\ 2 & -1 & 0 & 1 & 0 & 0 \\ 3 & 1 & 0 & 0 & 1 & 0 \\ 4 & 2 & 1 & 0 & 0 & 0 \end{pmatrix}$$

produces the basis $\{v_1, v_2, e_4, e_2\}$.

Exercises

In Questions 1–4, do the given vectors form a basis for the vector space V?

1. $\begin{pmatrix} 5 \\ 8 \\ -7 \end{pmatrix}, \begin{pmatrix} -1 \\ 2 \\ 0 \end{pmatrix}, \begin{pmatrix} 2 \\ 0 \\ 1 \end{pmatrix}, V = R^3$.

2. $\begin{pmatrix} 5 \\ 2 \\ -8 \end{pmatrix}, \begin{pmatrix} -1 \\ 2 \\ -11 \end{pmatrix}, \begin{pmatrix} 2 \\ 0 \\ 1 \end{pmatrix}, V = R^3$.

3. $6x^2 - 1$, $26x^2 + 12x + 9$, $14x^2 + 3x + 1$, $V = $ all polynomials of degree ≤ 2.

4. $\begin{pmatrix} 1 & 2 \\ 3 & 4 \end{pmatrix}, \begin{pmatrix} 4 & 3 \\ 2 & 1 \end{pmatrix}, \begin{pmatrix} 1 & 3 \\ 2 & 4 \end{pmatrix}, \begin{pmatrix} 4 & 3 \\ 3 & 1 \end{pmatrix}, V = $ all 2×2 matrices.

5. Find two bases for R^3 containing $\begin{pmatrix} 1 \\ 3 \\ -1 \end{pmatrix}, \begin{pmatrix} 0 \\ 3 \\ 0 \end{pmatrix}$.

6. Find a basis for the subspace of R^4 spanned by

$$\begin{pmatrix} 1 \\ 9 \\ -2 \\ 2 \end{pmatrix}, \begin{pmatrix} -25 \\ 1 \\ 2 \\ 1 \end{pmatrix}, \begin{pmatrix} 2 \\ 0 \\ 0 \\ 1 \end{pmatrix}, \begin{pmatrix} 1 \\ 1 \\ 0 \\ 2 \end{pmatrix}.$$

7. a) Show that the polynomials

$$x^3 + 4x^2 + 3x + 3,$$
$$2x^3 + 6x^2 + x - 1,$$
$$x^3 + x^2 + 2x + 2,$$
$$2x^3 - 2x^2 + 9x + 1$$

form a basis for the vector space of all polynomials of degree ≤ 3.

b) Express $x^3 + x^2 + x + 1$ as a linear combination of the polynomials in this basis, i.e., find its coordinates with respect to the basis.

8. a) Find a basis for the vector space of all polynomials of degree ≤ 3 that contains $x^2 + x$ and $x - 1$.

b) Express $ax^3 + bx^2 + cx + d$ as a linear combination of the basis in a).

■ 29.3 Null, Row, and Column Spaces

29.3.1 Example: Using the Reduced Row-Echelon Form

This example shows how to use the reduced row-echelon form (*RREF*) of a matrix to determine its null, row, and column spaces. We'll take the example

$$A = \begin{pmatrix} 1 & 3 & 6 & -6 & 1 \\ 4 & 0 & 0 & 12 & 3 \\ 1 & -2 & -4 & 7 & 0 \\ 3 & 2 & 4 & 5 & 3 \end{pmatrix}.$$

*   Find the *RREF* of the above matrix. (**Matrices ▷ Row Reduce**)

$$\triangleright \text{ Row reduce the matrix } \begin{pmatrix} 1 & 3 & 6 & -6 & 1 \\ 4 & 0 & 0 & 12 & 3 \\ 1 & -2 & -4 & 7 & 0 \\ 3 & 2 & 4 & 5 & 3 \end{pmatrix}$$

$$Out[2]= \begin{pmatrix} 1 & 0 & 0 & 0 & -\frac{1}{8} \\ 0 & 1 & 2 & 0 & \frac{23}{24} \\ 0 & 0 & 0 & 1 & \frac{7}{24} \\ 0 & 0 & 0 & 0 & 0 \end{pmatrix}$$

We can read off the following information about A from its *RREF*:

The null space of A is the set of solutions to $A\boldsymbol{x} = \boldsymbol{0}$. These vectors are in R^5 and can be expressed in terms of two *basic* or *free variables* x_3, x_5:

$$\begin{pmatrix} x_1 \\ x_2 \\ x_3 \\ x_4 \\ x_5 \end{pmatrix} = \begin{pmatrix} \frac{1}{8} x_5 \\ -2 x_3 - \frac{23}{24} x_5 \\ x_3 \\ -\frac{7}{24} x_5 \\ x_5 \end{pmatrix} = x_3 \begin{pmatrix} 0 \\ -2 \\ 1 \\ 0 \\ 0 \end{pmatrix} + x_5 \begin{pmatrix} \frac{1}{8} \\ -\frac{23}{24} \\ 0 \\ -\frac{7}{24} \\ 1 \end{pmatrix}.$$

This is true because elementary row operations do not change the solutions to a matrix equation $A x = b$.

The null space of A has dimension 2, with a basis of two vectors:

$$\begin{pmatrix} 0 \\ -2 \\ 1 \\ 0 \\ 0 \end{pmatrix}, \begin{pmatrix} \frac{1}{8} \\ -\frac{23}{24} \\ 0 \\ -\frac{7}{24} \\ 1 \end{pmatrix}.$$

This follows because the two vectors span the null space and are linearly independent.

The row space of A is the span of its rows. It is the subspace of R^5 whose basis consists of the three *nonzero rows* of the *RREF*, and so has dimension 3:

$$(1, 0, 0, 0, -\tfrac{1}{8}), \ (0, 1, 2, 0, \tfrac{23}{24}), \ (0, 0, 0, 1, \tfrac{7}{24}).$$

This follows because elementary row operations do not change the linear combinations of the rows of A, although they may change the rows themselves. So A and its *RREF* have the same row space.

The column space of A is the span of its columns. It is the subspace of R^4 whose basis consists of the three columns of A that produce the three *pivot columns* of the *RREF*. The pivot columns are those containing the leading 1's in the nonzero rows. So the column space of A has dimension 3, with this basis:

$$\begin{pmatrix} 1 \\ 4 \\ 1 \\ 3 \end{pmatrix}, \begin{pmatrix} 3 \\ 0 \\ -2 \\ 2 \end{pmatrix}, \begin{pmatrix} -6 \\ 12 \\ 7 \\ 5 \end{pmatrix}.$$

This is true because elementary row operations do not change the number of linearly independent columns, although they may change the column space. Here A and its *RREF* have different column spaces, but the two column spaces each have dimension 3.

The *nullity* of A is the dimension of its null space, and is 2. It can be proved that the row and column spaces of a matrix always have the same dimension, called the *rank*. So rank $A = 3$. Since every column contributes to a basis either for the null space or for the column space, *the sum of the rank and nullity is always the number of columns.*

Exercises

1. Find the rank and nullity of $\begin{pmatrix} 1 & -25 & 2 & 1 \\ 9 & 1 & 0 & 1 \\ -2 & 2 & 0 & 0 \\ 2 & 1 & 1 & 2 \end{pmatrix}$.

2. Find a basis for the row space of $\begin{pmatrix} 1 & 0 & 0 & 0 \\ 3 & 3 & 0 & 1 \\ -1 & 0 & 1 & 0 \end{pmatrix}$.

3. Find a basis for the column space of $\begin{pmatrix} 2 & 3 & 1 \\ -2 & -1 & 1 \\ -4 & 0 & 4 \\ 8 & -2 & -10 \end{pmatrix}$.

4. Find a basis for the null space of the matrix in Question 3.

5. Find a basis for the null space of the matrix in Question 1.

6. Find a basis for R^4 that contains a basis for the null space of the matrix in Question 3. (Hint: see Example 29.2.1.)

■ 29.4 Eigenvalues

Additional Prerequisite:
  Secs. 3.3–5   Plotting points, combining graphs, parameterized curves
  Sec. 3.9   How to Animate Graphs
  Sec. 16.15   How to Find Eigenvalues and Eigenvectors

29.4.1 Example: Visualizing Eigenvalues

This example shows how to visualize the eigenvalues and eigenvectors of a 2×2 matrix by using animation. We'll use the example

$$A = \begin{pmatrix} 1 & -1 \\ 2 & 4 \end{pmatrix}.$$

- Click in your *Mathematica* notebook beneath the last output, if any, so that a horizontal line appears.
- Type **A** = and click the **Matrix** button on the *Joy* palette.

- Fill in 2 for the number of rows and columns and click **Paste**.
- Fill in the preceding matrix and press ⌞SHIFT⌟ − ⌞RET⌟.

The result is:

In[2]:=   **A** = $\begin{pmatrix} 1 & -1 \\ 2 & 4 \end{pmatrix}$

Out[2]= $\begin{pmatrix} 1 & -1 \\ 2 & 4 \end{pmatrix}$

The *eigenvectors* of A are the nonzero vectors v for which Av is a multiple of v, i.e., $Av = \lambda v$ for some scalar λ, which is called an *eigenvalue* of A. We'll follow the path of a vector $v = (\cos t, \sin t)$ and of its image Av as v moves around the unit circle $x^2 + y^2 = 1$.

First we'll plot the circle and its image and then combine the two graphs.

- From the *Joy* menus, choose **Graph ▷ Parametric Plot**.
- Fill in x = Cos[t], y = Sin[t].
- Fill in $0 \le t \le 2\pi$ and check **Equal scales**.

   Enter π with the π button on the *Joy* palette. Choosing equal scales makes the curve appear as a circle instead of an ellipse.

- Check **Assign name** and fill in circle.
- Click **OK**.

When we multiply v on the circle by the matrix A, we apply a linear transformation that results in

$$Av = \begin{pmatrix} 1 & -1 \\ 2 & 4 \end{pmatrix}\begin{pmatrix} \cos t \\ \sin t \end{pmatrix} = \begin{pmatrix} \cos t - \sin t \\ 2\cos t + 4\sin t \end{pmatrix}.$$

These vectors form the image of the circle under the linear transformation. It can be proved that the image is an ellipse, which we can now plot and then combine with the circle

- Click **Last Dialog** on the *Joy* palette to reopen the **Parametric Plot** dialog.
- Fill in x = Cos[t] − Sin[t], y = 2 Cos[t] + 4 Sin[t].
- Change the name to ellipse and click **OK**.
- Combine *circle* and *ellipse*, and call the result combo. (**Graph ▷ Combine Graphs, Assign name**)

▷ Combining the graphs {circle, ellipse} gives

combo =

Out[5]=   - Graphics -

We'll use *combo* as the background for an animation of the vectors v, Av as v moves around the unit circle. The animation will have 15 frames, and we will divide the circle into arcs of $2\pi/15$ radians. Section 3.9 explains more about animation.

- If necessary, click in your notebook so that a horizontal line appears.
- Type v = {Cos [2 π⌴ $\frac{k}{15}$], Sin [2 π⌴ $\frac{k}{15}$]}.

  The symbol ⌴ means to leave a space to indicate multiplication. Be sure to use curly brackets { } to specify that v is a vector.

- Press SHIFT – RET.

In[6]:=   v = $\left\{ Cos\left[2\pi\dfrac{k}{15}\right], Sin\left[2\pi\dfrac{k}{15}\right] \right\}$

Out[6]= $\left\{ Cos\left[\dfrac{2k\pi}{15}\right], Sin\left[\dfrac{2k\pi}{15}\right] \right\}$

- Choose **Graph ▷ Animation**.
- From the popup menu, choose **Points** and click **OK**.
- Fill in v, A . v in the input field.

  This plots the two vectors v and Av together as points in the plane.

- In the **Animation Info** window, fill in $1 \le k \le 15$.
- Fill in $-2 \le x \le 2$, $-5 \le y \le 5$ as the common plot frame.

  The graph *combo* shows that this frame will contain all the points.

- Check **Equal scales**.

This makes the path of *v* appear as a circle.

- Check **Combine with** and fill in combo.
- In the **Plot Points** window, click **OK**.

Mathematica graphs the frames, collapses them, and starts the animation.

▷ Plot of the points {v, A.v}   Animation for k = 1 to 15.

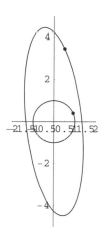

You can control the animation with the buttons that appear at the bottom of the *Mathematica* window. Clicking anywhere in the notebook will stop the motion and double-clicking on the graphs will restart it. You can change the speed with the buttons or by pressing a number between 1 (slowest) and 9 (fastest). Slowing down the animation will help you see the relation between *v* and *Av* better.

 In some situations, the moving points may be too small to be seen well. Here's how you can enlarge them. Click on the triangle ▷ in the *Joy* paraphrase to open the *Mathematica* command. Find `PointSize[0.02]` and change the size to `0.04`. Then, while the insertion point is still in the cell containing the command, press ⎙SHIFT⎙ − ⎙RET⎙. You may need to experiment with the point size to find one that you like.

The animation contains four frames where *Av* is approximately a multiple of *v*. They correspond to two eigenvalues, each with two unit eigenvectors that point in opposite directions. Each eigenvalue determines a line of eigenvectors. Here's how to see them in more detail.

- Open the animation frames by double-clicking on the cell bracket that contains them.

  The following frame shows a vector *v* on the unit circle and its image *Av* on the ellipse. Here *Av* is approximately a multiple of *v*, and *v* is close to an eigenvector of *A*. Including more frames in the animation can give a better approximation.

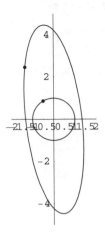

We can use *Joy* to find the eigenvalues and eigenvectors of A.

- Choose **Matrices ▷ Eigenvalues**.
- Fill in **A**.
- Make sure the boxes for **eigenvalues** and **eigenvectors** are both checked.
- Click **Algebraically** and click **OK**.

▷ The eigenvalues and eigenvectors of A are

Out[10]//MatrixForm=

$$\begin{pmatrix} 2 & \{-1, 1\} \\ 3 & \{-1, 2\} \end{pmatrix}$$

The eigenvalues are $\lambda = 2, 3$. The eigenvectors for $\lambda = 3$ are the nonzero vectors on the line $y = -2\,x$, and the vectors in the frame shown above are approximately on that line.

It can be proved that the image of the unit circle under A is an ellipse and that the axes of the ellipse are determined by the eigenvectors of $A^T A$. In fact, let w_1, w_2 be linearly independent unit eigenvectors of $A^T A$. Then w_1, w_2 are orthogonal, the vectors Aw_1, Aw_2 are orthogonal, and Aw_1, Aw_2 lie at the ends of the two axes of the ellipse. This result is related to the singular value decomposition of A. (In order for the vectors to appear orthogonal in a graph, the axes must have equal scales.)

Exercises

1. a) Find the characteristic polynomial of $\begin{pmatrix} 1 & 1 & 1 \\ 1 & 1 & 0 \\ 1 & 0 & 0 \end{pmatrix}$. Section 16.14 shows one way to do this.

    (Matrices ▷ Determinant)

b) Use a) to find the eigenvalues of this matrix. (**Algebra** ▷ **Solve**)

2. a) Find a basis for R^3 of eigenvectors for $A = \begin{pmatrix} 0 & 1 & 1 \\ 1 & 0 & 1 \\ 1 & 1 & 0 \end{pmatrix}$.

    b) Use the basis in a) to diagonalize A, i.e., find an invertible matrix P for which $P^{-1} A P$ is diagonal. Calculate the product to verify it is diagonal.

3. a) Plot the ellipse in the example of this section, Example 29.4.1. (**Graph** ▷ **Parametric Plot**)

    b) Find *unit* eigenvectors w_1, w_2 for the two eigenvalues of $A^T A$, where A is the matrix in the example. Verify that they are orthogonal by taking their dot product. (Enter $w_1 \cdot w_2$ and simplify by choosing **Algebra** ▷ **Simplify**.)

    c) Similarly, show that Aw_1, Aw_2 are orthogonal.

    d) It can be proved that Aw_1, Aw_2 lie at the ends of the two axes of the ellipse. Use the **Numeric** button on the *Joy* palette to express Aw_1 and Aw_2 in numeric form and then plot them on a printout of the graph in a). Do Aw_1 and Aw_2 appear to be at the ends of the axes of the ellipse?

4. a) Follow the procedure of the example of this section, Example 29.4.1, to create an animation using the matrix $A = \begin{pmatrix} 1 & 2 \\ 2 & 1 \end{pmatrix}$.

    b) Where do the eigenvectors of A appear to be relative to the ellipse, according to the animation? Then use *Joy* to find the eigenvectors and check your answer.

    c) Note that A is a *symmetric* matrix ($A^T = A$). By hand, find the relationship between the eigenvalues of A and of $A^T A$, and prove that they have the same eigenvectors. This, together with the last paragraph of Example 29.4.1, page 556, explains the connection between th eigenvectors of A and the ellipse.

5. Square matrices with 1's above the main diagonal and 0's elsewhere play a role in the Jordan canonical form. An example of this is the matrix

    $$A = \begin{pmatrix} 0 & 1 & 0 & 0 & 0 \\ 0 & 0 & 1 & 0 & 0 \\ 0 & 0 & 0 & 1 & 0 \\ 0 & 0 & 0 & 0 & 1 \\ 0 & 0 & 0 & 0 & 0 \end{pmatrix}.$$

    a) Raise A to powers 2, 3, 4, 5 and describe the pattern that the powers take. What will the sixth power be? (**Matrices** ▷ **Powers**, *do not* enter these in the form A^n)

    b) Find the eigenvalues of A **numerically**. (**Matrices** ▷ **Eigenvalues**)

    c) Use a) to prove that the only eigenvalue of A is $\lambda = 0$ and to find (by hand) its eigenvectors.

Lab 29.1 Rotations

Prerequisites:
Ch. 1   A Brief Tour of *Joy*
Sec. 4.9   How to Simplify Expressions in General
Secs. 16.4,8 Creating and multiplying matrices

In this lab, you will explore various properties of rotations. You will see how each rotation corresponds to a matrix, what happens when you combine rotations, and how to find the inverse matrix for a rotation.

■ Before the Lab (Do by Hand)

Rotating a vector around the origin through some angle is a linear transformation, so it can be accomplished by multiplying the vector by some matrix. Before the lab, you will set up these matrices, and during the lab you will use them to study how rotations behave.

Rotating a vector v counterclockwise around the origin through an angle θ produces a vector v'. The goal in this section is to find a matrix $R(\alpha)$ that depends on α and such that $v' = R(\alpha)\,v$, for all v in R^2.

Let e_1, e_2 be the standard basis vectors for R^2,

$$e_1 = \begin{pmatrix} 1 \\ 0 \end{pmatrix} \text{ and } e_2 = \begin{pmatrix} 0 \\ 1 \end{pmatrix}.$$

Since rotation is linear, the columns of $R(\alpha)$ will be e_1', e_2'.

1. Suppose $\alpha = \pi/2$. Sketch e_1', e_2', find their coordinates, and find the matrix $R(\pi/2)$.

2. Find $v' = R(\pi/2)\,v$ when $v = \begin{pmatrix} 1 \\ 1 \end{pmatrix}$. Then sketch v' and see if your result appears to give the correct rotation of v.

3. Find a formula for $v' = R(\pi/2)\,v$ when $v = \begin{pmatrix} x \\ y \end{pmatrix}$.

4. Repeat Questions 1–3 for $\alpha = \pi/4$.

5. Use trigonometry to find e_1', e_2', and $R(\alpha)$ for an arbitrary angle α. It will help to recall that every vector v can be written in the form

$$v = \begin{pmatrix} r\cos\theta \\ r\sin\theta \end{pmatrix},$$

where r is the length of v and θ is the angle it makes with the positive x-axis. Check to see that your matrix $R(\alpha)$ agrees with what you found for $\alpha = \pi/2$ and $\pi/4$.

6. Find a formula for $v' = R(\alpha)\, v$ when $v = \begin{pmatrix} x \\ y \end{pmatrix}$.

■ In the Lab

1. Create a function $R(\alpha)$ whose value is the matrix you found in Question 5 (Before the Lab). Use the **Greek** button on the *Joy* palette to enter α and the **Matrix** button to enter the matrix. (**Create ▷ Function**)

   When you are done, close the Greek palette by clicking the close box in its upper-left corner (Macintosh) or the close button $\boxed{\textbf{X}}$ in its upper-right corner (Windows).

2. Find the product $R\,(\pi/6)\,R\,(\pi/3)$. (**Matrices ▷ Multiply**)

   You can enter π with either the π or **Greek** button on the *Joy* palette.

3. The inverse of a rotation should also be a rotation, by some angle. What angle do you think corresponds to $R(\pi/6)^{-1}$? Test your answer by using *Joy* to multiply $R(\pi/6)$ by $R(\ldots)$, filling in the angle that you think is correct. If the result is not the identity, try again.

4. Find the product $R(\alpha)\,R(\beta)$, where α and β do not have specific values.

5. Simplify the last output. (**Algebra ▷ Simplify,** choose the **Simplify** button)

6. Print your work.

■ After the Lab

1. Explain why the product $R(\pi/6)\,R(\pi/3)$ must turn out to be a rotation matrix. What is the angle of rotation?

2. In the lab, you computed the product $R(\alpha)\,R(\beta)$. To what single geometric operation does this correspond? Use trigonometric identities to verify that the matrix of this single operation is the same as $R\,(\alpha)\,R\,(\beta)$.

3. In Question 2, should it matter if you compute $R(\beta)\,R(\alpha)$ instead? Why?

4. What rotation matrix is the inverse of $R(\alpha)$? Test your answer by substituting $\alpha = \pi/6$ and comparing it with your answer to Question 3 (In the Lab).

Lab 29.2 Reflections

Prerequisites:
    Ch. 1   A Brief Tour of *Joy*
    Sec. 4.9   How to Simplify Expressions in General
    Secs. 16.4,8 Creating and multiplying matrices

In this lab, you will explore various properties of reflections. You will see how each reflection corresponds to a matrix, what the inverse matrix of a reflection is, and what happens when you combine reflections.

■ Before the Lab (Do by Hand)

Reflecting a vector across a line through the origin is a linear transformation, so it corresponds to multiplying the vector by some matrix. Before the lab, you will set up these matrices, and during the lab you will use them to study how reflections behave.

Reflecting a vector v across a line through the origin produces a vector v'. The line of reflection bisects the angle between v and v', and the two vectors have the same length. Suppose the angle between the line and the positive direction of the x-axis is α, that is, the line has the equation $y = (\tan \alpha)\, x$. The goal in this section is to find a matrix $S(\alpha)$ that depends on α and such that $v' = S(\alpha)\, v$ for all v in R^2.

Let e_1, e_2 be the standard basis vectors for R^2,

$$e_1 = \begin{pmatrix} 1 \\ 0 \end{pmatrix} \text{ and } e_2 = \begin{pmatrix} 0 \\ 1 \end{pmatrix}.$$

Since reflection is linear, the columns of $S(\alpha)$ will be $e_1{}'$, $e_2{}'$.

1. Suppose $\alpha = \pi/4$, so that the reflection is across the line $y = x$. Sketch $e_1{}'$, $e_2{}'$, find their coordinates, and find the matrix $S(\pi/4)$.

2. Find $v' = S(\pi/4)\, v$ when $v = \begin{pmatrix} -1 \\ 1 \end{pmatrix}$. Then sketch v' and see if your result appears to give the correct reflection of v.

3. Find a formula for $v' = S(\pi/4)\, v$ when $v = \begin{pmatrix} x \\ y \end{pmatrix}$.

4. Repeat Questions 1–3 for $\alpha = 3\pi/4$.

5. Use trigonometry to find $e_1{}'$, $e_2{}'$, and $S(\alpha)$ for an arbitrary angle α. It will help to recall that every vector v can be written in the form

$$v = \begin{pmatrix} r\cos\theta \\ r\sin\theta \end{pmatrix},$$

where r is the length of v and θ is the angle it makes with the positive x-axis. Check to see that your matrix $S(\alpha)$ agrees with what you found for $\alpha = \pi/4$ and $3\pi/4$.

6. Find a formula for $v' = S(\alpha)\,v$ when $v = \begin{pmatrix} x \\ y \end{pmatrix}$.

■ In the Lab

1. Create a function $S(\alpha)$ whose value is the matrix you found in Question 5 (Before the Lab). Use the **Greek** button on the *Joy* palette to enter α and the **Matrix** button to enter the matrix. **(Create ▷ Function)**

   When you are done, close the Greek palette by clicking the close box in its upper-left corner (Macintosh) or the close button **X** in its upper-right corner (Windows).

2. The inverse of a reflection should also be a reflection, across a line making some angle with the x-axis. What angle do you think corresponds to $S(\pi/3)^{-1}$? Test your answer by using *Joy* to multiply $S(\pi/3)$ by $S(\ldots)$, filling in the angle that you think is correct. If the result is not the identity, try again.

   Use the π button or the **Greek** button on Joy *palette* to enter π.

3. Find the product $S(\pi/4)\,S(\pi/3)$. Then compute $S(\pi/3)\,S(\pi/4)$. **(Matrices ▷ Multiply)**

4. Use *Joy* to multiply $S(\alpha)\,S(\beta)$, where α and β do not have specific values.

5. Simplify the last output. **(Algebra ▷ Simplify,** choose the **Simplify** button)

6. Print your work.

■ After the Lab

1. What reflection matrix is the inverse of $S(\alpha)$? Test your answer by substituting $\alpha = \pi/3$ and comparing it with your answer to Question 2 (In the Lab).

2. The matrix of a counterclockwise *rotation* through an angle θ is
   $$\begin{pmatrix} \cos\theta & -\sin\theta \\ \sin\theta & \cos\theta \end{pmatrix}.$$

   Your result in Question 5 (In the Lab) should show that $S(\alpha)\,S(\beta)$ is a rotation. For what angle θ is it a rotation?

3. Since $(A\,B)\,v = A(B\ v)$, multiplying two matrices together corresponds to a composition. For $S(\pi/4)\,S(\pi/3)$, sketch the result of applying the two reflections consecutively to the vectors e_1, e_2. Then sketch the rotation to which this corresponds and check that they are the same.

4. Repeat Question 3 for $S(\pi/3)\,S(\pi/4)$.

■ For Further Exploration

What is the product of a reflection and a rotation? You can explore this topic further by considering products of the form $R(\alpha)\,S(\beta)$ and $S(\beta)\,R(\alpha)$, where $R(\alpha)$ is the matrix for a counterclockwise rotation through an angle of α radians.

1. a) Do rotations and reflections commute, i.e., is $R(\alpha)\,S(\beta) = S(\beta)\,R(\alpha)$? Try some examples using specific values of α and β and make a conjecture.

    b) Multiply the matrices without specific values and check your conjecture.

2. a) Is the product of a reflection and a rotation another reflection? a rotation? Try some specific examples and make a conjecture.

    b) Use *Joy* to check your conjecture for the general case, without using specific numerical values.

Lab 29.3 Markov Chains and Market Share

Prerequisites:
      Ch. 1   A Brief Tour of *Joy*
      Secs. 16.4,13,15   Creating matrices, powers, eigenvalues

The breakfast cereals Crispy Bran and Shredded Bran appeal to the same customers. Market research tells us that each week, a certain percentage of each brand's customers will switch to the other brand. If the same percentages apply each week, can we predict whether one brand of bran will take over the market in the long run? In this lab, you can explore using Markov chains and eigenvalues to answer this question.

■ Before the Lab (Do by Hand)

Since the cereals are so similar to each other, their customers don't have much brand loyalty and aggressive pricing can cause many customers to switch brands of bran. Suppose half the people who buy Crispy Bran this week will buy it again next week, but the other half will switch to Shredded Bran. Similarly, 75% of those who buy Shredded Bran this week will be loyal to it next week but 25% will switch to Crispy Bran.

Let the vector $x_0 = (c, s)$ give the numbers of customers who buy these two cereals during some week, where c stands for Crispy Bran and s for Shredded Bran.

1. How many will buy Crispy Bran during the next week? How many will buy Shredded Bran? Use your answers to write an equation $x_1 = P x_0$ that expresses next week's vector in terms of this week's vector x_0 and a matrix P.

When the percentage of people who switch from each brand to the other is the same each week, the system is called a *Markov chain*. The matrix P is called the *transition matrix* of the chain. We can rewrite $x_k = P x_{k-1}$ as $x_k = P^k x_0$ and express the vector for week k in terms of the vector for the initial week.

A vector whose coordinates are nonnegative and add up to 1 is called a *probability vector*. A matrix whose columns are probability vectors is called a *stochastic* (or *column-stochastic*) *matrix*. Your matrix P should be a stochastic matrix.

2. What probability vector gives the market shares of the two cereals, i.e., the proportion of the total customer pool that buys each cereal?

An *equilibrium vector* for P is a vector x for which $P x = x$, i.e., x doesn't change from week to week. It is therefore an eigenvector for P with eigenvalue $\lambda = 1$.

3. Find an equilibrium vector for P that corresponds to a total of 3 million customers. Is there more than one?

4. Find an equilibrium probability vector for P. Is there more than one?

5. Suppose the fraction of one week's Crispy Bran buyers who will buy it the following week is a, $0 < a < 1$. Suppose the fraction of one week's Shredded Bran buyers who will buy it the following week is b, $0 < b < 1$. What is the transition matrix Q for these fractions? It should be a stochastic matrix.

■ In the Lab

In the lab, you will use *Joy* to create the matrices and vectors you found before the lab, and to carry out calculations with them to predict how the two brands will fare against each other.

1. Create the matrix P as follows:

   - Click in your *Mathematica* notebook below the last output, if any, so that a horizontal line appears.
   - Type P = and click the **Matrix** button on the *Joy* palette.
   - Fill in 2 for the number of rows and columns and click **Paste**.
   - Fill in the matrix P and press ⏑SHIFT⏑ − ⏑RET⏑.

2. Suppose 2,500,000 people buy Crispy Bran this week and only 500,000 buy Shredded Bran. Using these figures, create the vectors x_0 as a column vector. (**Matrix** button, **Subscript** button)

3. Calculate $x_1 = P\,x_0$, $x_2 = P\,x_1$, and $x_3 = P\,x_2$. (**Matrices** ▷ **Multiply**)

4. Find $P^5\,x_0$ by first finding P^5 and then multiplying the result by x_0. Find $P^{10}\,x_0$ in the same way. (**Matrices** ▷ **Powers**, *do not* enter these in the form P^n)

5. Create any other initial vector x_0 whose coordinates are nonnegative and add up to 3 million. Compute powers $P^k\,x_0$ for large enough k so that you think you have found the value of $\lim_{k\to\infty} P^k\,x_0$. Do it for a third vector x_0.

6. Create the matrix Q that you found in Question 5 (Before the Lab). Here a and b will not have specific values.

7. Find the eigenvalues and eigenvectors of Q. (**Matrices** ▷ **Eigenvalues,** choose **Algebraically**)

8. Find Q^n without specifying a numerical value for n. Then find $Q^n\,v_0$, where $v_0 = \begin{pmatrix} r \\ 1 - r \end{pmatrix}$ is a probability vector and r is not specified.

9. Print your work.

■ After the Lab

1. Even though 2,500,000 people buy Crispy Bran and only 500,000 buy Shredded Bran, the latter quickly dominates the market. How could you have predicted this from the weekly transition percentages?

2. What effect does changing the initial distribution of the 3 million customers appear to have on the long-run distribution?

3. How does the long-run distribution of the 3 million customers compare to the equilibrium you found before the lab?

4. What do the long-run market shares for the two brands appear to be, and how does this compare to the equilibrium probability vector you found before the lab?

5. Based on Question 7 (In the Lab), what are the market shares for the two cereals in equilibrium when the transition matrix is Q? Your answers should be in terms of a and b.

6. a) Since a and b are between 0 and 1, what is $\lim_{n \to \infty} (-1 + a + b)^n$?

   b) Use a) and your output for Question 8 (In the Lab) to find the long-run market shares when the transition rates are a and b. How do these compare to the equilibrium market shares in Question 5 (After the Lab)?

The phenomena you observed in this lab can be shown to be a general property of stochastic matrices: If every entry of P (or of some power of P) is positive, then $\lim_{k \to \infty} P^k x_0$ exists and is the same for every x_0 with nonnegative coordinates that have the same sum. Also, this limit is an eigenvector of P with eigenvalue $\lambda = 1$ and the dimension of the corresponding eigenspace is 1. That is, the long-run vector is also an equilibrium vector. Moreover, if x_0 is a probability vector, then so is the limit, and P has only one equilibrium probability vector.

Lab 29.4 Systems of Linear Differential Equations: An Employment Model

---

Prerequisites:

Ch. 1	A Brief Tour of *Joy*
Secs. 16.4,8,9	Creating, multiplying, inverting matrices
Secs. 16.12,15	Transpose, eigenvalues

In this lab, you can explore how to use Joy to solve a homogeneous system $x' = Ax$ of linear differential equations with constant coefficients, where the coordinates of the vector x are functions of some variable t. The solution will use the eigenvalues and eigenvectors of the matrix A and the exponential matrix e^{At} of the system. Such systems arise in many situations, including the following model of a population that doesn't change in size but is redistributed continuously among three classes over time. Variations of this model can account for additions to and subtractions from the total population. A similar analysis can be used to study the diffusion of a substance, such as a chemical compound, through a system of several compartments, such as organs of the body. You can also use *Joy* to construct a discrete model, based on fixed time steps rather than on a continuous time variable t.

■ Before the Lab (Do by Hand)

In this section, you can construct a model for changes in the level of unemployment based on the following assumption: Each week, fixed percentages of those working, looking for work, and not looking change from one status to another.

According to the U.S. Bureau of Labor Statistics (*Monthly Labor Review*, Sept. 1992) there were about 118 million people over the age of 16 employed in the civilian work force in 1992. About 10 million more were unemployed and looking for work, and about 64 million were not looking for work and therefore not counted in the labor force.

Suppose that each week, 0.26% of all workers lose their jobs and start looking for work. Also suppose that 3.0% of those who are looking for work find jobs, while another 2.0% give up and stop looking. Finally, suppose that 0.40% of those who haven't been looking enter the labor force and begin a job search. How would unemployment vary under these conditions? To simplify the question, ignore any changes in the total working age population over the period of time in question.

1. a) Based on the number of people given above for each sector, how many people would lose their jobs in the first week? How many would find work? How many would drop out of the labor force, and how many would enter it (and be counted among the unemployed)?

    b) Would the level of unemployment begin to increase or decrease?

Rather than tracking employment changes week by week, you can use a continuous model based on differential equations. Let $w = w(t)$ be the number of people who are working at time t, let

$u = u(t)$ be the number who are unemployed and looking for work, and let $v = v(t)$ be the number who aren't looking. Here t measures time in weeks.

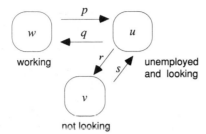

working

not looking

unemployed and looking

The assumptions made earlier suggest the differential equations

$$
\begin{aligned}
w'(t) &= -pw + qu \\
u'(t) &= pw - (q+r)u + sv \ , \\
v'(t) &= ru - sv
\end{aligned}
$$

where $p = 0.0026$, $q = 0.03$, $r = 0.02$, $s = 0.004$.

2. Explain why these equations give a reasonable model for the rate at which the size of each group changes from week to week.

The equations can be expressed in matrix form $x' = Ax$, where

$$
A = \begin{pmatrix} -p & q & 0 \\ p & -(q+r) & s \\ 0 & r & -s \end{pmatrix}, \quad x = x(t) = \begin{pmatrix} w \\ u \\ v \end{pmatrix}.
$$

It can be shown that the general solution to this equation is $x = e^{At} k$, where $k = x(0)$ is a vector of constants containing the initial conditions and e^{At} is the *exponential matrix function*. This is analogous to the case of the scalar differential equation $x' = ax$, whose solution is $x = x(t) = k e^{at}$, where a and k are scalars, $k = x(0)$. The matrix function e^{At} shares similar properties with the scalar function e^{at}. It is defined by a power series of matrices, $e^{At} = \sum_{n=0}^{\infty} (At)^n / n!$, that is analogous to the scalar function $e^{at} = \sum_{n=0}^{\infty} (at)^n / n!$.

If A can be diagonalized, then there is a way to calculate e^{At} without using power series. In this case, there is an invertible matrix S such that $L = S^{-1} A S$ is diagonal,

$$
L = \begin{pmatrix} \lambda_1 & 0 & 0 \\ 0 & \lambda_2 & 0 \\ 0 & 0 & \lambda_3 \end{pmatrix}.
$$

The diagonal entries in L are the eigenvalues of A and the columns of S form a basis of eigenvectors for R^3 corresponding to the eigenvalues. One can prove that in this case

$$e^{At} = S\, e^{Lt}\, S^{-1} \text{ and } e^{Lt} = \begin{pmatrix} e^{\lambda_1 t} & 0 & 0 \\ 0 & e^{\lambda_2 t} & 0 \\ 0 & 0 & e^{\lambda_3 t} \end{pmatrix}.$$

So in this case e^{At} can be found using the eigenvalues and eigenvectors of A.

> *Mathematica* has a built-in function for the exponential function matrix, `MatrixExp[A t]`. This can also be used as a "black box" technique for finding e^{At} instead of following the diagonalization method shown here.

3. The preceding description of e^{At} applies in any dimension. Calculate e^{At} for this example in R^2:

$$A = \begin{pmatrix} 1 & 2 \\ 2 & 1 \end{pmatrix}, L = \begin{pmatrix} -1 & 0 \\ 0 & 3 \end{pmatrix}, S = \begin{pmatrix} -1 & 1 \\ 1 & 1 \end{pmatrix}, S^{-1} = \begin{pmatrix} -\frac{1}{2} & \frac{1}{2} \\ \frac{1}{2} & \frac{1}{2} \end{pmatrix}.$$

■ In the Lab

1. Create the matrix A as follows:

   - Click in your *Mathematica* notebook below the last output, if any, so that a horizontal line appears.
   - Type **A** = and click the **Matrix** button on the *Joy* palette.
   - Fill in 3 for the number of rows and columns and click **Paste**.
   - Fill in the matrix $\begin{pmatrix} -p & q & 0 \\ p & -(q+r) & s \\ 0 & r & -s \end{pmatrix}$ and press $\boxed{\text{SHIFT}} - \boxed{\text{RET}}$.

2. Enter the values for p, q, r, s. For example:

   - If necessary, click below the last output so that a horizontal line appears.
   - Type **p** = **0.0026** and press $\boxed{\text{SHIFT}} - \boxed{\text{RET}}$.
   - Repeat for q, r, s.

3. a) Find the eigenvalues and eigenvectors of A **numerically**. (**Matrices ▷ Eigenvalues**)

   You should find that one of the eigenvalues is zero.

   b) Use the eigenvalues to create a matrix *expL* equal to e^{Lt} as defined before the lab. That is, *expL* will be a diagonal matrix with exponential functions along its diagonal. Enter e as **E** and use the **Superscript** button on the *Joy* palette.

   Mathematica doesn't permit using e^{Lt} as a name, so we'll call the matrix `expL`.

4. You can use the eigenvectors of A to determine S. Rather than copy and paste the eigenvectors, take this shortcut:

- Click **Last Dialog** to reopen the **Eigenvalues** dialog.
- *Uncheck* **Eigenvalues** and make sure **Numerically** is checked.

    This will produce a matrix whose *rows* are the eigenvectors.

- Click **OK**.

To find *S*, we transpose this matrix.

- Take the transpose of the **Last Output** and assign the name S. (**Matrices ▷ Transpose, Assign name**)

5. Take the inverse of *S* and assign the name Sinverse. (**Matrices ▷ Inverse, Assign name**)

6. Now create a matrix *expA* equal to $e^{At} = S\, e^{Lt}\, S^{-1}$. That is, enter
expA = S . expL . Sinverse in your notebook.

    Mathematica may warn of a possible spelling error. Also, the matrix may be so wide that it's hard to see on your monitor.

7. a) Create a column vector *k* whose entries are the initial values of *w*, *u*, *v* given for 1992 before the lab.

    b) Enter x = expA . k in your notebook. This is $x(t) = e^{At}\, k$, the solution to the system of differential equations.

    c) Simplify x. (**Algebra ▷ Simplify,** click the **Simplify** button)

8. a) From the preceding result, read off the vector $x_\infty = \lim_{t\to\infty} x(t)$,

    b) In your notebook, calculate the product Ax_∞.

9. Print your work. Save your notebook if you will be doing the section **For Further Exploration**.

■ After the Lab

1. What do the coordinates of x_∞ mean in terms of the labor force?

2. The *unemployment rate* is the proportion of people in the labor force who are unemployed. Those who are not looking for work are not included in this calculation. What is the long-run unemployment rate for this model?

3. a) When *t* is very large, $x(t) \approx x_\infty$. What would you expect $x'(t)$ to be for large values of *t*?

    b) What would you expect the coordinates of Ax_∞ to equal?

4. a) In Question 3 (In the Lab), your output should contain an eigenvector for the eigenvalue $\lambda = 0$. Show that x_∞ is a multiple of that eigenvector by showing that their three coordinates have a common ratio.

b) Why does a) imply the value you found for $A\boldsymbol{x}_\infty$?

■ For Further Exploration

1. a) Create $\boldsymbol{k} = \begin{pmatrix} a \\ b \\ c \end{pmatrix}$, i.e., redefine the initial population distribution to be arbitrary.

    b) Enter $\mathbf{x} = \mathbf{expA}$. $\mathbf{k}$ to find the solution $\boldsymbol{x}(t) = e^{At} \boldsymbol{k}$ for this arbitrary $\boldsymbol{k}$.

2. It may be hard to see the coordinates of the vector $\boldsymbol{x}$ on your monitor. It's easier to take the coordinates separately.

    - Type $\mathbf{x}[\,[\,1\,]\,]$ and press $\boxed{\text{SHIFT}} - \boxed{\text{RET}}$.

      This displays only the first coordinate of $\boldsymbol{x}$.

    - Repeat for the second and third coordinates of $\boldsymbol{x}$.

3. a) What does $a + b + c$ equal?

    b) What is $\boldsymbol{x}_\infty = \lim_{t\to\infty} \boldsymbol{x}(t)$ in this case?

    c) Show that $\boldsymbol{x}_\infty$ is an eigenvector of A for eigenvalue $\lambda = 0$.

4. What is the long-run unemployment rate for this model? Does it depend on the initial distribution of the population among employed, unemployed, and not looking?

Index